AS/A2

PHYSICS

D1330188

Chris Mee   Brian Arnold
Mike Crundell   Wendy Brown

Hodder Murray

www.hod

How to link this book with your specification

The Physics AS/A2 Specifications offered by the five Awarding Bodies in England, Wales and Northern Ireland do not present the core material of the subject in precisely the same order. According to the rules laid down by the Qualifications and Curriculum Authority, some topics must be presented in the AS course, and others in the A2 course. Other topics may appear in either the AS or the A2 course, or may be introduced in AS and re-visited in A2. The Table below gives a broad indication of how the various chapters of the book are linked to the AS and A2 courses of the Specifications offered by each board.

Chapter	AQA Physics A 5451/6451	AQA Physics B 5456/6456	Edexcel Physics 8540/9540	Edexcel Salters' Physics 8552/9552	NICCEA Physics 1210	OCR Physics A 3883/7883	OCR Advancing Physics	WJEC/CBAC Physics 54080/54090
1 Physical quantities and units	AS	AS	AS	AS	AS	AS	AS	AS
2 Kinematics	AS	AS	AS	AS	AS	AS	AS	AS
3 Work, energy and power	AS	AS & A2	AS	AS	AS & A2	AS & A2	AS & A2	A2
4 Force and collisions	AS	AS & A2	AS	AS & A2	AS & A2	AS & A2	AS & A2	AS & A2
5 Thermal physics	AS	A2	AS	A2	A2	A2	A2	A2
6 Electricity	AS	AS	AS	AS	AS	AS	AS	AS
7 Waves	A2	AS	A2	AS	AS	AS	AS	AS
8 Photons, electrons and atoms	AS	A2	A2	AS	AS	AS	AS	AS
9 Circular motion and oscillations	A2	A2	A2	A2	A2	A2	A2	A2
10 Fields	A2	A2	A2	A2	A2	A2	A2	A2
11 Electromagnetism	A2	A2	A2	A2	A2	A2	A2	A2
12 Nuclear physics	A2	AS & A2	AS	AS & A2	A2	A2	A2	AS & A2
13 Synoptic assessment	A2	A2	A2	A2	A2	A2	A2	A2

CONTENTS

Preface

New Specifications (syllabuses) for GCE Advanced Level Physics examinations were published by the five Awarding Bodies in England, Wales and Northern Ireland early in 2000. In the new courses, the full A-level qualification is taken in two stages. The first stage is the Advanced Subsidiary (AS) examination, which is assessed at a standard appropriate for candidates who have completed half of the full A-level course. The AS accounts for 50% of the full A level in terms of teaching time and content. The second half of the course is referred to as A2. This is assessed at a higher standard than the AS. An important feature is that the AS can be taken as a stand-alone qualification, without progression to A2. All students seeking a full A level must be assessed at both AS and A2 stages. The A2 assessment includes a substantial weighting (40% of the total marks for A2) of synoptic assessment, in which the ability of students to bring together principles and concepts from different areas of physics, and to use physics skills in contexts which bring together different areas of the subject, are tested. The full A-level award is based on an equally-weighted aggregation of marks from the AS and the A2.

All the new Specifications are based on the Subject Criteria for Physics, published by the Qualifications and Curriculum Authority. The Criteria comprise a core of skills and topics. The subject matter laid down in the Criteria is intended to occupy about 60% of the total subject matter of the A-level course. Certain topics must be taken in the AS part of the course, and others in the A2. Other topics may appear either in the AS or the A2 course. All the new Specifications are presented in modular form.

These substantial changes in both the organisation of the A-level course and the method of assessment require a textbook written closely to the Subject Criteria for Physics.

This book covers all the subject matter of the Criteria, together with a number of additional topics which are common to most of the specifications. The contents have been arranged to follow closely the order of the Criteria. Topics which must be studied in the AS course are presented in a way which should be accessible, in terms of language and mathematical demands, to students who have very recently made the progression from GCSE work. Areas which must appear in the A2 course are treated more rigorously. The way in which the various chapters are linked to the AS and A2 material of the Specification is shown in a Table on page ii. The book covers about 80% of the subject matter of any given Specification. The remaining 20% is likely to differ from one Specification to another, and will probably be presented in the form of Options or Topics. To keep the size of the book manageable, no attempt has been made to cover this optional material.

Key points, definitions and equations are highlighted. Material which, according to the Criteria, must be known, appears in yellow boxes with a red star. After each Section, there is a brief summary of the ground covered. Mathematical derivations have been kept to the minimum required by the Criteria. Where some guidance in mathematics may be helpful, this is provided in a *Maths Note* box. To achieve familiarity through practice, worked examples are provided after each topic. These are followed by similar questions for students to attempt, under the heading *Now it's your turn*. In addition, there are groups of problems and questions after each Section, and selections of examination questions at the end of each Chapter. The final Chapter deals with the important matter of synoptic assessment, with examples of different types of question which may be set.

Chris Mee
Mike Crundell
Brian Arnold
Wendy Brown

March 2000

CHAPTER ONE
Physical quantities and units

Physics is the study of how the world behaves and how the laws of nature operate. Accurate measurement is very important in the development of any science, and this is particularly true of physics. Throughout history, as methods of taking measurements have improved, then new ideas have developed.

In general, physicists begin by observing, measuring and collecting data. These data are then analysed to discover whether they fit into a pattern. If there is a pattern and this pattern can be used to explain other events, it becomes a theory. If the theory predicts events successfully, the theory becomes a law. The process is known as **scientific method** (see Figure 1.1).

This chapter looks at the enormous range of numbers which are met in physics, the units in which quantities are measured and how these are expressed.

A galaxy contains several hundred billion stars with a total mass of 10^{42} kg. It could have a diameter of about 1.2×10^{21} m and be 6×10^{23} m away from Earth. Light would take about 65 million years to reach Earth – dinosaurs were roaming Earth when the light began its journey!

CHAPTER 1

Figure 1.1 *Block diagram to illustrate the process of Scientific Method.*

```
        OBSERVE and MEASURE
                 │
                 ▼
              RECORD
                 │
                 ▼
         DEVELOP THEORY
                 │
                 ▼
MODIFY THEORY ◄── TEST ──► REJECT
                 │
                 ▼
              ACCEPT
```

Brahe (1546–1601) measured the elevations of stars; these days a modern theodolite is used for measuring angular elevation.

1.1 | Quantities and units

A **physical quantity** is a feature of something which can be measured, for example, length, weight or time of fall. Every physical quantity has a numerical value and a unit. If someone says they have a waist measurement of 50, they could be very slim or very fat depending on whether the measurement is in centimetres or inches! Take care – it is vital to give the unit of measurement whenever a quantity is measured or written down.

Large and small quantities are usually expressed in scientific notation i.e. as a simple number multiplied by a power of ten. For example, 0.00034 would be written as 3.4×10^{-4} and 154 000 000 as 1.54×10^{8}. There is far less chance of making a mistake with the number of noughts!

The elephant is large in comparison with the boy but small compared with the jumbo-jet.

The SI System of Units

In very much the same way that languages have developed in various parts of the world, many different systems of measurement have evolved. Just as languages can be translated from one to another, units of measurement can also be converted between systems. Although some conversion factors are easy to remember, some are very difficult. It is much better to have just one system of units. For this reason, scientists around the world use the **Systeme International** (SI System). The SI System of Units is based on the metric system of measurement. Each quantity has just one unit and this unit can have **multiples** and **sub-multiples** to cater for larger or smaller values. The unit is given a **prefix** to denote the multiple or submultiple (see Figure 1.2). For example, one thousandth of a metre is known as a millimetre (mm) and 1.0 millimetre equals 1.0×10^{-3} metres (m).

The mass of this jewel could be measured in kilograms, pounds, carats, grains etc.

Figure 1.2 *The more commonly used prefixes.*

prefix	symbol	multiplying factor
peta	P	10^{15}
tera	T	10^{12}
giga	G	10^{9}
mega	M	10^{6}
kilo	k	10^{3}
deci	d	10^{-1}
centi	c	10^{-2}
milli	m	10^{-3}
micro	μ	10^{-6}
nano	n	10^{-9}
pico	p	10^{-12}
femto	f	10^{-15}
atto	a	10^{-18}

Beware when converting units for areas and volumes!

$$1\,\text{mm} = 10^{-3}\,\text{m}$$

Squaring both sides, $\quad 1\,\text{mm}^2 = (10^{-3})^2\,\text{m}^2 = 10^{-6}\,\text{m}^2$

and $\quad 1\,\text{mm}^3 = (10^{-3})^3\,\text{m}^3 = 10^{-9}\,\text{m}^3.$

Note also that $\quad 1\,\text{cm}^2 = (10^{-2})^2\,\text{m}^2 = 10^{-4}\,\text{m}^2$

and $\quad 1\,\text{cm}^3 = (10^{-2})^3\,\text{m}^3 = 10^{-6}\,\text{m}^3.$

Note: A distance of thirty metres should be written as 30 m and not 30 ms or 30 m s. The letter s is *never* included in a unit for the plural. If a space is left between two letters, the letters denote different units. So, 30 m s would mean thirty metre seconds and 30 ms means 30 milliseconds.

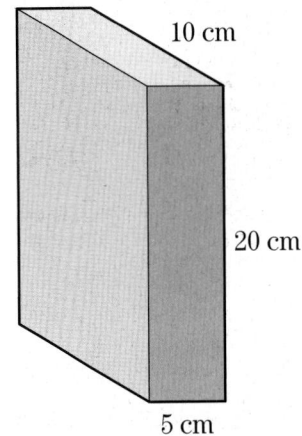

10 cm

20 cm

5 cm

Figure 1.3 *This box has a volume of 1.0×10^3 cm^3 or 1.0×10^6 mm^3 or 1.0×10^{-3} m^3.*

Example

Calculate the number of micrograms in 1.0 milligram.

$$1.0 \text{ g} = 1.0 \times 10^3 \text{ mg}$$
$$\text{and } 1.0 \text{ g} = 1.0 \times 10^6 \text{ micrograms (µg)}.$$
$$\text{So, } 1.0 \times 10^3 \text{ mg} = 1.0 \times 10^6 \text{ µg}$$
$$\text{and } 1.0 \text{ mg} = (1.0 \times 10^6)/(1.0 \times 10^3) = 1.0 \times 10^3 \text{ µg}$$

Now it's your turn

1 Calculate the area, in cm², of the top of a table with sides of 1.2 m and 0.9 m.
Ans: 10 800 cm² = 1.08×10^4 cm²

2 Determine the number of cubic metres in one cubic kilometre.
Ans: 1.0×10^9

Base units

If a quantity is to be measured accurately, the unit in which it is measured must be defined as precisely as possible.

★ The SI System of Units is founded upon seven fundamental or **base units**.

The base quantities and the units with which they are measured are listed in Figure 1.4. For completeness, the candela has been included, but this unit will not be used in the A Level course.

Figure 1.4 *The base quantities.*

quantity	unit	symbol
mass	kilogram	kg
length	metre	m
time	second	s
electric current	ampere (amp)	A
thermodynamic temperature	kelvin	K
amount of substance	mole	mol
luminous intensity	candela	cd

Derived units

All quantities, apart from the base quantities, can be measured using **derived units**.

Derived units consist of some combination of the base units. The base units may be multiplied together or divided into one another, but never added or subtracted.

See Figure 1.5 for examples of derived units. Some quantities have a named unit. For example, the unit of force is the newton, symbol N, but the newton can be expressed in terms of base units (see Figure 1.5). Quantities which do not have a named unit are expressed in terms of other units. For example, specific heat capacity (Chapter 3) is measured in joules per kilogram per kelvin ($J\,kg^{-1}\,K^{-1}$).

quantity	unit	derived unit
frequency	hertz (Hz)	s^{-1}
speed	$m\,s^{-1}$	$m\,s^{-1}$
acceleration	$m\,s^{-2}$	$m\,s^{-2}$
force	newton (N)	$kg\,m\,s^{-2}$
energy	joule (J)	$kg\,m^2\,s^{-2}$
power	watt (W)	$kg\,m^2\,s^{-3}$
electric charge	coulomb (C)	$A\,s$
potential difference	volt (V)	$kg\,m^2\,s^{-3}\,A^{-1}$
electrical resistance	ohm (Ω)	$kg\,m^2\,s^{-3}\,A^{-2}$
specific latent heat	$J\,kg^{-1}\,K^{-1}$	$m^2\,s^{-2}\,K^{-1}$

Figure 1.5 *Some examples of derived units which may be used in the AS course.*

Example

What are the base units of speed?

$$\text{Speed is defined as } \frac{distance}{time} \text{ and the unit is } \frac{m}{s}.$$

Division by a unit is shown using a negative index, that is s^{-1}. The base units of speed are **$m\,s^{-1}$**.

Now it's your turn

Use the information in Figures 1.4 and 1.5 to determine the base units of the following quantities.

1 Density $\left(= \dfrac{mass}{volume} \right)$.

 Ans: $kg\,m^{-3}$

2 Force ($= mass \times acceleration$).

 Ans: $kg\,m\,s^{-2}$

CHAPTER 1

Equations

It is possible to work out the total number of apples in two bags if one bag contains four and the other five (the answer is nine!). This exercise would, of course, be nonsense if one bag contains three apples and the other four pears. In the same way, for any equation to make sense, each term involved in the equation must have the same base units. A term in an equation is a group of numbers and symbols, and each of these terms (or groups) is added to, or subtracted from, other terms. For example, in the equation

$$v = u + at,$$

the terms are v, u, and at.

> In any equation where each term has the same base units, the equation is said to be **homogeneous** or 'balanced'.

In the example above, each term has the base units m s^{-1}. If the equation is not homogeneous, then it is incorrect and is not valid. When an equation is known to be homogeneous, then the balancing of base units provides a means of finding the units of an unknown quantity.

Example

Use base units to show that the following equation is homogeneous.

work done = gain in kinetic energy + gain in gravitational potential energy

The terms in the equation are work, (gain in) kinetic energy, and (gain in) gravitational potential energy.

work done = force $\times$ distance moved

and so the base units are kg m s^{-2} $\times$ m = kg m^2 s^{-2}.

kinetic energy = $\frac{1}{2} \times$ mass $\times$ (speed)^2 and since any pure number such as $\frac{1}{2}$ has no unit, the base units are kg $\times$ (m s^{-1})^2 = kg m^2 s^{-2}.

potential energy = mass $\times$ gravitational field strength g $\times$ distance

Base units are kg $\times$ m s^{-2} $\times$ m = kg m^2 s^{-2}

Conclusion: All terms have the same base units and the equation is homogeneous.

Now it's your turn

1 Use base units to check whether the following equations are balanced:
   (a) *pressure = depth × density × gravitational field strength*
   Ans: yes
   (b) *energy = mass × (speed of light)²*
   Ans: yes

2 The thermal energy Q needed to melt a solid of mass m without any change of temperature is given by the equation

$$Q = mL,$$

   where L is a constant. Find the base units of L.
   Ans: $m^2 s^{-2}$

Section 1.1 summary

★ All quantities have a magnitude (size) and a unit.

★ The SI base units of mass, length, time, electric current, thermodynamic temperature and amount of substance are the kilogram, metre, second, ampere, kelvin and mole respectively.

★ Units of all mechanical, electrical, magnetic and thermal quantities may be derived in terms of these base units.

★ Physical equations must be homogeneous (balanced). Each term in an equation must have the same base units.

Section 1.1 questions

1 Write down, using scientific notation, the values of the following quantities:
   (a) 6.8 pF
   (b) 32 µC
   (c) 60 GW

2 How many electric fires, each rated at 2.5 kW, can be powered from a generator providing 2.0 MW of electric power?

3 An atom of gold, Figure 1.6, has a diameter of 0.26 nm and the diameter of its nucleus is 5.6 fm. Calculate the ratio of the diameter of the atom to that of the nucleus.

4 Determine the base units of the following quantities:
   (a) Energy (= *force × distance*).
   (b) Specific heat capacity
   (*thermal energy change = mass × specific heat capacity × temperature change*)

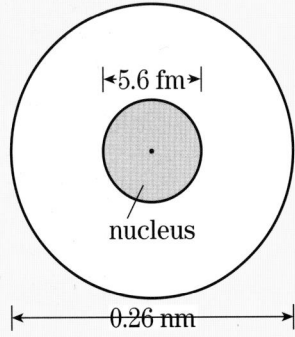
|◄5.6 fm►|
nucleus
|◄——— 0.26 nm ———►|

Figure 1.6 *Simple Bohr model of atom with dimensions as shown in question.*

5   Show that the equation:

$$pressure + \tfrac{1}{2}\,density \times (speed)^2 = constant$$

is homogeneous and find the base units of the constant.

6   The period T of a pendulum of mass M is given by the expression

$$T = 2\pi \sqrt{\frac{I}{Mgh}} \, ,$$

where g is the gravitational field strength and h is a length.

Determine the base units of the constant I.

1.2   Order of magnitude of quantities

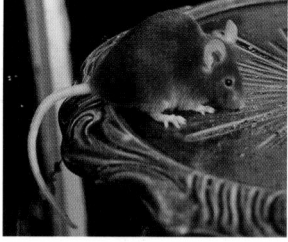

The ratio of their masses is about 10^4. That is minute when the mass of a galaxy is compared with the mass of nucleus (10^{68})!!

It is often useful to be able to estimate the size, or **order of magnitude**, of a quantity. Strictly speaking, the order of magnitude is the power of ten to which the number is raised. The need to be able to estimate is particularly true in a subject like physics where quantities have such widely different values. A *short* distance for an astrophysicist is a light-year (about 9.5×10^{15} m) whereas a *long* distance for a nuclear physicist is 6×10^{-15} m (the approximate diameter of a nucleus)! Figure 1.7 gives some values of distance which may be met in the A Level physics course.

The ability to estimate orders of magnitude is valuable when planning and carrying out experiments or when suggesting theories. Having an idea of the expected result provides a useful check that a gross error has not been made. This is also true when using a calculator. For example, the acceleration of free fall at the Earth's surface is about 10 m s^{-2}. If a value of 9800 m s^{-2} is calculated, then this is obviously wrong and a simple error in the power of ten is likely to be the cause. Similarly, a calculation in which the cost of boiling a kettle-full of water is found to be about £9, rather than about 1 p, may indicate that the energy had been measured in watt-hours rather than kilowatt-hours.

Figure 1.7 *Some values of distance.*

distance from Earth to edge of observable Universe	1.4×10^{26} m
diameter of a galaxy	1.2×10^{21} m
distance from Earth to the Sun	1.5×10^{11} m
distance from London to Paris	3.5×10^5 m
length of a car	4 m
diameter of a hair	5×10^{-4} m
diameter of an atom	3×10^{-10} m
diameter of a nucleus	6×10^{-15} m

Example

It is worthwhile to remember the sizes of some common objects so that comparisons can be made. For example, a jar of jam has a mass of about 500 g and a carton of orange juice has a volume of 1000 cm^3 (1 litre).

Now it's your turn

Estimate the following quantities:

(a) The mass of an apple
(b) The mass of an adult human
(c) The height of a room in a house
(d) The diameter of a pencil
(e) The volume of a pea
(f) The volume of a human head
(g) The speed of a jumbo jet
(h) The temperature of the human body

Ans: (a) 100–150 g; (b) 50–100 kg; (c) 2–3 m; (d) 0.5–1 cm; (e) 0.5 cm^3; (f) 4×10^{-3} m^3; (g) 220 m s^{-1}; (h) 310 K

1.3 Accuracy and precision

Whenever an experimenter makes a measurement, he uses a measuring instrument which will give a reading as close as possible to the true value. This does require skill, and these skills have to be practised, whether it involves the use of something simple such as a metre rule or more complex equipment such as a travelling microscope. If the measurement is repeated, or another person makes the measurement, it is quite likely that different results will be obtained. This is because, in any reading, there is some **uncertainty** in the value. Physical quantities cannot be measured exactly, no matter what measuring device is used. It is possible to measure time intervals to less than one millionth of a second using a cathode-ray oscilloscope, but even then the measurements are not exact.

When a student measures the length of the page of a book, the most likely measuring instrument to be used is a metre rule (or a 30 cm rule). Using the rule, the length may be measured to the nearest millimetre and so the length may be stated to be, for example, 297 ± 1 mm. This means that the student thinks that the most likely length of the page is 297 mm, but it could be as low as 296 mm or as high as 298 mm. The ±1 mm is the uncertainty in the length. Some people refer to this uncertainty as the **error** reading. However, an error implies some sort of a mistake. There is no mistake here, it is just that no measurement can be made with 100% certainty.

A digital meter in use.

It is easy to visualise an uncertainty in the reading of a meter with a scale, but at first sight, digital meters would appear to have little or no uncertainty. However, manufacturers of such instruments usually provide details of the uncertainty in any reading. For example, the uncertainty of a digital milliammeter may be quoted as $\pm 1\% \pm 1$ digit. For a reading of 400 mA, the uncertainty would be:

$$1\% \text{ of } 400 \text{ (i.e. } \pm 4\text{)} \pm 1 = \pm 5,$$

and the reading would be 400 ± 5 mA.

This uncertainty is in the meter. It does not take into account any other uncertainties such as fluctuations in the meter reading.

There are two different types of uncertainty, **systematic** and **random**.

A systematic uncertainty (or 'error') will result in all readings being either too large or too small. This uncertainty cannot be eliminated by taking an average of several values.

Examples of systematic errors include:

- non-zero reading on a meter (a zero error)
- incorrectly calibrated scale
- reaction time of experimenter

Systematic uncertainty is sometimes thought to be found only in measuring instruments but, as can be seen above, it is also associated with poor experimental technique.

Random uncertainty (or 'error') gives rise to a scatter of readings about the true value. The uncertainty can be reduced by repeating readings and taking an average.

Examples of random errors are:

- reading a scale
- taking a reading which changes with time
- counting oscillations
- measuring out a certain volume of liquid.
- reading a scale from the wrong position (a parallax error)

Figure 1.8 *If you look at a clock face at an angle, it appears to give a slightly different time. This is known as a parallax error.*

1.3 Accuracy and precision

Since random errors can be reduced by repeating readings and taking an average, it is good experimental technique to repeat readings whenever possible.

Example

The current in a resistor is to be measured. State one source of uncertainty which is:
(a) systematic,
(b) random.

In both cases suggest how the uncertainty may be reduced.

(a) Systematic uncertainty could be a zero error on the meter (or a wrongly calibrated scale). This can be reduced by checking for a 'zero reading' before starting the experiment (or use two ammeters in series and check that the readings agree).
(b) Random uncertainty could be a parallax error when taking a reading. This can be reduced by the use of a mirror behind the scale and viewing normally.

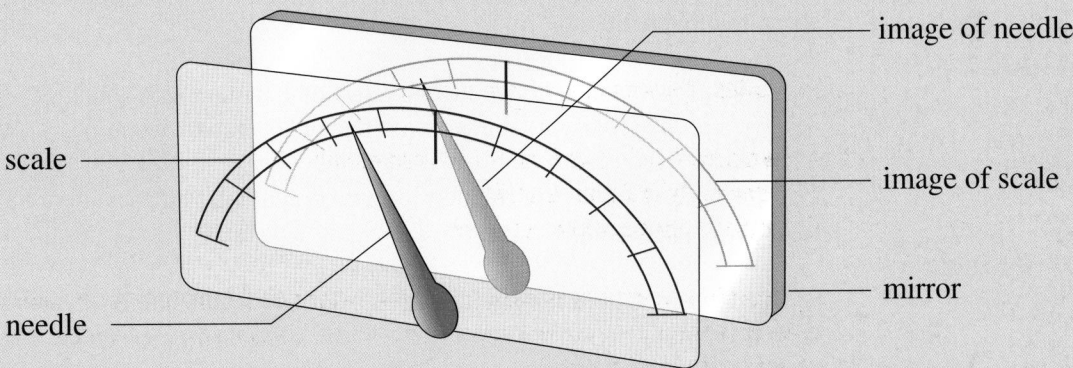

image of needle

scale

image of scale

mirror

needle

Figure 1.9 *Meter with a mirror next to the scale. To avoid parallax errors, the reading is taken from a position such that the image of the pointer is hidden by the pointer itself.*

Now it's your turn

1 The length of a pencil is measured with a 30 cm rule. Suggest one possible source of:
(a) a systematic uncertainty,
(b) a random uncertainty.
In each case, suggest how the uncertainty may be reduced.
Ans: (a) End of ruler damaged. Measure length from 10 cm mark; (b) Parallax error. Place pencil on top of scale.

2 The diameter of a wire is to be measured using a micrometer screw gauge.
(a) Suggest two sources of systematic uncertainty.
(b) Suggest why measurements are taken spirally along the whole length of the wire.
Ans: (a) zero error on drum; dirt between faces gripping the wire; (b) to allow for a non-circular cross-section, moving along its length to allow for tapering.

In question 1, the length of the pencil could have been 136 ± 1 mm. The answer has been given to three significant figures (3 sig. fig.). The number of significant figures quoted for a measurement, or the answer to a question, gives an indication of the uncertainty in that result. To have given the answer to four significant figures (4 sig. fig.), for example, 136.2 mm, would have been pointless because the measurement could be made only to ±1 mm.

> Measurements and numerical answers to questions should always be given to an appropriate number of significant figures, so that the uncertainty in the numerical value may be judged.

A common fault of students is to believe that, because a calculator will give the answer with eight digits, the answer may be written down with eight significant figures. This is untrue. The result can never be more reliable than the data from which it is obtained. The result should be quoted with the same number of significant figures as the data.

There is a problem when writing down a reading such as 470 mm. This could mean that the result is

> either 470 mm, with a possible range of 469 mm to 471 mm,
> or 470 mm, with a possible range of 460 mm to 480 mm.

For this reason, values are frequently given in scientific notation in order to avoid any confusion. So,

4.70×10^2 mm implies a range of values from 4.69×10^2 mm to 4.71×10^2 mm,

and

4.7×10^2 mm implies a range of values from 4.6×10^2 mm to 4.8×10^2 mm.

Example

A marble rolls a distance of 43.7 cm in 4.7 s. Calculate the average speed of the marble.

$$average\ speed = \frac{distance\ travelled}{time\ taken}$$

Using a calculator, $\dfrac{43.7}{4.7}$ is equal to 9.2978723.

The distance is measured to 3 sig. fig. but the time is measured to 2 sig. fig. and so the answer should be quoted to 2 sig. fig.

$$\therefore average\ speed = \mathbf{9.3\ cm\ s^{-1}}$$

Now it's your turn

Calculate the following, giving your answers to an appropriate number of significant figures.

1   The resistance R in ohms (Ω) of a resistor, given that $R = \dfrac{V}{I}$ where $V = 2.1$ V and $I = 0.72$ A.

    Ans: 2.9 Ω

2   The mass M of an aluminium block, given that the density ρ of aluminium is 2.7 g cm^{-3} and the volume V is 16.2 cm^3 ($M = \rho V$).

    Ans: 44 g

3   Using some kitchen scales, the mass of a book is found to be 732 g. A bookmark of mass 2.3 g is put in the book. What is the new reading on the scales?

    Ans: 734 g

The number of significant figures quoted for a measurement gives an indication of the **precision** with which the measurement can be made. A greater number of significant figures implies greater precision. Precision and **accuracy** are often taken to mean the same although, in physics, they are very different. Accuracy is concerned with how close a measurement is to its true value. Thus, the diameter of a wire may be measured with precision using a micrometer screw gauge but, if there is a zero error on the gauge, the result will not be accurate.

Example

The time for 50 oscillations of a pendulum was found to be 57.62 s. Explain why the reading is precise, but the result for the period of the pendulum may not be accurate.

The time has been measured to the nearest one hundredth of a second, which is precise. The result may not be accurate owing to the following:

★   the number of oscillations may not have been counted correctly
★   poor reaction times of the experimenter
★   starting and stopping the timer at the incorrect position in the oscillation

Note: It would be considered to be poor experimental technique to quote the time with such precision knowing that the number of significant figures is not justified by the level of accuracy of the result.

Now it's your turn

Two distances are measured using a metre rule. The distances are 975 mm and 964 mm. Explain why, although the readings are precise, the result when the distances are subtracted may not be accurate.

Ans: Each distance is measured to about 0.1% and may be thought to be precise. Result is 11 mm with uncertainty likely to be ±2 mm i.e. 18%. Probably inaccurate.

Section 1.3 summary

★ No measurement can be made with absolute accuracy; all measurements have an uncertainty.

★ Uncertainties may be either systematic or random.

★ A systematic uncertainty gives rise to all readings being either too high or too low. It cannot be eliminated by averaging.

★ Random uncertainty gives rise to a scatter of readings about the true value. It may be reduced by averaging.

★ The number of significant figures gives an indication of the uncertainty in any particular value.

★ The precision of a reading increases with the number of significant figures with which the result may be quoted.

★ Accuracy is concerned with how close a measurement is to the true value.

Section 1.3 questions

1 The distance of the image of an object from the centre of a convex lens is measured with a metre rule.

State two possible sources of systematic uncertainty and two of random uncertainty and suggest how they might be reduced.

2 A digital meter is quoted as having an uncertainty of $\pm1.5\%$ ±2 digits. The reading on the meter is 56.72. Write down the reading with an appropriate number of significant figures.

3 In an experiment, the time for a ball to fall from a particular height was measured. The measurement was repeated many times and the number of occasions n on which a particular time t was obtained was plotted against t. For each of the graphs in Figure 1.10, state whether the measurements are accurate or precise or both. The true value of the time is T.

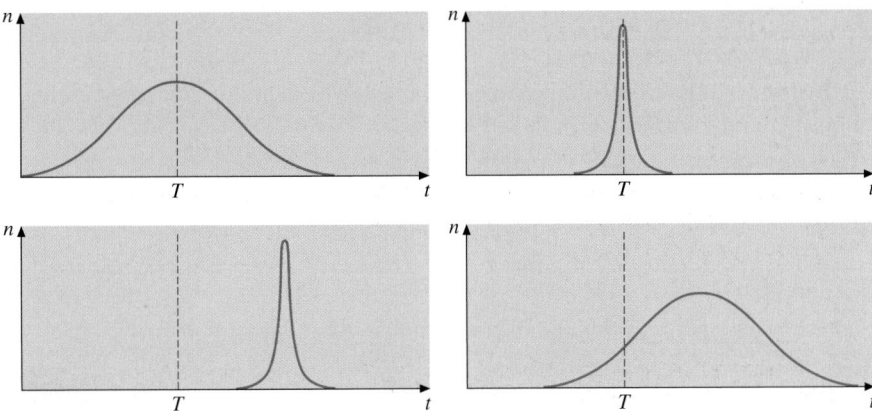

Figure 1.10

1.4 Scalars and vectors

All physical quantities have a magnitude and a unit. For some quantities, magnitude and units do not give us enough information to fully describe the quantity. For example, if we are given the time for which a car travels at a certain speed, then we can calculate the distance travelled. However, we cannot find out how far the car is from its starting point unless we are told the direction of travel. In this case, the speed and direction must be specified.

A quantity which can be described fully by giving its magnitude is known as a **scalar quantity**. A **vector quantity** has magnitude and direction.

Some examples of scalar and vector quantities are given in Figure 1.11.

Although the athlete runs 10 km in the race, his final distance from the starting point may well be zero!

quantity	scalar	vector
mass	✓	
weight		✓
speed	✓	
velocity		✓
force		✓
pressure	✓	
electric current		✓
temperature	✓	

Figure 1.11 Some scalars and vectors.

Example

A 'big wheel' at a fairground has a diameter of 14 m and people on the ride complete one revolution in 24 s. Calculate:
(a) the distance a rider moves in 3.0 minutes,
(b) the distance of the rider from the starting position.

(a) In 3.0 minutes, the rider completes $\dfrac{3.0 \times 60}{24} = 7.5$ revolutions

$$\begin{aligned} \text{distance travelled} &= 7.5 \times \text{circumference of wheel} \\ &= 7.5 \times 2\pi \times 7.0 \\ &= \textbf{330 m} \end{aligned}$$

(b) 7.5 revolutions completed. Rider is $\frac{1}{2}$ revolution from starting point. The rider is at the opposite end of a diameter of the 'big wheel'. So, the distance from starting position = **14 m**

Now it's your turn

1 State whether the following quantities are scalars or vectors:
   (a) time of departure of a train,
   (b) temperature change in a greenhouse,
   (c) gravitational field strength,
   (d) density of a liquid.
   Ans: (a) scalar; (b) vector; (c) vector; (d) scalar

2 A student has a part-time job, earning $3.50 per hour. He works for four hours and then spends $12. Calculate:
   (a) the total amount of money involved (a scalar quantity)
   (b) the balance (a vector).
   Ans: (a) $26; (b) $2.0 credit

Vector representation

When you hit a tennis ball, you have to judge the direction you want it to move in, as well as how hard to hit it. The force you exert is therefore a vector quantity and cannot be represented by a number alone. One way to represent a vector is by means of an arrow. The direction of the arrow is the direction of the vector quantity. The length of the arrow, drawn to scale, represents its magnitude. This is illustrated below.

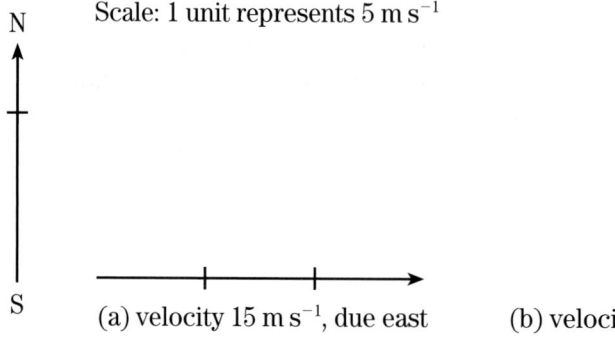

Scale: 1 unit represents 5 m s^{-1}

(a) velocity 15 m s^{-1}, due east   (b) velocity 10 m s^{-1}, due south

Section 1.4 summary

★ A scalar quantity has magnitude only.

★ A vector has magnitude and direction.

★ A vector quantity may be represented by an arrow, with the length of the arrow drawn to scale to give the magnitude.

Section 1.4 questions

1 State whether the following quantities are scalars or vectors:

   (a) movement of the hands of a clock,
   (b) frequency of vibration,
   (c) flow of water in a pipe.

2 Speed and velocity have the same units. Explain why speed is a scalar quantity whereas velocity is a vector quantity.

3 A student states that a bag of sugar has a weight of 10 N and that this weight is a vector quantity. Suggest whether the student is correct when stating that weight is a vector.

4 Explain how an arrow may be used to represent a vector quantity.

1.5 Addition of vectors

The addition of two scalar quantities which have the same unit is no problem. The quantities are added using the normal rules of addition. For example, a beaker of volume 250 cm^3 and a bucket of volume 9.0 litres have a total volume of 9250 cm^3.

Adding together two vectors is more difficult because they have direction as well as magnitude. If the two vectors are in the same direction, then they can simply be added together. Two objects of weight 50 N and 40 N have a combined weight of 90 N because both weights act in the same direction (vertically downwards). Figure 1.12 shows the effect of adding two forces of magnitudes 30 N and 20 N which act in the same direction or in opposite directions. The angle between the forces is 0° when they act in the same direction and 180° when they are in opposite directions. For all other angles between the directions of the forces, the combined effect, or **resultant**, is some value between 10 N and 50 N.

Figure 1.12 Vector addition.

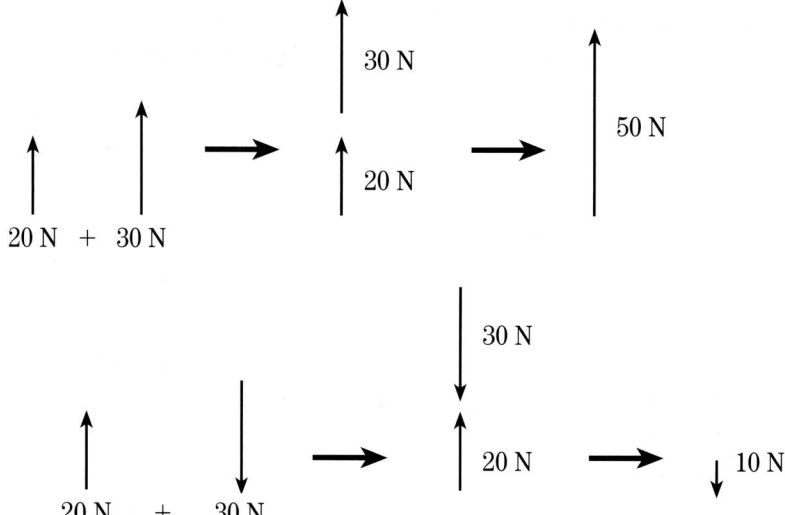

In cases where the vectors do not act in the same or opposite directions, the resultant is found by means of a **vector triangle**. Each one of the two vectors V_1 and V_2 is represented in magnitude and direction by the side of a triangle. Note that both vectors must be in either a clockwise or an anticlockwise direction (see Figure 1.13). The combined effect, or resultant R, is given in magnitude and direction by the third side of the triangle. It is important to remember that, if V_1 and V_2 are drawn clockwise, then R is anticlockwise and *vice versa*.

The resultant may be found by means of a scale diagram. Alternatively, having drawn a sketch of the vector triangle, the problem may be solved using trigonometry.

See Maths Note on page 23

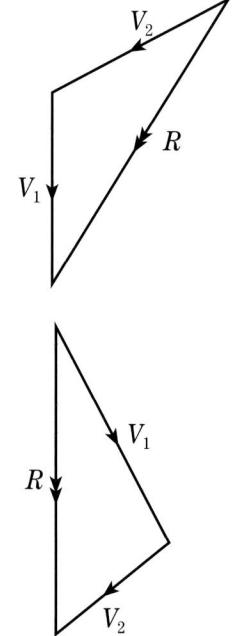

Figure 1.13 Vector triangles.

Example

A ship is travelling due north with a speed of 12 km h^{-1} relative to the water. There is a current in the water flowing at 4.0 km h^{-1} in an easterly direction relative to the shore. Determine the velocity of the ship relative to the shore by:
(a) scale drawing,
(b) calculation.

(a) By scale drawing:

Scale: 1 cm represents 2 km h^{-1}

resultant R

The velocity relative to the shore is:

$6.3 \times 2 = $ **12.6 km h^{-1}** in a direction **18° east of north**

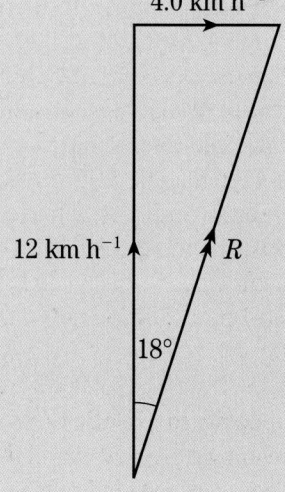

1.5 Addition of vectors

(b) By calculation:
Referring to the diagram and using Pythagoras' theorem,

$$R^2 = 12^2 + 4^2 = 160$$

$$R = \sqrt{160} = 12.6$$

$$\tan \alpha = \frac{4}{12} = 0.33$$

$$\alpha = 18.4°$$

The velocity of the ship relative to the shore is **12.6 km h^{-1}** in a direction **18.4° east of north**

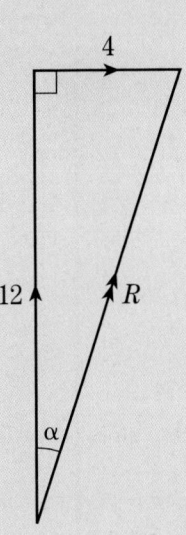

Now it's your turn

1 A swimmer who can swim in still water at a speed of 4 km h^{-1} is swimming in a river. The river flows at a speed of 3 km h^{-1}. Calculate the speed of the swimmer relative to the river bank when she swims
   (a) downstream,
   (b) upstream.
   Ans: (a) 7 km h^{-1}; (b) 1 km h^{-1}

2 For the following diagram, draw to scale a vector triangle to determine the resultant of the two forces.
   Check your answer by calculating the resultant.
   Ans: 11 N at an angle to the 6.0 N force of 56°, in an anticlockwise direction.

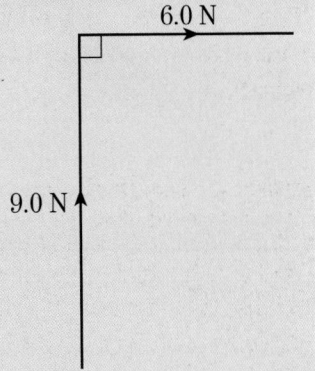

The use of a vector triangle for finding the resultant can be demonstrated by means of a simple laboratory experiment. A weight is attached to each end of a flexible thread and the thread is then suspended over two pulleys, as shown in Figure 1.14. A third weight is attached to a point P near the centre of the thread. The string moves over the pulleys and then comes to rest. The positions of the threads are marked on a piece of paper held on a board behind the threads. This is easy to do if light from a small lamp is shone at the board. Having noted the sizes W_1 and W_2 of the weights on the ends of the thread, a vector triangle can then be drawn on the paper, as shown in Figure 1.15. The resultant of W_1 and W_2 is found to be equal in magnitude but opposite in direction to the weight W_3. If this were not so, there would be a resultant force at P and the thread and weights would move. The use of a vector triangle is justified.

Figure 1.14 *Apparatus to check the use of vector triangle.*

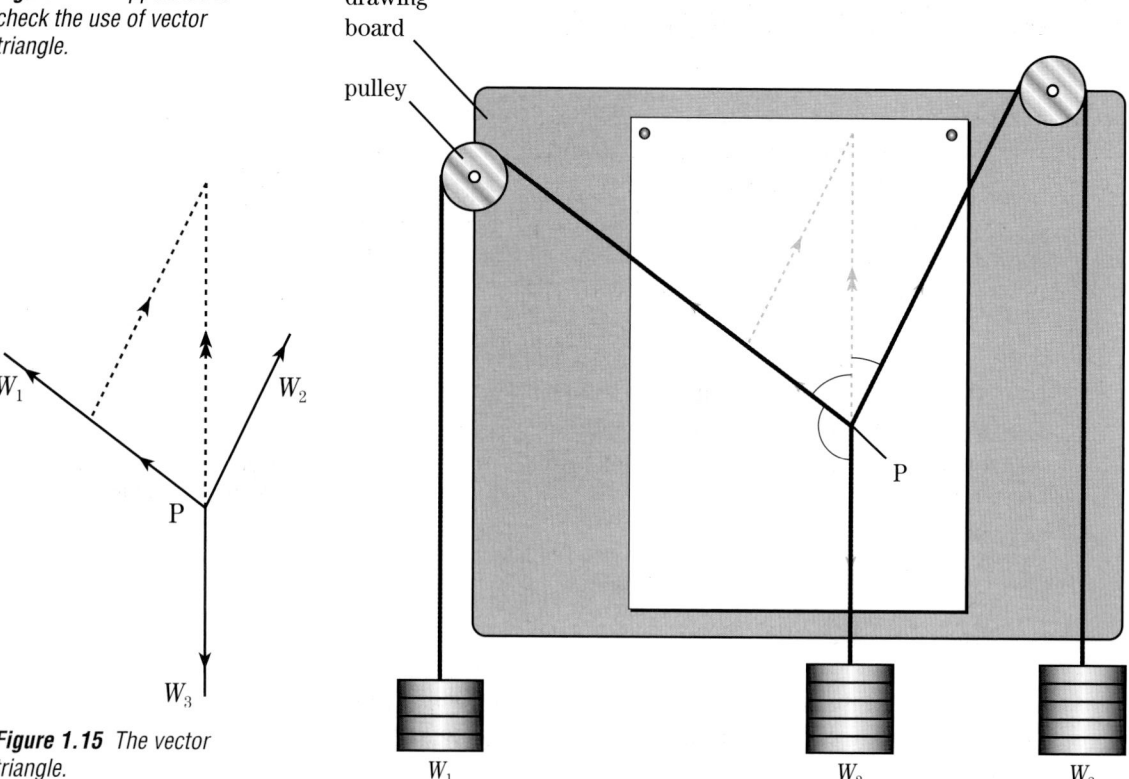

Figure 1.15 *The vector triangle.*

Section 1.5 summary

★ The combined effect of two (or more) vectors is known as the resultant.

★ Two vectors may be added using a vector triangle.

★ The resultant may be found using a scale diagram of the vector triangle or by calculation.

Section 1.5 questions

1 Two forces are of magnitude 450 N and 240 N respectively.
   Determine:

   (a) the maximum magnitude of resultant force,
   (b) the minimum magnitude of the resultant force,
   (c) the resultant force when the forces act at right-angles to each other.

   Use a vector diagram and then check your result by calculation.

2 A boat can be rowed at a speed of 7.0 km h^{-1} in still water. A river flows at a constant speed of 1.5 km h^{-1}. Use a scale diagram to determine the angle to the bank at which the boat must be rowed in order that the boat travels directly across the river.

3 Two forces act at a point P as shown below. Draw a vector diagram, to scale, to determine the resultant force. Check your work by calculation.

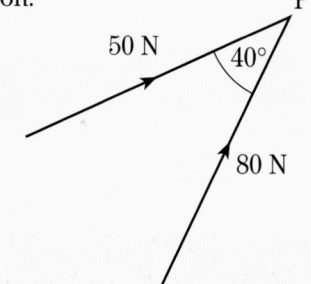

1.6 Resolution of vectors

In section 1.5, we saw that two vectors may be added together to produce a single resultant. This resultant behaves in the same way as the two individual vectors. It follows that a single vector may be split up, or **resolved**, into two vectors, or **components**. The combined effect of the components is the same as the original vector. In later chapters, we will see that resolution of a vector into two perpendicular components is a very useful means of solving certain types of problem.

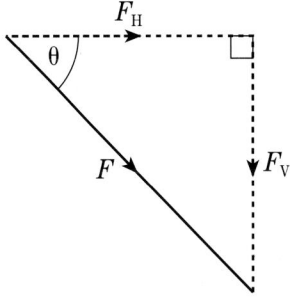

Consider a force of magnitude F acting at an angle of θ below the horizontal (see Figure 1.16). A vector triangle can be drawn with a component F_H in the horizontal direction and a component F_V acting vertically. Remembering that F, F_H and F_V form a right-angled triangle, then

Figure 1.16 *The resolving of a vector into components.*

$$F_H = F \cos \theta$$

and $\quad F_V = F \sin \theta.$

The force F has been resolved into two perpendicular components, F_H and F_V. The example chosen is concerned with forces, but the method applies to all types of vector quantity.

Example

A glider is launched by an aircraft with a cable, as shown in the photograph. At one particular moment, the tension in the cable is 620 N and the cable makes an angle of 25° with the horizontal. Calculate:
(a) the force pulling the glider horizontally,
(b) the vertical force exerted by the cable on the nose of the glider.

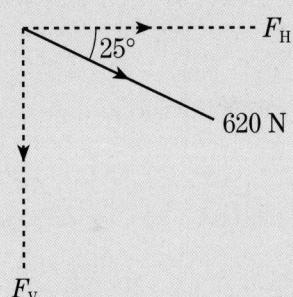

A glider being launched by means of a plane.

(a) horizontal component $F_H = 620 \cos 25 = \mathbf{560\ N}$
(b) vertical component $F_V = 620 \sin 25 = \mathbf{260\ N}$

Now it's your turn

1 An aircraft is travelling $35°$ east of north at a speed of 310 km h^{-1}. Calculate the speed of the aircraft in:
   (a) the northerly direction,
   (b) the easterly direction.
   Ans: (a) 250 km h^{-1}; (b) 180 km h^{-1}

2 A cyclist is travelling down a hill at a speed of 9.2 m s^{-1}. The hillside makes an angle of $6.3°$ with the horizontal. Calculate, for the cyclist:
   (a) the vertical speed,
   (b) the horizontal speed.
   Ans: (a) 1.0 m s^{-1}; (b) 9.1 m s^{-1}

Section 1.6 summary

★ A single vector may be divided into two separate components.

★ The dividing of a vector into components is known as the resolution of a vector.

★ In general, a vector is resolved into two components at right-angles to each other.

Exam Questions

1 (a) i) Explain what is meant by a *base unit*.
     ii) Give four examples of base units.
   (b) State what is meant by a *derived* unit.
   (c) i) For any equation to be valid, it must be homogeneous. Explain what is meant by a *homogeneous* equation.
     ii) The pressure p of an ideal gas of density ρ is given by the equation

$$p = \frac{1}{3}\rho\langle c^2 \rangle$$

   where $\langle c^2 \rangle$ is the mean square speed (i.e. it is a quantity measured as $[\text{speed}]^2$). Use base units to show that the equation is homogeneous.

2 (a) Determine the base units of:
     i) work done,
     ii) the moment of a force.
   (b) Explain why your answers to (a) mean that caution is required when the homogeneity of an equation is being tested.

3 (a) Distinguish between a *scalar* and a *vector* quantity.
   (b) A mass of weight 120 N is hung from two strings, as shown below.

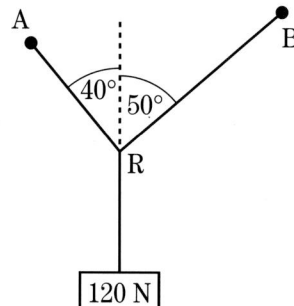

   Determine, by scale drawing or by calculation, the tension in:
     i) RA,   ii) RB.
   (c) Use your answers in (b) to determine the horizontal component of the tension in:
     i) RA,   ii) RB.
   Comment on your answer.

4 A fielder in a cricket match throws the ball to the wicket-keeper. At one moment of time, the ball has a horizontal velocity of 16 m s^{-1} and a velocity in the vertically upward direction of 8.9 m s^{-1}.

(a) Determine, for the ball:
   i)   its resultant speed,
   ii) the direction in which it is travelling relative to the horizontal.

(b) During the flight of the ball to the wicket-keeper, the horizontal velocity remains unchanged. The speed of the ball at the moment when the wicket-keeper catches it is 19 m s^{-1}. Calculate, for the ball just as it is caught:
   i)   its vertical speed,
   ii) the angle that the path of the ball makes with the horizontal.

(c) Suggest with a reason whether the ball, at the moment it is caught, is rising or falling.

For any triangle,

$$\frac{a}{\sin A} = \frac{b}{\sin B} = \frac{c}{\sin C}$$

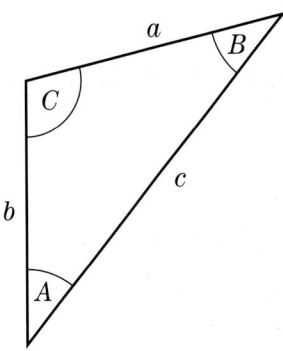

sine rule

and $a^2 = b^2 + c^2 - 2bc \cos A$

$b^2 = a^2 + c^2 - 2ac \cos B$

$c^2 = a^2 + b^2 - 2ab \cos C$

cosine rule

For a right-angled triangle

$$h^2 = o^2 + a^2$$

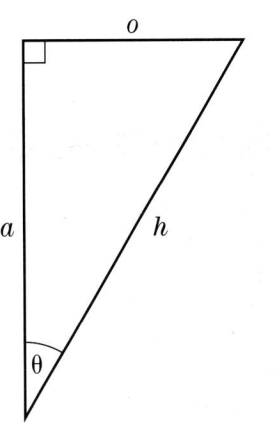

Pythagoras' theorem

$$\sin \theta = \frac{o}{h}$$

$$\cos \theta = \frac{a}{h}$$

$$\tan \theta = \frac{o}{a}$$

CHAPTER TWO
Kinematics

Free-fall parachutists fall through distances of many hundreds of metres before opening their parachutes. Their motion starts off with an acceleration of nearly 10 m s^{-2}, but air resistance progressively reduces this. When they open their parachutes, there is a sudden deceleration and they move with almost constant speed by the time they approach the ground.

The aim of this chapter is to describe motion in terms of quantities such as position, speed, velocity and acceleration. To make matters easy, we shall talk about the motion of a particle, rather than of a real body. The simplification of dealing with a particle – a body with no size – is one that is adopted in many branches of physics.

2.1 Average speed

When talking about motion we shall discuss the way in which the position of a particle varies with time. Think about a particle moving along a straight line. In a certain time, the particle will cover a certain distance. The average speed of the particle is defined as the distance covered divided by the time taken. Written as a word equation, this is:

$$\bigstar \quad average\ speed = \frac{distance\ covered}{time\ taken}$$

The unit of speed is the metre per second (m s^{-1}).

One of the most fundamental of physical constants is the speed of light in a vacuum. It is important because it is used in the definition of the metre, and because, according to the theory of Relativity, it defines an upper limit to attainable speeds. The range of speeds that you are likely to come across is enormous; some are summarised in Figure 2.1.

Light	$3.0 \times 10^8 \text{ m s}^{-1}$
Electron round nucleus	$2.2 \times 10^6 \text{ m s}^{-1}$
Earth round Sun	$3.0 \times 10^3 \text{ m s}^{-1}$
Jet airliner	$2.5 \times 10^2 \text{ m s}^{-1}$
Motorway speed limit (70 m.p.h.)	31 m s^{-1}
Sprinter	10 m s^{-1}
Walking speed	1.5 m s^{-1}
Snail	$1 \times 10^{-3} \text{ m s}^{-1}$

Figure 2.1 *Examples of speeds.*

It is important to recognise that speed has a meaning only if it is quoted relative to a fixed reference. In most cases, speeds are quoted relative to the surface of the Earth, which, although is moving relative to the solar system, is often taken to be fixed. Thus, when we say that a bird can fly at a certain average speed, we are relating its speed to the Earth. However, a passenger on a ferry may see that a seagull, flying parallel to the boat, appears to be practically stationary. If this is the case, the seagull's speed relative to the boat is zero. However, if the speed of the boat through the water is 8 m s^{-1}, then the speed of the seagull relative to Earth is also 8 m s^{-1}. When talking about relative speeds we must be careful about directions. It is easy if the motions are in the same direction, like in the example of the ferry and the seagull. You have already met this in Section 1.5, the addition of vectors.

Examples

1 The radius of the Earth is 6.4×10^6 m; one revolution about its axis takes 24 hours (8.6×10^4 s). Calculate the average speed of a point on the Equator relative to the centre of the Earth.

   In 24 hours, the point on the Equator completes one revolution and travels a distance of $2\pi \times$ the Earth's radius, that is $2\pi \times 6.4 \times 10^6 = 4.0 \times 10^7$ m. The average speed is (distance covered)/(time taken), or $4.0 \times 10^7/8.6 \times 10^4 = \mathbf{4.7 \times 10^2 \text{ m s}^{-1}}$.

2 How far does a cyclist travel in 11 minutes if his average speed is 22 km hr^{-1}?

First convert the average speed in km hr^{-1} to a value in m s^{-1}. 22 km (2.2×10^4 m) in 1 hour (3.6×10^3 s) is an average speed of 6.1 m s^{-1}. 11 minutes is 660 s. Since average speed is (distance covered)/(time taken), the distance covered is given by (average speed) × (time taken), or $6.1 \times 660 = \textbf{4000 m}$.
Note the importance of working in consistent units: this is why the average speed and the time were converted to m s^{-1} and s respectively.

3 A train is travelling at a speed of 25 m s^{-1} along a straight track. A boy walks along the aisle, towards the rear of the train, at a speed of 1 m s^{-1} relative to the train. What is his speed relative to Earth?

In one second, the train travels 25 m forwards along the track. In the same time the boy moves 1 m towards the rear of the train, so he has moved 24 m along the track. His speed relative to Earth is thus $25 - 1 = \textbf{24 m s}^{-1}$.

Now it's your turn

1 The speed of an electron in orbit about the nucleus of a hydrogen atom is 2.2×10^6 m s^{-1}. It takes 1.5×10^{-16} s for the electron to complete one orbit. Find the radius of the orbit.
Ans: 5.3×10^{-11} m

2 The average speed of an airliner on a domestic flight is 220 m s^{-1}. How long will it take to fly between two airports on a flightpath 700 km long?
Ans: 3200 s or 53 min

3 Two cars are travelling in the same direction on a long, straight road. The one in front has an average speed of 25 m s^{-1} relative to Earth; the other's is 31 m s^{-1}, also relative to Earth. What is the speed of the second car relative to the first when it is overtaking?
Ans: 6 m s^{-1}

2.2 Speed and velocity

In ordinary language, there is no difference between the terms *speed* and *velocity*. However, in physics there is an important distinction between the two. The term *velocity* is used to represent a vector quantity: the magnitude of how fast a particle is moving, and the direction in which it is moving. *Speed* does not have an associated direction. It is a scalar quantity (see Section 1.4).

So far, we have talked about the total distance travelled by a body. Like speed, distance is a scalar quantity, because we do not have to specify the direction in which the distance is travelled. However, in defining velocity we introduce a quantity called *displacement*. Displacement of a particle is its change of position. Consider a cyclist travelling 500 m along a straight road, and then turning round and coming back 300 m. The total distance travelled is 800 m, but the displacement is only 200 m, since the cyclist has ended up 200 m from the starting point.

The average velocity is defined as the displacement divided by the time taken.

$$\bigstar \quad average \; velocity = \frac{displacement}{time \; taken}$$

Because distance and displacement are different quantities, the average speed of motion will sometimes be different from the magnitude of the average velocity. If the time taken for the cyclist's trip in the example above is 120 s, the average speed is $800/120 = 6.7$ m s^{-1}, whereas the magnitude of the average velocity is $200/120 = 1.7$ m s^{-1}. This may seem confusing, but the difficulty arises only when the motion involves a change of direction and we take an average value. If we are interested in describing the motion of a particle at a particular moment in time, the speed at that moment is the same as the magnitude of the velocity at that moment.

We now need to define average velocity more precisely, in terms of a mathematical equation, instead of our previous word equation. Suppose that at time t_1 a particle is at a point x_1 on the x-axis (Figure 2.2). At a later time t_2, the particle has moved to x_2. The displacement (the change in position) is $(x_2 - x_1)$, and the time taken is $(t_2 - t_1)$. The average velocity $\bar{v}$ is then

Figure 2.2

$$\bigstar \quad \bar{v} = \frac{x_2 - x_1}{t_2 - t_1}.$$

The bar over v is the symbol meaning 'average'. As a shorthand, we can write $(x_2 - x_1)$ as Δx, where Δ (the Greek capital letter delta) means 'the change in'. Similarly, $t_2 - t_1$ is written as Δt. This gives us

$$\bigstar \quad \bar{v} = \frac{\Delta x}{\Delta t}.$$

If x_2 were less than x_1, $(x_2 - x_1)$ and Δx would be negative. This would mean that the particle had moved to the left, instead of to the right as in Figure 2.2. The sign of the displacement gives the direction of particle motion. If Δx is negative, then the average velocity v is also negative. The sign of the velocity, as well as the sign of the displacement, indicates the direction of the particle's motion. This is because both displacement and velocity are vector quantities.

2.3 Describing motion by graphs

Position-time graphs

Figure 2.3 is a graph of position x against time t for a particle moving in a straight line. This curve gives a complete description of the motion of the particle. We can see from the graph that the particle starts at the origin O (at which $x = 0$) at time $t = 0$. From O to A the graph is a straight line: the particle is covering equal distances in equal periods of time. This represents a period of *uniform velocity*. The average velocity during this time is just $(x_1 - 0)/(t_1 - 0)$. Clearly, this is just the gradient of the straight-line part of the graph between O and A. Between A and B the particle is slowing down, because the distances travelled in equal periods of time are getting smaller. The average velocity during this period is $(x_2 - x_1)/(t_2 - t_1)$. On the graph, this is represented by the gradient of the straight line joining A and B. At B, for a moment, the particle is at rest, and after B it has reversed its direction and is heading back towards the origin. Between B and C the average velocity is $(x_3 - x_2)/(t_3 - t_2)$. Because x_3 is less than x_2, this is a negative quantity, indicating the reversal of direction.

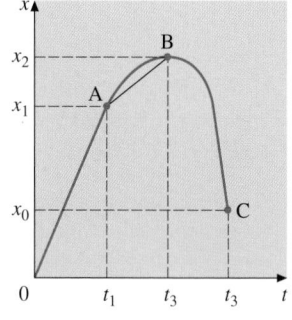

Figure 2.3

Calculating the average velocity of the particle over the relatively long intervals t_1, $(t_2 - t_1)$ and $(t_3 - t_2)$ will not, in fact, give us the complete description of the motion that was claimed at the beginning of this section. To describe the motion exactly, we need to know the particle's velocity at every instant. We introduce the idea of **instantaneous velocity**. To define instantaneous velocity we make the intervals of time over which we measure the average velocity shorter and shorter. This has the effect of approximating the curved displacement-time graph by a series of short straight-line segments. The approximation becomes better the shorter the time interval, as illustrated in Figure 2.4. Eventually, in the case of extremely small time intervals (mathematically we would say 'infinitesimally small'), the straight-line segment has the same direction as the tangent to the curve. This limiting case gives the instantaneous velocity as the slope of the tangent to the displacement-time curve.

Figure 2.4

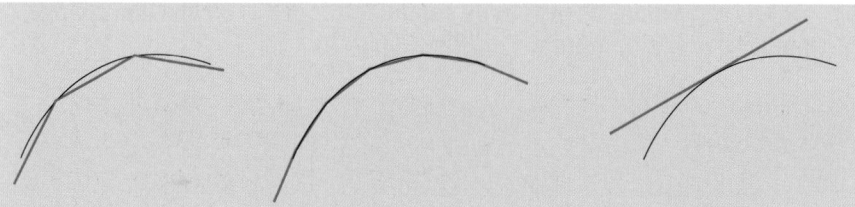

Displacement-time and velocity-time graphs

Figure 2.5 is a sketch graph showing how the displacement of a car, travelling along a straight test track, varies with time. We interpret this graph in a descriptive way by noting that between O and A the distances travelled

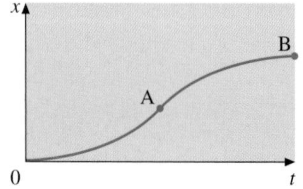

Figure 2.5

2.3 Describing motion by graphs

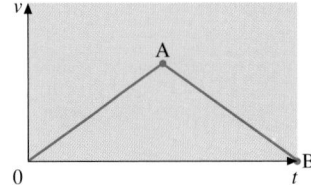
in equal intervals of time are progressively increasing: that is, the velocity is increasing as the car is accelerating. Between A and B the distances for equal time intervals are decreasing; the car is slowing down. Finally, there is no change in position, even though time passes, so the car must be at rest. We can use Figure 2.5 to deduce the details of the way in which the car's instantaneous velocity v varies with time. To do this, we draw tangents to the curve in Figure 2.5 at regular intervals of time, and measure the slope of each tangent to obtain values of v. The plot of v against t gives the graph in Figure 2.6. This confirms our descriptive interpretation: the velocity increases from zero to a maximum value, and then decreases to zero again. We will look at this example in more detail on page 37.

Figure 2.6

2.4 Acceleration

We have used the word *accelerating* in describing the increase in velocity of the car in the previous section. Acceleration is a measure of the rate at which the velocity of a particle is changing. Average acceleration is defined by the word equation

$$\star \quad average\ acceleration = \frac{change\ in\ velocity}{time\ taken}\ .$$

The unit of acceleration is the unit of velocity (the metre per second) divided by the unit of time (the second), giving the metre per second per second (m s^{-2}). In symbols, this equation is:

$$\star \quad \bar{a} = \frac{v_2 - v_1}{t_2 - t_1} = \frac{\Delta v}{\Delta t},$$

where v_1 and v_2 are the velocities at times t_1 and t_2 respectively. To obtain the *instantaneous acceleration*, we take extremely small time intervals, just as we did when defining instantaneous velocity. Because it involves a change in velocity (a vector quantity), acceleration is also a vector quantity: we need to specify both its magnitude and its direction.

We can deduce the acceleration of a particle from its velocity-time graph by drawing a tangent to the curve and finding the slope of the tangent. Figure 2.7 shows the result of doing this for the car's motion described by Figures 2.5 (the displacement-time graph) and 2.6 (the velocity-time graph). The car accelerates at a constant rate between O and A, and then decelerates (that is, slows down) uniformly between A and B.

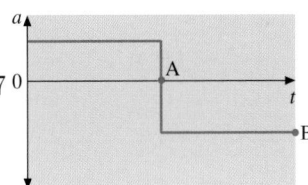

Figure 2.7

An acceleration with a very familiar value is the acceleration of free fall near the Earth's surface: this is $9.8\ \text{m s}^{-2}$, often approximated to $10\ \text{m s}^{-2}$. To illustrate the range of values you may come across, some accelerations are summarised in Figure 2.8.

Figure 2.8 *Examples of accelerations.*

Due to circular motion of electron around nucleus	9×10^{26} m s^{-2}
Car crash	1×10^{3} m s^{-2}
Free fall on Earth	10 m s^{-2}
Family car	2 m s^{-2}
Free fall on Moon	2 m s^{-2}
At equator, due to rotation of Earth	3×10^{-2} m s^{-2}
Due to circular motion of Earth around Sun	6×10^{-5} m s^{-2}

Examples

1   A sports car accelerates along a straight test track from rest to 70 km h^{-1} in 6.3 s. What is its average acceleration?

First convert the data into consistent units. 70 km (7.0×10^4 m) in 1 hour (3.6×10^3 s) is 19 m s^{-1}. Since average acceleration is (change of velocity)/(time taken), the acceleration is 19/6.3 = **3.0 m s^{-2}**.

2   A railway train, travelling along a straight track, takes 1.5 minutes to come to rest from a speed of 115 km h^{-1}. What is its average acceleration while braking?

115 km hr^{-1} is 31.9 m s^{-1}, and 1.5 minutes is 90 s. The average acceleration is (change of velocity)/(time taken) = $-31.9/90$ = **-0.35 m s^{-2}**. Note that the acceleration is a negative quantity because the change of velocity is negative – the final velocity is less than the initial. A negative acceleration is often called a deceleration.

Now it's your turn

1   A sprinter, starting from the blocks, reaches his full speed of 9.0 m s^{-1} in 1.5 s. What is his average acceleration?
Ans: 6.0 m s^{-2}

2   A car is travelling at a speed of 25 m s^{-1} on the motorway. At this speed, it is capable of accelerating at 1.8 m s^{-2}. How long would it take to accelerate from 25 m s^{-1} to the motorway speed limit of 31 m s^{-1}?
Ans: 3.3 s

Sections 2.1–2.4 summary

★   Kinematics is the description of how objects move.

★   Average speed is defined by (*distance covered*)/(*time taken*).

★ Speed is a scalar quantity, and is described by magnitude only. Velocity is a vector, and requires magnitude and direction. Speed is the magnitude of velocity. Displacement is the change in position of a particle, and is a vector quantity.

★ Average velocity is defined by (*displacement*)/(*time taken*) or $\Delta x/\Delta t$.

★ The instantaneous velocity is the average velocity measured over an infinitesimally short time interval.

★ Average acceleration is defined by (*change in velocity*)/(*time taken*) or $\Delta v/\Delta t$.

★ Acceleration is a vector. Instantaneous acceleration is the average acceleration measured over an infinitesimally short time interval.

★ Velocity may be deduced from the gradient of a displacement-time graph, and acceleration from the gradient of a velocity-time graph.

Sections 2.1–2.4 questions

1   At an average speed of 24 km h^{-1}, how many kilometres will a cyclist travel in 75 minutes?

2   An aircraft travels 1600 km in 2.5 hours. What is its average speed, in m s^{-1}?

3   Does a car speedometer register speed or velocity? Explain.

4   An aircraft travels 1400 km at a speed of 700 km h^{-1}, and then runs into a headwind that reduces its speed over the ground to 500 km h^{-1} for the next 800 km. What is the total time for the flight? What is the average speed of the aircraft?

5   A sports car can stop in 6.1 s from a speed of 110 km h^{-1}. What is its acceleration?

6   Can the velocity of a particle change if its speed is constant? Can the speed of a particle change if its velocity is constant? If the answer to either question is 'yes', give examples.

2.5 Uniformly accelerated motion

Having defined displacement, velocity and acceleration, we shall use the definitions to derive a series of equations, called the *kinematic equations*, which can be used to give a complete description of the motion of a particle in a straight line. The mathematics will be simplified if we deal with situations in which the acceleration does not vary with time, that is, the acceleration is uniform (or constant). This approximation applies for many practical cases. However, there are two important types of motion for which the kinematic equations do not apply: circular motion, and the oscillatory motion called simple harmonic motion. We shall deal with these separately in Chapter 9.

Think about a particle moving along a straight line with constant acceleration a. Suppose that its initial velocity, at time $t = 0$, is u. After a further time t its velocity has increased to v. From the definition of acceleration as (change in velocity)/(time taken), we have $a = (v - u)/t$, or, re-arranging

$$v = u + at.$$

From the definition of average velocity $\bar{v}$ as (distance travelled)/(time taken), over the time t the distance travelled will be given by the average velocity multiplied by the time taken, or

$$s = \bar{v}t.$$

The average velocity $\bar{v}$ is written in terms of the initial velocity u and final velocity v as

$$\bar{v} = \frac{u + v}{2}$$

and, using the previous equation for v,

$$\bar{v} = (u + u + at)/2 = u + at/2.$$

Substituting this we have

$$s = ut + \tfrac{1}{2}at^2.$$

The right-hand side of this equation is the sum of two terms. The ut term is the distance the particle would have travelled in time t if it had been travelling with a constant speed u, and the $\tfrac{1}{2}at^2$ term is the additional distance travelled as a result of the acceleration.

The equation relating the final velocity v, the initial velocity u, the acceleration a and the distance travelled s is

$$v^2 = u^2 + 2as.$$

If you wish to see how this is obtained from previous equations, see the Maths Note.

2.5 Uniformly accelerated motion

From $v = u + at$, $t = (v - u)/a$.

Substitute this in $s = ut + \frac{1}{2}at^2$: $s = u(v - u)/a + \frac{1}{2}a(v - u)^2/a^2$

Multiplying both sides by $2a$ and expanding the terms,

$$2as = 2uv - 2u^2 + v^2 - 2uv + u^2$$

or

$$v^2 = u^2 + 2as.$$

The four equations relating the various quantities which define the motion of the particle in a straight line in uniformly-accelerated motion are:

★ $v = u + at$

$s = ut + \frac{1}{2}at^2$

$v^2 = u^2 + 2as$

$\bar{v} = (u + v)/2.$

In these equations u is the initial velocity, v is the final velocity, $\bar{v}$ is the average velocity, a is the acceleration, s is the distance travelled, and t is the time taken.

In solving problems involving kinematics, it is important to understand the situation before you try to substitute numerical values into an equation. Identify the quantity you want to know, and then make a list of the quantities you know already. This should make it obvious which equation is to be used.

Free fall acceleration

A very common example of uniformly accelerated motion is when a body falls freely near the Earth's surface. Because of the gravitational attraction of the Earth, all objects fall with the same uniform acceleration. This acceleration is called the *acceleration of free fall*, and is represented by the symbol g. It has a value of 9.81 m s^{-2}, and is directed downwards. Strictly speaking, we ought to qualify this statement by saying that the fall must be in the absence of air resistance, but in most situations this is taken for granted.

Until the sixteenth century the idea of the acceleration of a falling body was not fully appreciated. It was commonly thought that heavier bodies fell faster than light ones. This idea was a consequence of the effect of air resistance on light objects with a large surface area, such as feathers. However, Galileo Galilei (1564–1642) suggested that, in the absence of resistance, all bodies would fall with the same constant acceleration. He showed mathematically

Strobo-flash photograph of objects in free fall.

Galileo in his study.

Leaning Tower of Pisa.

that, for a body falling from rest, the distance travelled is proportional to the square of the time. Galileo tested the relation experimentally by timing the fall of objects from various levels of the Leaning Tower of Pisa. This is the relation we have derived as $s = ut + \frac{1}{2}at^2$. For a body starting from rest, $v = 0$ and $s = \frac{1}{2}at^2$, that is, the distance is proportional to time squared.

We have mentioned that in most situations air resistance can be neglected. In fact, there are some applications in which this resistance is most important. One is the case of the fall of a parachutist, where air resistance plays a vital part. The velocity of a body falling through a resistive fluid (a liquid or a gas) does not increase indefinitely, but eventually reaches a maximum velocity, called the **terminal velocity**. The force due to air resistance increases with speed. When this resistive force has reached a value equal and opposite to the weight of the falling body, the body no longer accelerates and continues at uniform velocity. This is a case of motion with non-uniform acceleration. The acceleration starts off with a value of g, but decreases to zero at the time when the terminal velocity is achieved. Thus, raindrops and parachutists are normally travelling at a constant speed by the time they approach the ground.

Parachutist about to land.

Examples

1 A car increases its speed from 25 m s^{-1} to 31 m s^{-1} with a uniform acceleration of 1.8 m s^{-2}. How far does it travel while accelerating?

   In this problem we want to know the distance s. We know the initial speed $u = 25$ m s^{-1}, the final speed $v = 31$ m s^{-1}, and the acceleration $a = 1.8$ m s^{-2}. We need an equation linking s with u, v and a: this is $v^2 = u^2 + 2as$. Substituting the values, we have $31^2 = 25^2 + 2 \times 1.8\,s$. Re-arranging, $s = (31^2 - 25^2)/2 \times 1.8 = $ **93 m.**

2 The average acceleration of a sprinter from the time of leaving the blocks to reaching his maximum speed of 9.0 m s^{-1} is 6.0 m s^{-2}. For how long does he accelerate? What distance does he cover in this time?

In the first part of this problem we want to know the time t. We know the initial speed $u = 0$, the final speed $v = 9.0$ m s^{-1}, and the acceleration $a = 6.0$ m s^{-2}. We need an equation linking t with u, v and a: this is $v = u + at$. Substituting the values, we have $9.0 = 0 + 6.0t$. Re-arranging, $t = 9.0/6.0 = \mathbf{1.5\ s}$.

For the second part of the problem, we want to know the distance s. We know the initial speed $u = 0$, the final speed $v = 9.0$ m s^{-1}, and the acceleration $a = 6.0$ m s^{-2}; we have also just found the time $t = 1.5$ s. There is a choice of equations linking s with u, v, a and t. We can use $s = ut + \frac{1}{2}at^2$. Substituting the values, $s = 0 + \frac{1}{2} \times 6.0 \times (1.5)^2 = \mathbf{6.8\ m}$. Another relevant equation is $v = \Delta x/\Delta t$. Here the average velocity $\bar{v}$ is given by $\bar{v} = (u + v) = 9.0/2 = 4.5$ m s^{-1}. $\Delta x/\Delta t$ is the same as s/t, so $4.5 = s/1.5$, and $s = 4.5 \times 1.5 = \mathbf{6.8\ m}$ as before.

3 A cricketer throws a ball vertically upward into the air with an initial velocity of 18.0 m s^{-1}. How high does the ball go? How long is it before it returns to the cricketer's hands?

In the first part of the problem we want to know the distance s. We know the initial velocity $u = 18.0$ m s^{-1} upwards and the acceleration $a = g = 9.81$ m s^{-2} downwards. At the highest point the ball is momentarily at rest, so the final velocity $v = 0$. The equation linking s with u, v and a is $v^2 = u^2 + 2as$. Substituting the values, $0 = (18.0)^2 + 2(-9.81)s$. Thus $s = -(18.0)^2/2(-9.81) = \mathbf{16.5\ m}$. Note that here the ball has an upward velocity but a downward acceleration, and that at the highest point the velocity is zero but the acceleration is not zero.

In the second part we want to know the time t for the ball's up-and-down flight. We know u and a, and also the overall displacement $s = 0$, as the ball returns to the same point at which it was thrown. The equation to use is $s = ut + \frac{1}{2}at^2$. Substituting the values, $0 = 18.0t + \frac{1}{2}(-9.81)t^2$. Doing some algebra, $t(36.0 - 9.81t) = 0$. There are two solutions, $t = 0$ and $t = 36.0/9.81 = 3.7$ s. The $t = 0$ value corresponds to the time when the displacement was zero when the ball was on the point of leaving the cricketer's hands. The answer required here is $\mathbf{3.7\ s}$.

Now it's your turn

1 An airliner must reach a speed of 110 m s^{-1} to take off. If the available length of the runway is 2.4 km and the aircraft accelerates uniformly from rest at one end, what minimum acceleration must be available if it is to take off?

Ans: 2.5 m s^{-2}

2   A speeding motorist passes a traffic policeman on a stationary motor-cycle. The policeman immediately gives chase: his uniform acceleration is 4.0 m s^{-2}, and by the time he draws level with the motorist he is travelling at 30 m s^{-1}. How long does it take for the motor-cyclist to catch the car? If the car continues to travel at a steady speed during the chase, what is that speed?

Ans: 7.5 s; 15 m s^{-1}

3   A cricket ball is thrown vertically upwards with a speed of 15.0 m s^{-1}. What is its velocity when it first passes through a point 8.0 m above the cricketer's hands?

Ans: 8.2 m s^{-1} upwards

2.6 | Graphs of the kinematic equations

It is often useful to represent the motion of a particle graphically, instead of by means of a series of equations. In this section we bring together the graphs which correspond to the equations we have already derived. We shall see that there are some important links between the graphs.

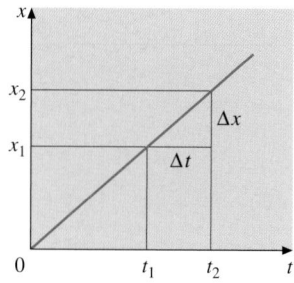

Figure 2.9

First, think about a particle moving in a straight line with constant velocity. Constant velocity means that the particle covers equal distances in equal intervals of time. A graph of displacement against time is thus a straight line, as in Figure 2.9. Here the particle has started at $x = 0$ and at time $t = 0$. The slope of the graph is equal to the magnitude of the velocity, since, from the definition of average velocity, $\bar{v} = (x_2 - x_1)/(t_2 - t_1) = \Delta x/\Delta t$. Because this graph is a straight line, the average velocity and the instantaneous velocity are the same. The equation describing the graph is $x = vt$.

Now think about a particle moving in a straight line with constant acceleration. The particle's velocity will change by equal amounts in equal intervals of time. A graph of the magnitude v of the velocity against time t will be a straight line, as in Figure 2.10. Here the particle has started with velocity u at time $t = 0$. The slope of the graph is equal to the magnitude of the acceleration. The graph is a straight line showing that the acceleration is a constant. The equation describing the graph is $v = u + at$.

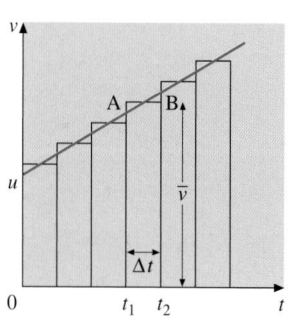

Figure 2.10

An important feature of the velocity-time graph is that we can deduce the displacement of the particle by calculating the area between the graph and the t-axis, between appropriate limits of time. Suppose we want to obtain the displacement of the particle between times t_1 and t_2 in Figure 2.10. Between these times the average velocity $\bar{v}$ is represented by the horizontal line AB. The area between the graph and the t-axis is equal to the area of the rectangle topped by the average velocity $\bar{v}$. This area is $\bar{v}\Delta t$. But, by the

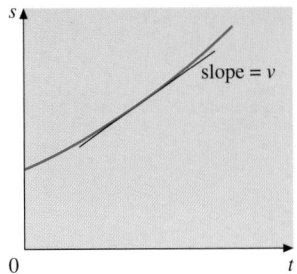
definition of average velocity ($\bar{v} = \Delta x/\Delta t$), $\bar{v}\Delta t$ is equal to the displacement Δx during the time interval Δt.

We can deduce the graph of displacement s against time t from the velocity-time graph by calculating the area between the graph and the t-axis for a succession of value of t. As shown in Figure 2.10, we can split the area up into a number of rectangles. The displacement at a certain time is then just the sum of the areas of the rectangles up to that time. Figure 2.11 shows the result of plotting the displacement s determined in this way against time t. It is a curve with a slope which increases the higher the value of t, indicating that the particle is accelerating. The slope at a particular time gives the magnitude of the instantaneous velocity. The equation describing Figure 2.11 is $s = ut + \frac{1}{2}at^2$.

Figure 2.11

Figure 2.12

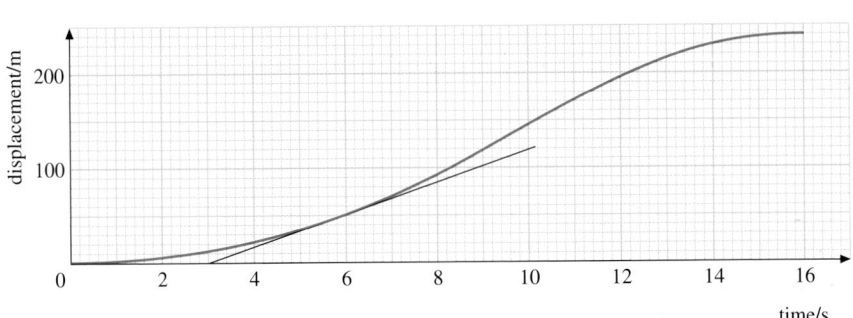

time/s

Example

The displacement-time graph for a car on a straight test track is shown in Figure 2.12.

Use this graph to draw velocity-time and acceleration-time graphs for the test run.

We have already met this graph when we discussed the concepts of velocity and acceleration. In Figure 2.12 it has been re-drawn to scale, and figures have been put on the displacement and time axes. We find the magnitude of the velocity by measuring the gradient of the displacement-time graph. As an example, a tangent to the graph has been drawn at $t = 6.0$ s. The slope of this tangent is 18 m s^{-1}. If the process is repeated at different times, the following velocities are determined:

t/s	2	4	6	8	10	12	14	16
v/m s^{-1}	6	12	18	24	30	20	10	0

These values are plotted on the velocity-time graph of Figure 2.13. Check some of the values by drawing tangents yourself! Hint: when drawing tangents, use a mirror or a transparent ruler.

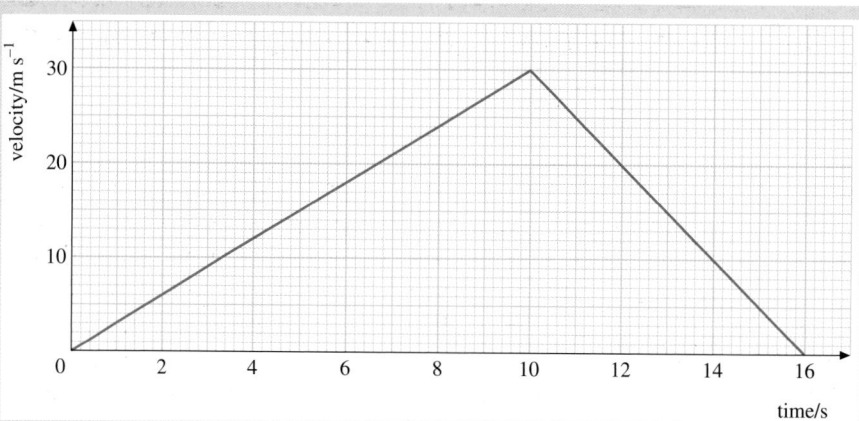

Figure 2.13

This graph shows two straight-line portions. Initially, from $t = 0$ to $t = 10$ s, the car is accelerating uniformly, and from $t = 10$ s to $t = 16$ s it is decelerating. The acceleration is given by $a = \Delta v/\Delta t = 30/10 = 3$ m s^{-2} up to $t = 10$ s. Beyond $t = 10$ s the acceleration is $-30/6 = -5$ m s^{-2}. (The minus sign shows that the car is decelerating). The acceleration-time graph is plotted in Figure 2.14.

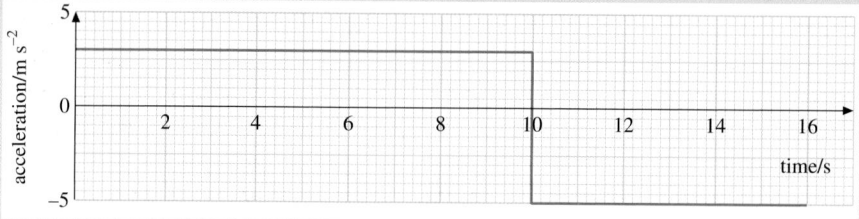

Figure 2.14

Finally, we can confirm that the area under a velocity-time graph gives the displacement. The area of the triangle in Figure 2.13 is $\frac{1}{2} \times 16 \times 30 = 240$ m, the value of s at $t = 16$ s on Figure 2.12.

Now it's your turn

In a test of a sports car on a straight track, the following readings of velocity v were obtained at the times t stated:

t/s	0	5	10	15	20	25	30	35
v/m s^{-1}	0	15	23	28	32	35	37	38

On graph paper, draw a velocity-time graph and use it to determine the acceleration of the car at time $t = 5$ s. Find also the total distance travelled between $t = 0$ and $t = 30$ s.
Ans: 1.8 m s^{-2}; 770 m

Note: these figures refer to a case of non-uniform acceleration, which is more realistic than the previous worked example. However, the same rules apply: the acceleration is given by the slope of the velocity-time graph at the relevant time, and the distance travelled can be found from the area under the graph.

2.7 Two-dimensional motion under a constant force

So far we have been dealing with motion along a straight line, that is, one-dimensional motion. We now think about the motion of particles moving in paths in two dimensions. We shall need to make use of ideas we have already met regarding vectors in Chapter 1. The particular example we shall take is where a particle moves in a plane under the action of a constant force. An example is the motion of a ball thrown at an angle to the vertical, or an electron moving at an angle to an electric field. In the case of the ball, the constant force acting on it is its weight. For the electron, the constant force is the force provided by the electric field.

Cricket fielder returning ball.

This topic is often called **projectile motion**. Galileo first gave an accurate analysis of this motion. He did so by splitting the motion up into its vertical and horizontal components, and considering these separately. The key is that the two components can be considered independently.

As an example, think about a particle sent off in a horizontal direction and subject to a vertical gravitational force (its weight). As before, air resistance will be neglected. We will analyse the motion in terms of the horizontal and vertical components of velocity. The particle is projected at time $t = 0$ at the origin of a system of x, y co-ordinates (Figure 2.15) with velocity u_x in the x-direction. Think first about the particle's vertical motion (in the y-direction). Throughout the motion it has an acceleration of g (the acceleration of free fall) in the y-direction. The initial value of the vertical component of velocity is $u_y = 0$. The vertical component increases continuously under the uniform acceleration g. Using $v = u + at$, its value v_y at time t is given by $v_y = gt$. Also at time t, the vertical displacement y downwards is given by $y = \frac{1}{2}gt^2$. Now for the horizontal motion (in the x-direction): here the acceleration is zero, so the horizontal component of velocity remains constant at u_x. At time t the horizontal displacement x is given by $x = u_x t$. To find the velocity of the particle at any time t, the two components v_x and v_y must be added vectorially. The direction of the resultant vector is the direction of motion of the particle. The curve traced out by a particle subject to a constant force in one direction is a *parabola*.

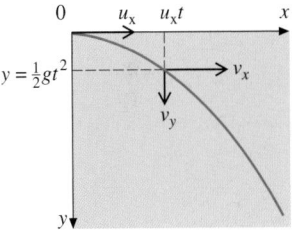

Figure 2.15

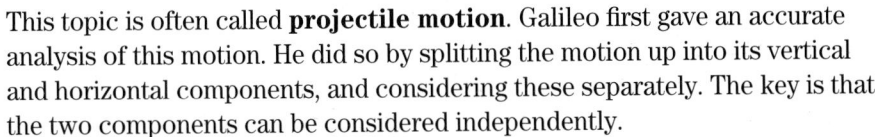

Water jets from a garden sprinkler showing a parabola.

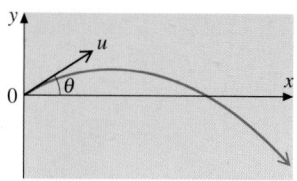

Figure 2.16

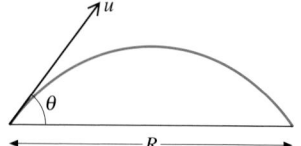

Figure 2.17

If the particle had been sent off with velocity u at an angle θ to the horizontal, as in Figure 2.16, the only difference to the analysis of the motion is that the initial y-component of velocity is $u_y \sin \theta$. In the example illustrated in Figure 2.16 this is upwards. Because of the downwards acceleration g, the y-component of velocity decreases to zero, at which time the particle is at the crest of its path, and then increases in magnitude again but this time in the opposite direction. The path is again a parabola.

For the particular case of a particle projected with velocity u at an angle θ to the horizontal from a point on level ground (Figure 2.17), the range R is defined as the distance from the point of projection to the point at which the particle reaches the ground again. We can show that R is given by

$$R = \frac{(u^2 \sin 2\theta)}{g}.$$

For details, see the Maths Note.

Maths Note

Suppose that the particle is projected from the origin ($x = 0$, $y = 0$). We can interpret the range R as being the horizontal distance x travelled at the time t when the value of y is again zero. The equation which links displacement, initial speed, acceleration and time is $s = ut + \frac{1}{2}at^2$. Adapting this for the vertical component of the motion, we have

$$0 = (u \sin \theta)t - \tfrac{1}{2}gt^2.$$

The two solutions of this equation are $t = 0$ and $t = (2u \sin \theta)/g$. The $t = 0$ case is when the particle was projected; the second is when it returns to the ground at $y = 0$. We use this second value of t with the horizontal component of velocity $u \cos \theta$ to find the distance x travelled (the range R). This is

$$x = R = (u \cos \theta)t = (2u^2 \sin \theta \cos \theta)/g.$$

There is a trigonometric relation $\sin 2\theta = 2 \sin \theta \cos \theta$, use of which puts the range expression in the required form

$$R = (u^2 \sin 2\theta)/g.$$

We can see that R will have its maximum value for a given speed of projection u when $\sin 2\theta = 1$, or $2\theta = 90°$, or $\theta = 45°$. The value of this maximum range is $R_{max} = u^2/g$.

Examples

1   A stone is thrown from the top of a vertical cliff, 45 m high above level
    ground, with an initial velocity of 15 m s^{-1} in a horizontal direction
    (Figure 2.18). How long does it take to reach the ground? How far
    from the base of the cliff is it when it reaches the ground?

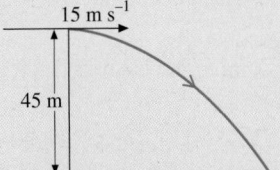

Figure 2.18

To find the time t for which the stone is in the air, work with the
vertical component of the motion, for which we know that the initial
component of velocity is zero, the displacement $y = 45$ m, and the
acceleration a is 9.81 m s^{-2}. The equation linking these is $y = \frac{1}{2}gt^2$.
Substituting the values, we have $45 = \frac{1}{2} \times 9.81t^2$. This gives
$t = (2 \times 45/9.81)^{1/2} = \mathbf{3.0}$ **s**.

For the second part of the question, we need to find the horizontal
distance x travelled in the time t. Because the horizontal component of
the motion is not accelerating, x is given simply by $x = u_x t$.
Substituting the values, we have $x = 15 \times 3.0 = \mathbf{45}$ **m**.

2   An electron, travelling with a velocity of 2.0×10^7 m s^{-1} in a horizontal
    direction, enters a uniform electric field. This field gives the electron a
    constant acceleration of 5.0×10^{15} m s^{-2} in a direction perpendicular
    to its original velocity (Figure 2.19). The field extends for a horizontal
    distance of 60 mm. What is the magnitude and direction of the velocity
    of the electron when it leaves the field?

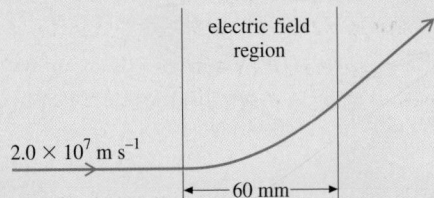

Figure 2.19

The horizontal motion of the electron is not accelerated. The time t
spent by the electron in the field is given by $t = x/u_x$
$= 60 \times 10^{-3}/2.0 \times 10^7 = 3.0 \times 10^{-9}$ s. When the electron enters the
field, its vertical component of velocity is zero; in time t, it has been
accelerated to $v_y = at = 5.0 \times 10^{15} \times 3.0 \times 10^{-9} = 1.5 \times 10^7$ m s^{-1}.
When the electron leaves the field, it has a horizontal component of
velocity $v_x = 2.0 \times 10^7$ m s^{-1}, unchanged from the initial value u_x. The
vertical component is $v_y = 1.5 \times 10^7$ m s^{-1}. The resultant velocity v is
given by $v = (v_x^2 + v_y^2)^{1/2} = [(2.0 \times 10^7)^2 + (1.5 \times 10^7)^2]^{1/2}$
$= \mathbf{2.5 \times 10^7}$ **m s**$^{-1}$. The direction of this resultant velocity makes an
angle θ to the horizontal, where θ is given by
$\tan θ = v_y/v_x = 1.5 \times 10^7/2.0 \times 10^7$. The angle θ is **36.9°**.

Now it's your turn

1   A ball is thrown horizontally from the top of a tower 30 m high and lands 15 m from its base (Figure 2.20). What is the ball's initial speed?
Ans: 6.1 m s^{-1}

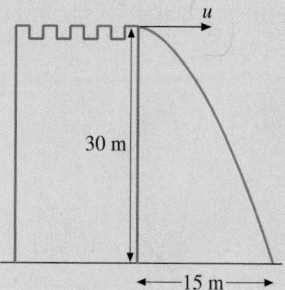

Figure 2.20

2   A football is kicked on level ground at a velocity of 15 m s^{-1} at an angle of $30°$ to the horizontal (Figure 2.21). How far away is the first bounce?
Ans: 20 m

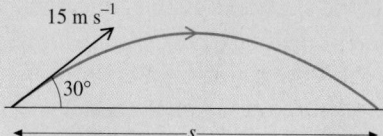

Figure 2.21

Sections 2.6–2.8 summary

★   The kinematic equations for a body moving in a straight line with uniform acceleration are:

$$v = u + at$$
$$s = ut + \tfrac{1}{2}at^2$$
$$v^2 = u^2 + 2as$$
$$\bar{v} = (u + v)/2$$

★   Objects falling freely near the surface of the Earth in the absence of air resistance, experience the same acceleration, the acceleration of free fall g, which has the value $g = 9.81 \text{ m s}^{-2}$.

★   The gradient of a displacement-time graph gives the velocity of a particle, and the gradient of a velocity-time graph gives its acceleration. The area between a velocity-time graph and the time axis gives the displacement.

★   The motion of projectiles is analysed in terms of two independent motions at right angles. The horizontal component of the motion is at a constant velocity, while the vertical motion is subject to a constant acceleration g.

Sections 2.5–2.7 questions

1 A car accelerates from 5.0 m s^{-1} to 20 m s^{-1} in 6.0 s. Assuming uniform acceleration, how far does it travel in this time?

2 If a raindrop were to fall from a height of 1 km, with what velocity would it hit the ground if there were no air resistance?

3 Traffic police can estimate the speed of vehicles involved in accidents by the length of the marks made by skidding tyres on the road surface. It is known that the maximum deceleration that a car can attain when braking on a normal road surface is about 9 m s^{-2}. In one motorway accident, the tyre-marks were found to be 125 m long. Estimate the speed of the vehicle before braking.

4 A lift is travelling upwards at constant speed. As it passes the second floor, a passenger drops coin A through an old-fashioned gate grille. At exactly the same time, a person standing on the second-floor landing drops coin B through the gate grille at the landing. Which coin, A or B (if either), reaches the bottom of the lift shaft first? Which (if either) has the greater speed on impact?

5 William Tell was faced with the agonising task of shooting an apple from his son Jemmy's head. Assume that William is placed 25 m from Jemmy; his crossbow fires a bolt with an initial speed of 45 m s^{-1}. The crossbow and apple are on the same horizontal line. At what angle to the horizontal should William aim so that the bolt hits the apple?

6 The position of a sports car on a straight test track is monitored by taking a series of photographs at fixed time intervals. The following record of position x was obtained at the stated times t:

t/s	0	0.5	1.0	1.5	2.0	2.5	3.0	3.5	4.0	4.5	5.0
x/m	0	0.4	1.8	4.2	7.7	12.4	18.3	25.5	33.9	43.5	54.3

On graph paper, draw a graph of x against t. Use your graph to obtain values of the velocity v of the car at a number of values of t. Draw a second graph of v against t. From this graph, what can you deduce about the acceleration of the car?

Exam Questions

1 In a driving manual, it is suggested that, when driving at 13 m s^{-1} (about 30 m.p.h.), a driver should always keep a minimum of two car-lengths between the driver's car and the one in front.

(a) Suggest a scientific justification for this safety tip, making reasonable assumptions about the magnitudes of any quantities you need.

(b) How would you expect the length of this 'exclusion zone' to depend on speed for speeds higher than 13 m s^{-1}?

2 A student, standing on the platform at Cambridge railway station, notices that the first two carriages of an arriving train pass her in 2.0 s, and the next two in 2.4 s. The train is decelerating uniformly. Each carriage is 20 m long. When the train stops, the student is opposite the last carriage. How many carriages are there in the train?

CHAPTER 2

3 A ball is to be kicked so that, at the highest point of its path, it just clears a horizontal cross-bar on a pair of goal-posts. The ground is level and the cross-bar is 2.5 m high. The ball is kicked from ground level with an initial velocity of 8.0 m s^{-1}.

   (a) Calculate the angle of projection of the ball and the distance of the point where the ball was kicked from the goal-line.

   (b) Also calculate the horizontal velocity of the ball as it passes over the cross-bar.

   (c) For how long is the ball in the air before it reaches the ground on the far side of the cross-bar?

4 An athlete competing in the long jump leaves the ground at an angle of 28° and makes a jump of 7.40 m.

   (a) Calculate the speed at which the athlete took off.

   (b) If the athlete had been able to increase this speed by 5%, what percentage difference would this have made to the length of the jump?

5 A hunter, armed with a bow and arrow, takes direct aim at a monkey hanging from the branch of a tree. At the instant that the hunter releases the arrow, the monkey takes avoiding action by releasing its hold on the branch. By setting up the relevant equations for the motion of the monkey and the motion of the arrow, show that the monkey was mistaken in its strategy.

CHAPTER THREE
Work, energy and power

Machines do work for us. They change energy from one form into some other more useful form. This chapter deals with the subject of energy in its various forms. Not only is the availability of useful forms of energy important, but also the rate at which it can be converted from one form to another. The rate of converting energy or using energy is known as power.

In developing and developed countries, the demand for energy is always increasing. Think what life would be like without any electrical energy – no radios, TV, electric lighting, washing machines etc. In the U.K. we take it for granted that electrical energy is always available on demand.

3.1 Work

The word 'work' is in use in everyday English language but it does have a variety of meanings. In physics, the word **work** has a definite meaning. The vagueness of the term 'work' in everyday speech causes problems for some students when they come to give a precise scientific definition of work.

'I'm going to work today.'

'Where do you work?'

'I've done some work in the garden.'

'Lots of work was done lifting the box.'

'I've done my homework.'

The weight-lifter uses a lot of energy to lift the weights but they can be rolled along the ground with little effort.

Work is done when a force moves the point at which it acts (the point of application) in the direction of the force.

★ *work done = force × distance moved by force in the direction of the force*

It is very important to include direction in the definition of work done. A car can be pushed horizontally quite easily but, if the car is to be lifted off its wheels, much more work has to be done and a machine, such as a car-jack, is used!

When a force moves its point of application in the direction of the force, the force does work and the work done *by* the force is said to be *positive*. Conversely, if the direction of the force is *opposite* to the direction of movement, work is done *on* the force. This work is then said to be *negative*. This is illustrated in Figure 3.1.

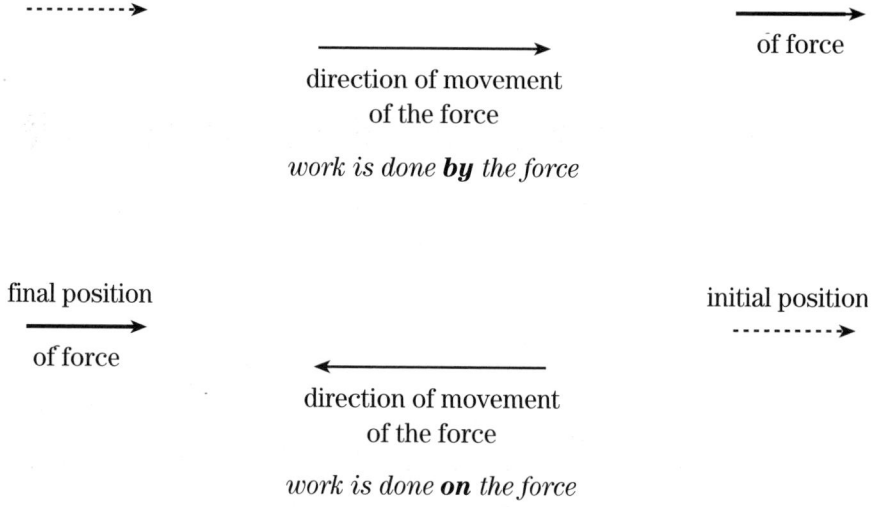

Figure 3.1

An alternative name for distance moved in a particular direction is **displacement**. Displacement is a vector quantity, as is force. However, work done has no direction, only size, and is a scalar quantity. It is measured in joules (J).

★ When a force of one newton moves its point of application by one metre in the direction of the force, one joule of work is done.

work done in joules = force in newtons × distance moved in metres in the direction of the force

It follows that a joule (J) may be said to be a newton-metre (N m). If the force and the displacement are not both in the same direction, then the component of the force in the direction of the displacement must be found. Consider a force F acting along a line at an angle θ to the displacement, as shown in Figure 3.2. The component of the force along the direction of the displacement is $F \cos \theta$.

$$\text{work done for displacement } x = F \cos \theta \times x$$
$$= Fx \cos \theta.$$

Note that the component $F \sin \theta$ of the force is at right angles to the displacement. Since there is no displacement in the direction of this component, no work is done.

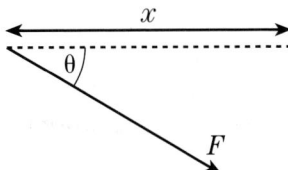

Figure 3.2

The useful work done by the horse is found using the component along the canal of the tension in the rope.

Example

A child tows a toy by means of a string as shown below.

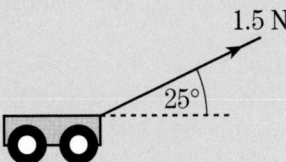

The tension in the string is 1.5 N and the string makes an angle of $25°$ with the horizontal. Calculate the work done in moving the toy horizontally through a distance of 265 cm.

$$work\ done = horizontal\ component\ of\ tension \times distance\ moved$$

$$= 1.5 \cos 25° \times \frac{265}{100}$$

$$= \textbf{3.6 J}$$

Now it's your turn

1 A box weighs 45 N. Calculate the work done in lifting the box through a vertical height of:
   (a) 4.0 m,
   (b) 67 cm.
   Ans: (a) 180 J; (b) 30 J

2 A force of 36 N acts at an angle of 55° to the vertical. The force moves its point of application by 64 cm in the direction of the force. Calculate the work done by:
   (a) the horizontal component of the force,
   (b) the vertical component of the force.
   Ans: (a) 15.5 J; (b) 7.6 J

Work done by an expanding gas

A building can be demolished with explosives. When the explosives are detonated, large quantities of gas at high pressure are produced. As the gas expands, it does work by breaking down the masonry. In this section, we will derive an equation for the work done when a gas changes its volume.

Explosives produce large quantities of high-pressure gas. When the gas expands, it does work in demolishing the building.

Consider a gas contained in a cylinder by means of a frictionless piston of area A, as shown in Figure 3.3. The pressure p of the gas in the cylinder is equal to the atmospheric pressure outside the cylinder. This pressure may be thought to be constant.

Since $pressure = \dfrac{force}{area}$, the gas produces a force F on the piston given by

$$F = pA$$

When the gas expands at constant pressure, the piston moves outwards through a distance x. So,

$$work\ done\ by\ the\ gas = force \times distance\ moved$$
$$W = pAx.$$

However, Ax is the increase in volume of the gas ΔV. Hence,

$$W = p\Delta V.$$

When the volume of a gas changes at constant pressure,

★ *work done = pressure × change in volume*

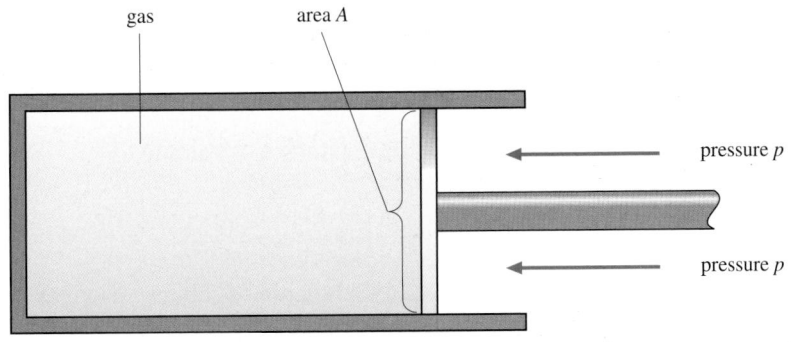

gas   area A

pressure p

pressure p

Figure 3.3

When the gas *expands*, work is done *by* the gas. If the gas *contracts*, then work is done *on* the gas.

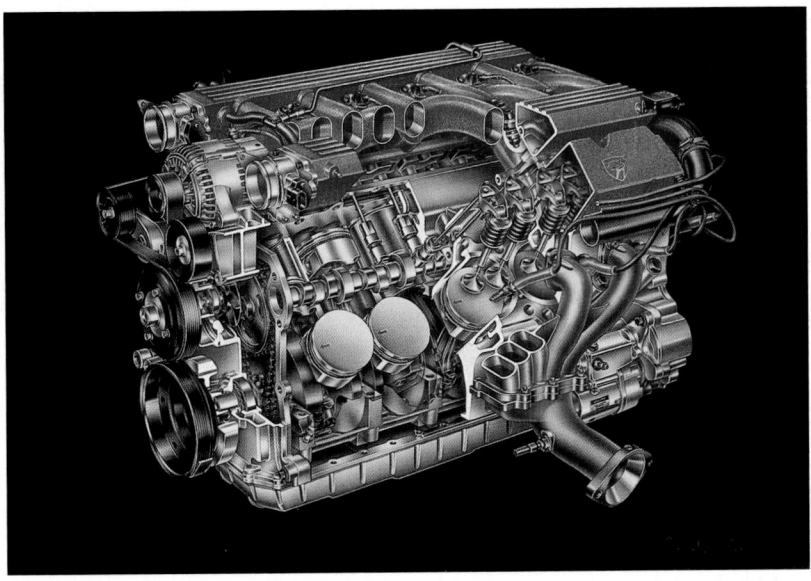

It is expanding gases pushing on the pistons which cause work to be done by the engine.

Remember that the unit of work done is the joule (J). The pressure must be in pascals (Pa) or newtons per metre squared (N m^{-2}) and the change in volume in metres cubed (m^3).

Example

A sample of gas has a volume of 750 cm^3. The gas expands at a constant pressure of 1.4×10^5 Pa so that its volume becomes 900 cm^3. Calculate the work done by the gas during the expansion.

$$
\begin{aligned}
\textit{change in volume } \Delta V &= (900 - 750) \\
&= 150 \text{ cm}^3 \\
&= 150 \times 10^{-6} \text{ m}^3
\end{aligned}
$$

$$
\begin{aligned}
\textit{work done by gas} &= p\Delta V \\
&= (1.4 \times 10^5) \times (150 \times 10^{-6}) \\
&= \mathbf{21 \ J}
\end{aligned}
$$

Now it's your turn

1 The volume of air in a tyre is $9.0 \times 10^{-3}\,\text{m}^3$. Atmospheric pressure is $1.0 \times 10^5\,\text{Pa}$. Calculate the work done against the atmosphere by the air when the tyre bursts and the air expands to a volume of $2.7 \times 10^{-2}\,\text{m}^3$.

Ans: 1800 J

2 High-pressure gas in a spray-can has a volume of $250\,\text{cm}^3$. The gas escapes into the atmosphere through a nozzle, so that its final volume is four times the volume of the can. Calculate the work done by the gas, given that atmospheric pressure is $1.0 \times 10^5\,\text{Pa}$.

Ans: 75 J

Section 3.1 summary

★ When a force moves its point of application in the direction of the force, work is done.

★ *Work done = Fx* cos θ, where θ is the angle between the direction of the force *F* and the displacement *x*.

★ When a gas expands at constant pressure,

$$work\ done = pressure \times change\ in\ volume$$

Section 3.1 questions

1 A force *F* moves its point of application by a distance *x* in a direction making an angle θ with the direction of the force, as shown below.

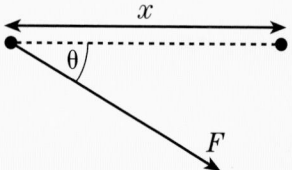

The force does an amount *W* of work. Copy and complete the following table.

F/N	x/m	θ/°	W/J
15	6.0	0	
15	6.0	90	
15	6.0	30	
46		23	6.4
2.4×10^3	1.6×10^2		3.1×10^5
	2.8	13	7.1×10^3

2   An elastic band is stretched so that its length increases by 2.4 cm. The force required to stretch the band increases linearly from 6.3 N to 9.5 N. Calculate:

   (a) the average force required to stretch the elastic band,

   (b) the work done in stretching the band.

3   When water boils at an atmospheric pressure of 101 kPa, 1.00 cm^3 of liquid becomes 1560 cm^3 of steam. Calculate the work done against the atmosphere when a saucepan containing 550 cm^3 of water is allowed to boil dry.

3.2   Energy

In order to wind up a spring, work has to be done because a force must be moved through a distance. When the spring is released, it can do work, for example, making a child's toy move. When the spring is wound, it stores the ability to do work.

The spring stores energy as it is stretched, releasing the energy as it returns to its original shape.

⭐   The ability to do work is called **energy**.

Since work done is measured in joules (J), energy is also measured in joules. Figure 3.4 lists some typical values of energy.

Figure 3.4 *Typical energy values.*

	order of magnitude of energy/J
radioactive decay of an atom	10^{-13}
sound of speech on ear for 1 second	10^{-8}
moonlight on face for 1 second	10^{-3}
beat of the heart	1
burning a match	10^{3}
large cream cake	10^{6}
energy released from 100 kg of coal	10^{10}
earthquake	10^{19}
energy received on Earth from the Sun in one year	10^{25}
rotational energy of the Milky Way Galaxy	10^{50}
estimated energy of formation of the Universe	10^{70}

Potential energy

 Potential energy is the ability of an object to do work as a result of its position or shape.

We have already seen that a wound-up spring stores energy. This energy is potential energy because the spring is strained. More specifically, the energy may be called **strain potential energy**. Strain potential energy or, for short, strain energy, is stored in objects which have had their shape changed elastically. Examples include stretched wires, twisted elastic bands and compressed gases.

A piece of wood stores energy in the sense that, when it is burned, energy is released. Carbon and hydrogen atoms in the wood combine with oxygen in the air to form carbon dioxide and water. A re-organisation of atoms has taken place. This reduces the potential energy of the atoms, releasing energy. The energy stored in the wood is referred to as **chemical potential energy** or, for short, chemical energy. Chemical energy is stored in batteries or cells. A 'charged' battery when connected to an electrical circuit can do work, for example, powering a radio or starting a car. When the battery is 'flat' it has no more useful chemical energy.

Another form of potential energy is **electric potential energy**. The law of charges – like charges repel, unlike charges attract – means that work has to be done when charges are moved relative to one another. If, for example, two positive charges are moved closer together, work is done and the electric potential energy of the charges increases. The electric potential energy stored is released when the charges move apart. Conversely, if a

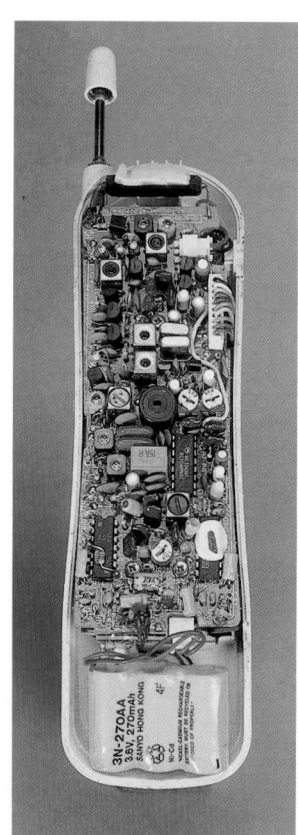

The batteries in the phone must be recharged regularly because they store a limited amount of chemical energy.

3.2 Energy

positive charge moves closer to a negative charge, energy is released because there is a force of attraction.

Newton's law of gravitation (see Chapter 10) tells us that all masses attract one another. We rely on the force of gravity to keep us on Earth! When two masses are pulled apart, work is done on them and so they gain **gravitational potential energy**. If the masses move closer together, they lose gravitational potential energy. Changes in gravitational potential energy are of particular importance for an object near to the Earth's surface because we frequently do work raising masses and, conversely, the energy stored is released when the mass is lowered again. An object of mass m near the Earth's surface has weight mg, where g is the acceleration of free fall. This weight is the force with which the Earth attracts the mass (and the mass attracts the Earth). If the mass moves a *vertical* distance Δh then,

$$work\ done\ = force \times distance\ moved$$
$$= mg\Delta h$$

When the mass is raised, the work done is stored as *gravitational potential energy* and this energy can be recovered when the mass falls.

★ Change in gravitational potential energy $\Delta E_{\mathrm{p}} = mg\Delta h$

It is important to remember that, for the energy to be measured in joules, the mass m must be in kilograms, the acceleration g in metre second^{-2} and the change in height Δh in metres.

Notice that a zero point of gravitational potential energy has not been stated. We are concerned with *changes* in potential energy when a mass rises or falls.

Example

A shop assistant stacks a shelf with 25 tins of beans, each of mass 460 g. Each tin has to be raised through a distance of 1.8 m. Calculate the gravitational potential energy gained by the tins of beans, given that the acceleration of free fall is 9.8 m s^{-2}.

$$\text{total mass raised} = 25 \times 460 = 11\ 500\ \text{g}$$
$$= 11.5\ \text{kg}$$
$$\text{increase in potential energy} = m \times g \times \Delta h$$
$$= 11.5 \times 9.8 \times 1.8$$
$$= \mathbf{203\ J}$$

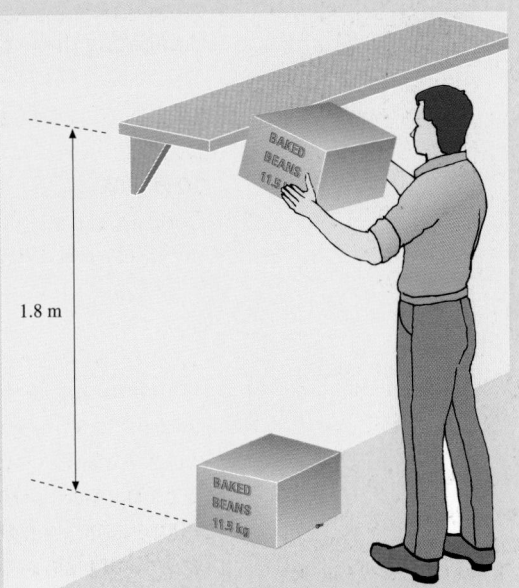

1.8 m

Figure 3.5

Kinetic energy

When the mass falls, it gains kinetic energy and drives the pile into the ground.

As an object falls, it loses gravitational potential energy and in so doing it speeds up. Energy is associated with a moving object. In fact, we know that a moving object can be made to do work as it slows down. For example, a moving hammer hits a nail and, as it stops, does work to drive the nail into a piece of wood.

★ Kinetic energy is energy due to motion.

Consider an object of mass m moving with a constant acceleration a. In a distance s, the object accelerates from velocity u to velocity v. Then, by referring to the equations of motion (see Chapter 2),

$$v^2 = u^2 + 2as.$$

By Newton's law (see Chapter 4), the force F giving rise to the acceleration a is given by

$$F = ma.$$

Combining these two equations,

$$v^2 = u^2 + 2\frac{F}{m}s.$$

Re-arranging,

$$mv^2 = mu^2 + 2Fs,$$
$$2Fs = mv^2 - mu^2,$$
$$Fs = \tfrac{1}{2}mv^2 - \tfrac{1}{2}mu^2.$$

By definition, the term Fs is the work done by the force moving a distance s. Therefore, since Fs represents work done, then the other terms in the equation, $\tfrac{1}{2}mv^2$ and $\tfrac{1}{2}mu^2$ must also have the units of work done, or energy (see Chapter 1). The magnitude of each of these terms depends on velocity squared and so $\tfrac{1}{2}mv^2$ and $\tfrac{1}{2}mu^2$ are terms representing energy which depends on velocity (or speed). The kinetic energy E_k of an object of mass m moving with speed v is given by

$$\bigstar \quad E_k = \tfrac{1}{2}mv^2$$

For the kinetic energy to be in joules, mass must be in kilograms and speed in metres per second.

The full name for the term $E_k = \tfrac{1}{2}mv^2$ is translational kinetic energy because it is energy due to an object moving in a straight line. It should be remembered that rotating objects also have kinetic energy and this form of energy is known as rotational kinetic energy.

Example

Calculate the kinetic energy of a car of mass 900 kg moving at a speed of 20 m s^{-1}. State the form of energy from which the kinetic energy is derived.

$$\begin{aligned} kinetic\ energy &= \tfrac{1}{2}mv^2 \\ &= \tfrac{1}{2} \times 900 \times 20^2 \\ &= \mathbf{1.8 \times 10^5\ J} \end{aligned}$$

This energy is derived from the chemical energy of the fuel.

Now it's your turn

1 Calculate the kinetic energy of a car of mass 800 kg moving at 100 kilometres per hour (about 60 m.p.h.).
   Ans: 3.1×10^5 J

2 A cycle and cyclist have a combined mass of 80 kg and are moving at 5.0 m s^{-1}. Calculate:
   (a) the kinetic energy of the cycle and cyclist,
   (b) the increase in kinetic energy for an increase in speed of 5.0 m s^{-1}.
   Ans: (a) 1000 J; (b) 3000 J

Hooke's law

A helical spring, attached to a fixed point, hangs vertically and has weights attached to its lower end, as shown in Figure 3.6. As the size of the weight is increased the spring becomes longer. The increase in length of the spring is called the **extension** of the spring and the weight attached to the spring is called the **load**.

If the load is increased greatly, the spring will permanently change its shape. However, for small loads, when the load is removed, the spring returns to its original length. The spring is said to have undergone an **elastic change**.

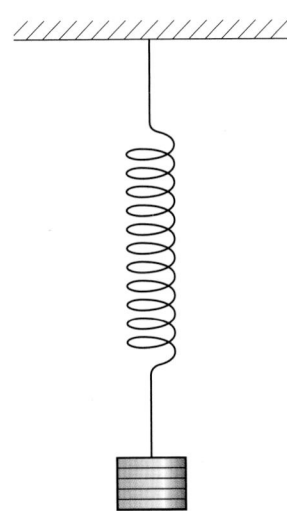

Figure 3.6 A loaded helical spring.

CHAPTER 3

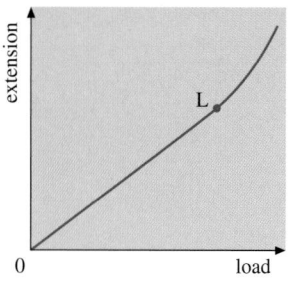

Figure 3.7 *Extension of a spring.*

In an elastic change a body returns to its original shape and size when the load on it is removed.

Figure 3.7 shows the variation with load of the extension of the spring. The section of the line from the origin to the point L is straight. In this region, the spring behaves elastically, and returns to its original length when the load is removed. The point L is referred to as the **elastic limit**. Beyond the point L, the spring is deformed permanently and the change is said to be **plastic**.

The fact that there is a straight line relationship between load and extension for the elastic change is expressed in Hooke's law. It should be appreciated that, although we have used a spring as an illustration, the law applies to any object, provided the elastic limit has not been exceeded.

> ★ Hooke's law states that, provided the elastic limit is not exceeded, the extension of a body is proportional to the applied load.

The law can be expressed in the form of an equation.

$$\text{Force } F \propto \text{extension } \Delta L$$

Removing the proportionality sign gives

$$F = k\Delta L,$$

where k is a constant, known as the **elastic constant** or **spring constant**.

> ★ The elastic constant (spring constant) is the force per unit extension.

The unit of the constant is newton per metre ($N\ m^{-1}$).

Example

An elastic cord has an unextended length of 25 cm. When the cord is extended by applying a force at each end, the length of the cord becomes 40 cm for forces of 0.75 N. Calculate the elastic constant of the cord.

$$\text{extension of cord} = 15 \text{ cm}$$
$$\text{elastic constant} = 0.75/0.15 \text{ (extension in metres)}$$
$$= \mathbf{5.0\ N\ m^{-1}}$$

Now it's your turn

1 Calculate the elastic constant for a spring which extends by a distance of 3.5 cm when a load of 14 N is hung from its end.

   Ans: 400 N m^{-1}

3.2 Energy

2 A steel wire extends by 1.5 mm when it is under a tension of 45 N. Calculate:

(a) the elastic constant of the wire,

(b) the tension required to produce an extension of 1.8 mm, assuming that the elastic limit is not exceeded.

Ans: (a) 3.0×10^4 N m^{-1}; (b) 54 N

Strain energy

When an object has its shape changed by forces acting on it, the object is said to be **strained**. Work has to be done by the forces to cause this strain. Provided that the elastic limit is not exceeded, the object can do work as it returns to its original shape when the forces are removed. Energy is stored in the body as potential energy when it is strained. This particular form of potential energy is called **strain energy**.

> ★ Strain energy is energy stored in a body due to change of shape.

Consider the spring shown in Figure 3.6. To produce an extension x, the force applied at the lower end of the spring increased linearly with extension from zero to a value F. The average force was $\frac{1}{2}F$ and the work done W by the force was

$$W = average\ force \times extension \qquad \text{(see Section 3.1)}$$
$$= \tfrac{1}{2}Fx.$$

However, the elastic constant k is given by the equation

$$F = k\Delta L$$

The value ΔL is equivalent to x. Therefore, substituting for F,

> ★ strain energy $W = \tfrac{1}{2}k(\Delta L)^2$

The energy is in joules if k is in newtons per metre and x (which is equal to ΔL) is in metres.

A graph of load (y-axis) against extension (x-axis) enables strain energy to be found even when the graph is not linear (see Figure 3.8). We have shown that strain energy is given by

$$strain\ energy\ W = \tfrac{1}{2}F\Delta L$$
$$= \tfrac{1}{2}Fx$$
$$= \tfrac{1}{2}kx^2.$$

The expression $\frac{1}{2}Fx$ represents the area between the straight line on Figure 3.8 and the x-axis. This means that strain energy is represented by the area under the line on a graph of load (y-axis) plotted against extension (x-axis).

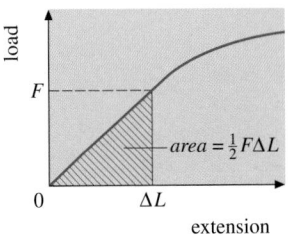

Figure 3.8 *Strain energy.*

Example

A spring has an elastic constant 65 N m^{-1} and is extended elastically by 1.2 cm. Calculate the strain energy stored in the spring.

$$strain\ energy\ W = \tfrac{1}{2}k(x)^2.$$
$$= \tfrac{1}{2} \times 65 \times (1.2 \times 10^{-2})^2$$
$$= \textbf{4.7} \times \textbf{10}^{-3}\ \textbf{J}$$

Now it's your turn

1 A wire has an elastic constant of 5.5×10^4 N m^{-1}. It is extended elastically by 1.4 mm. Calculate the strain energy stored in the wire.
Ans: 5.4×10^{-2} J

2 A rubber band has an elastic constant of 180 N m^{-1}. The work done in extending the band is 0.16 J. Calculate the extension of the band.
Ans: 4.2 cm

The Young modulus

The difficulty with using the elastic constant is that the constant is different for each specimen of a material having a different shape. It would be far more convenient if we had a constant for a particular material which would enable us to find extensions knowing the constant and the dimensions of the specimen. This is possible using the Young modulus.

We have already mentioned the term *strain*. When an object of original length L is extended by an amount ΔL, the strain ϵ is defined as

$$\bigstar\quad strain = \frac{extension}{original\ length}$$

$$\epsilon = \Delta L/L.$$

Strain is the ratio of two lengths and does not have a unit.

The strain produced within an object is caused by a **stress**. In our case, we are dealing with changes in length and so the stress is referred to as a **tensile stress**. When a tensile force F acts normally to an area A, the stress σ is given by

$$\bigstar\quad stress = \frac{force}{area\ normal\ to\ the\ force}$$

$$\sigma = F/A.$$

The unit of tensile stress is newtons per metre squared ($N\,m^{-2}$). This unit is also the unit of pressure and so an alternative unit for stress is the pascal (Pa).

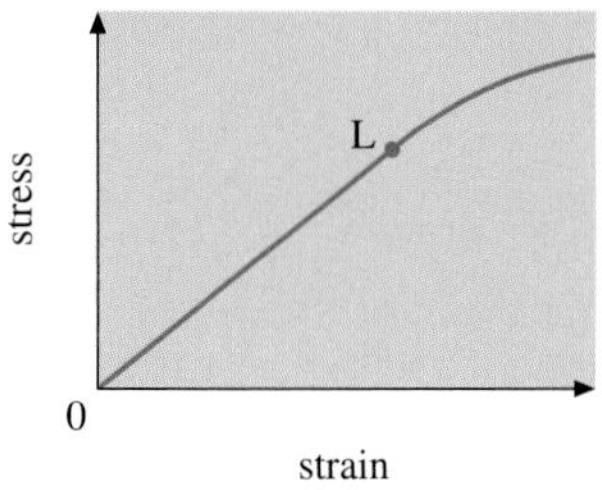

When discussing Hooke's law, we plotted a graph of load against extension. Since load is related to stress and extension is related to strain, a graph of stress plotted against strain would have the same basic shape, as shown in Figure 3.9. Once again, there is a straight line region between the origin and L, the elastic limit. In this region, changes of strain with stress are elastic.

Figure 3.9 *Stress/strain graph.*

In the region where the changes are elastic, it can be seen that

$$stress \propto strain,$$

or, removing the proportionality sign,

$$stress = E \times strain$$

The constant E is known as the Young modulus of the material.

> ★ *Young modulus* $E = \dfrac{stress}{strain}$

The unit of the Young modulus is the same as that for stress because strain is a ratio and has no unit. Some values of Young modulus for different materials are shown in Figure 3.10.

material	Young modulus E/Pa
aluminium	0.70×10^{11}
copper	1.1×10^{11}
steel	2.1×10^{11}
glass	0.41×10^{11}

Figure 3.10

Example

A steel wire of diameter 1.0 mm and length 2.5 m is suspended from a fixed point and a mass of weight 45 N is suspended from its free end. The Young modulus of the material of the wire is 2.1×10^{11} Pa. Assuming that the elastic limit of the wire is not exceeded, calculate:
(a) the applied stress,
(b) the strain,
(c) the extension of the wire.

(a)   area $= \pi \times (0.5 \times 10^{-3})^2$
           $= 7.9 \times 10^{-7}\,\text{m}^2$
    stress $=$ force/area
           $= 45/7.9 \times 10^{-7}$
           $= \mathbf{5.7 \times 10^7\,Pa}$

(b)   strain $=$ stress/Young modulus
           $= 5.7 \times 10^7/2.1 \times 10^{11}$
           $= \mathbf{2.7 \times 10^{-4}}$

(c) extension $=$ strain $\times$ length
           $2.7 \times 10^{-4} \times 2.5$
           $= 6.8 \times 10^{-4}\,\text{m}$
           $= \mathbf{0.68\,mm}$

Now it's your turn

1   A copper wire of diameter 1.78 mm and length 1.4 m is suspended
    from a fixed point and a mass of weight 32 N is suspended from its free
    end. The Young modulus of the material of the wire is 1.1×10^{11} Pa.
    Assuming that the elastic limit of the wire is not exceeded, calculate:
    (a) the applied stress,
    (b) the strain,
    (c) the extension of the wire.
    Ans: (a) 1.29×10^7 Pa; (b) 1.17×10^{-4}; (c) 0.16 mm

2   An elastic band of area of cross-section 2.0 mm^2 has an unextended
    length of 8.0 cm. When stretched by a force of 0.4 N, its length
    becomes 8.3 cm. Calculate the Young modulus of the elastic.
    Ans: 5.3×10^6 Pa

Energy conversion and conservation

Newspapers sometimes refer to a 'Global Energy Crisis'. What crisis? In the
near future there may well be a shortage of fossil fuels. Fossil fuels are
sources of chemical energy. It would be far more accurate to refer to a 'Fuel
Crisis'. When chemical energy is used, the energy is transformed into other
forms of energy, some of which are useful and some of which are not.
Eventually, all the chemical energy is likely to end up as energy which is no
longer useful to us. For example, when petrol is burned in a car engine, some
of the chemical energy is converted into the kinetic energy of the car and
some is wasted as heat (thermal) energy. When the car stops, its kinetic
energy is converted into heat energy in the brakes. The outcome is that the
chemical energy has been converted into heat energy which dissipates in the
atmosphere and is of no further use. However, the total energy present in the
Universe has remained constant. All energy changes are governed by the **law
of conservation of energy**. This law states that:

3.2 Energy

★ Energy cannot be created or destroyed. It can only be converted from one form to another.

There are many different forms of energy and you will meet a number of these during your AS/A2 level studies. Some of the more common forms are listed below.

energy	notes
potential energy	energy due to position
kinetic energy	energy due to motion
strain energy	energy due to stretching an object
electrical energy	energy associated with moving electric charge
sound energy	really, this is a mixture of potential and kinetic energy of the particles in the wave
wind energy	a particular type of kinetic energy
light energy	energy of electromagnetic waves
solar energy	a particular type of light energy, energy from the Sun
chemical energy	energy released during chemical reactions
nuclear energy	energy associated with particles in the nuclei of atoms
thermal energy	sometimes called heat energy

Example

Map out the energy changes taking place when a battery is connected to a light bulb.

Chemical energy in battery → electrical energy in wires
→ light energy and heat energy in bulb

Now it's your turn

Map out the following energy changes.

(a) a child swinging on a swing

Ans: kinetic at lowest point → potential at highest point → kinetic at lowest point etc.

(b) an aerosol can producing hair spray

Ans: potential energy of compressed gas → kinetic of spray droplets → heat when droplets have stopped moving

(c) a lump of clay thrown into the air which subsequently hits the ground

Ans: kinetic energy when thrown → potential and kinetic at highest point of motion → heat and sound energy when clay hits ground

Efficiency

In many examples of energy changes some energy is 'lost' as heat (thermal) energy. For example, when a ball rolls down a slope, the total change in gravitational potential energy is not equal to the gain in kinetic energy because heat (thermal) energy has been produced as a result of frictional forces.

Efficiency gives a measure of how much of the total energy may be used and is not 'lost'.

$$\bigstar \quad efficiency = \frac{useful\ work\ done}{total\ energy\ input}$$

Efficiency may be given either as a ratio or as a percentage. Since energy cannot be created, efficiency can never be greater than 100% and a 'perpetual motion' machine is not possible.

An attempt to design a machine to get something for nothing by breaking the law of conservation of energy.

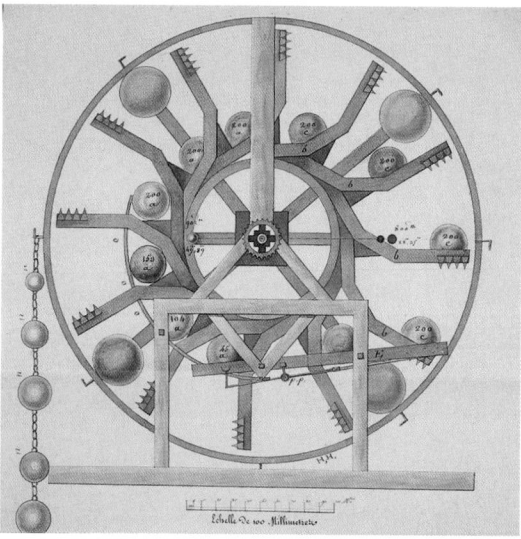

Example

A man lifts a weight of 480 N through a vertical distance of 3.5 m using a rope and some pulleys. The man pulls on the rope with a force of 200 N and a length of 10.5 m of rope passes through his hands. Calculate the efficiency of the pulley system.

$$
\begin{aligned}
work\ done\ by\ man &= force \times distance\ moved \\
&= 200 \times 10.5 \\
&= 2100\ \text{J} \\
work\ done\ lifting\ load &= 480 \times 3.5 \\
&= 1680\ \text{J} \\
efficiency &= work\ got\ out/work\ put\ in \\
&= 1680/2100 \\
&= \textbf{0.80 or 80\%}
\end{aligned}
$$

Now it's your turn

1   An electric heater converts electrical energy into heat energy. Suggest why this process may be 100% efficient.

2   The electric motor of a lift uses 630 kJ of electrical energy when raising the lift and passengers, of total weight 12 500 N through a vertical height of 29 m. Calculate the efficiency of the lift.

Ans: 58%

Section 3.2 summary

★   An elastic change occurs when an object returns to its original shape and size when the load is removed from it.

★   Hooke's law states that, for an elastic change, extension is proportional to load.

★   The elastic constant (spring constant) k is the ratio of force to extension.

★   *Strain energy* $= \frac{1}{2}kx^2$

★   *Tensile strain = extension/original length*

★   *Tensile stress = force/cross-sectional area*

★   *Young modulus = stress/strain*

★   Energy is needed to do work; energy is the ability to do work.

★   Potential energy is the energy stored in a body due to its position.

★   When an object of mass m moves vertically through a distance Δh, then the change in gravitational potential energy ΔE_p is given by

$$\Delta E_p = mg\Delta h,$$

where g is the acceleration of free fall.

★   Kinetic energy is the energy stored in a body due to its motion.

★   For an object of mass m moving with speed v, the kinetic energy E_k is given by the expression

$$E_k = \tfrac{1}{2}mv^2.$$

★   Energy cannot be created or destroyed. It can only be converted from one form to another.

$$efficiency = \frac{useful\ work\ done}{total\ energy\ input}$$

Section 3.2 questions

1. A spring has an unextended length of 12.4 cm. When a load of 4.5 N is suspended from the spring, its length becomes 13.3 cm. Calculate:

   (a) the elastic constant of the spring,

   (b) the length of the spring for a load of 3.5 N.

2. The elastic cord of a catapult has an elastic constant of 700 N m^{-1}. Calculate the energy stored in the elastic cord when it is extended by 15 cm.

3. Two wires each have length 1.8 m and diameter 1.2 mm. One wire has a Young modulus of 1.1×10^{11} Pa and the other 2.2×10^{11} Pa. One end of each wire is attached to the same fixed point and the other end of each wire is attached to the same load of 75 N so that each has the same extension. Assuming that the elastic limit of the wires is not exceeded, calculate the extension of the wires.

4. Name each of the following types of energy:

   (a) energy used in muscles,
   (b) energy stored in the Sun,
   (c) energy of water in a mountain lake,
   (d) energy captured by a wind turbine,
   (e) energy produced when a firework explodes,
   (f) energy of a compressed gas.

5. A child of mass 35 kg moves down a sloping path on a skate board. The sloping path makes an angle of 4.5 ° with the horizontal. The constant speed of the child along the path is 6.5 m s^{-1}. Calculate:

   (a) the vertical distance through which the child moves in 1.0 s,
   (b) the rate at which potential energy is being lost ($g = 10$ m s^{-2}).

6. A stone of mass 120 g is dropped down a well. The surface of the water in the well is 9.5 m below ground level. The acceleration of free fall of the stone is 9.8 m s^{-2}. Calculate, for the stone falling from ground level to the water surface:

   (a) the loss of potential energy,
   (b) its speed as it hits the water, assuming all the potential energy has been converted into kinetic energy.

7. An aircraft of mass 3.2×10^5 kg accelerates along a runway. Calculate the change in kinetic energy, in MJ, when the aircraft accelerates

   (a) from zero to 10 m s^{-1},
   (b) from 30 m s^{-1} to 40 m s^{-1},
   (c) from 60 m s^{-1} to 70 m s^{-1}.

8. In order to strengthen her legs, an athlete steps up on to a box and then down again thirty times per minute. The girl has mass 50 kg and the box is 35 cm high. The exercise lasts 4.0 minutes and as a result of the exercise, her leg muscles generate 120 kJ of heat energy. Calculate the efficiency of the leg muscles. ($g = 10$ m s^{-2}).

9. By accident, the door of a refrigerator is left open. Use the law of conservation of energy to explain whether the temperature of the room will rise, stay constant or fall after the refrigerator has been working for a few hours.

3.3 | Thermal (heat) energy

In this section, we will be looking at the effect of heating on temperature.

Specific heat capacity

When a solid, a liquid or a gas is heated, its temperature rises. Plotting a graph of thermal (heat) energy supplied against temperature rise (see Figure 3.11), it is seen that the temperature rise $\Delta\theta$ is proportional to the heat energy ΔQ supplied. Hence,

$$\Delta Q \propto \Delta\theta$$

for a particular mass of a particular substance.

Similarly, the heat energy required to produce a particular temperature rise is proportional to the mass m of substance being heated (see Figure 3.12). Therefore,

$$\Delta Q \propto m$$

for a particular temperature rise.

Combining these two relations gives

$$\Delta Q \propto m\Delta\theta,$$

★ $\Delta Q = mc\Delta\theta$

where c is the constant of proportionality known as the **specific heat capacity** of the substance. In this case, *specific* means *per unit mass*.

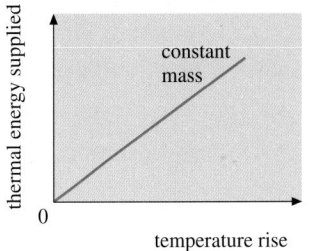

Figure 3.11

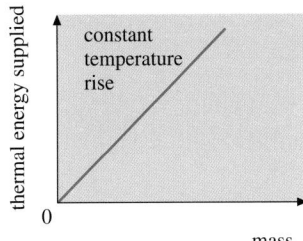

Figure 3.12

> ★ The numerical value of the specific heat capacity of a substance is the quantity of heat energy required to raise the temperature of unit mass of the substance by one degree.

In the SI system of units, the unit of specific heat capacity is J kg^{-1} K^{-1}. The unit of specific heat capacity is *not* the joule and this is why, in the definition of specific heat capacity, it is important to make reference to the *numerical value*. Specific heat capacity is different for different substances. Some values are given in Figure 3.13.

Figure 3.13

material	specific heat capacity/J kg^{-1} K^{-1}
ethanol	2500
glycerol	2420
ice	2100
mercury	140
water	4200
aluminium	913
copper	390
glass	640

It should be noted that, for relatively small changes in temperature, specific heat capacity is approximately constant. However, over a wide range of temperature, the value for a substance may vary considerably (see Figure 3.14). Unless stated otherwise, specific heat capacity is assumed to be constant.

Figure 3.14

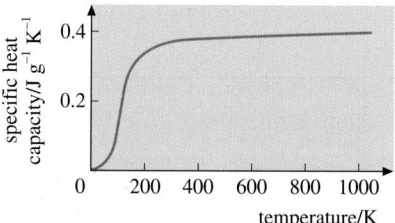

Example

Calculate the quantity of heat energy required to raise the temperature of a mass of 810 g of aluminium from 20 °C to 75 °C. The specific heat capacity of aluminium is 910 J kg^{-1} K^{-1}.

$$heat\ energy\ required = m \times c \times \Delta\theta$$

$$= \frac{810}{1000} \times 910 \times (75 - 20)$$

$$= 4.1 \times 10^4\ J$$

Now it's your turn

1 Calculate the heat energy gained or lost for the following temperature changes. Use Figure 3.13 to obtain values for specific heat capacity.

(a) 45 g of copper heated from 10 °C to 90 °C

(b) 1.3 kg of ice at 0 °C cooled to −15 °C

Ans: (a) 1400 J gain; (b) 41 kJ loss

2 Calculate the specific heat capacity of water given that 0.20 MJ of energy are required to raise the temperature of a mass of 600 g of water by 80 K.

Ans: 4200 J kg^{-1} K^{-1}

Thermal (heat) capacity

Specific heat capacity may be used for a single substance but often objects are made of several different materials. In order to find the change in heat energy, the heat capacity and mass of every substance in the object would need to be known. This can be avoided using the **thermal (heat) capacity** of the object.

> The numerical value of the thermal (heat) capacity of a body is the quantity of heat energy required to raise the temperature of the whole body by one degree.

For a body of thermal capacity C, the heat energy ΔQ supplied is related to the rise in temperature $\Delta\theta$ by the expression

$$\Delta Q = C\Delta\theta.$$

Heat capacity has the SI unit J K^{-1}.

Example

A room has a thermal capacity of 1.2 MJ K^{-1}. Calculate the heat energy required to raise the temperature of the room from 18 °C to 20 °C.

$$\begin{aligned} \Delta Q &= C \times \Delta\theta \\ &= 1.2 \times 10^6 \times (20 - 18) \\ &= 2.4 \times 10^6 \text{ J} \end{aligned}$$

Now it's your turn

1 A kettle contains 700 g of water. The specific heat capacity of water is $4200 \text{ J kg}^{-1} \text{ K}^{-1}$ and the kettle itself has heat capacity 540 J K^{-1}. The kettle and its contents are heated from 15 °C to the boiling point of water (100 °C).
   (a) Calculate the heat energy gained by:
      i)   the water,
      ii) the kettle.
   (b) What fraction of the total energy was used in heating the water?
   Ans: (a) 2.5×10^5 J, 4.6×10^4 J; (b) 0.84

2 An aluminium saucepan has a copper base. Use the information below to determine the thermal capacity of the saucepan.

metal	mass/g	specific heat capacity/J kg^{-1} K^{-1}
aluminium	1160	900
copper	160	390

Ans: 1100 J K^{-1}

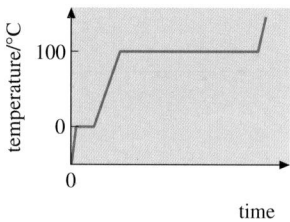

Figure 3.15

Specific latent heat

Figure 3.15 illustrates how the temperature of a mass of water varies with time when it is heated at a constant rate. At times when the substance is changing phase (ice to water or water to steam), heat energy is being supplied without any change of temperature. Because the heat energy does not change the temperature of the substance and is said to be **latent** (i.e. hidden). The latent heat required to melt (fuse) a solid is known as **latent heat of fusion**.

> ★ The numerical value of the specific latent heat of fusion is the quantity of heat energy required to convert unit mass of solid to liquid without any change in temperature.

The SI unit of latent heat of fusion is J kg^{-1}. For a substance with latent heat of fusion L_f, then the quantity of heat energy ΔQ required to fuse (melt) a mass m of solid is given by

$$★ \quad \Delta Q = mL_f.$$

The latent heat required to vaporise a liquid is referred to as **latent heat of vaporisation**.

> ★ The numerical value of the specific latent heat of vaporisation is the quantity of heat energy required to convert a unit mass of liquid to vapour without any change in temperature.

3.3 Thermal (heat) energy

The SI unit of latent heat of vaporisation is the same as that for fusion i.e. $J\,kg^{-1}$. For a substance with latent heat of vaporisation L_v, then the quantity of heat energy ΔQ required to vaporise a mass m of liquid is given by

$$\Delta Q = mL_v.$$

When a vapour condenses (vapour becomes liquid), the latent heat of vaporisation is released. Similarly, when a liquid solidifies (liquid becomes solid), the latent heat of fusion is released. Some values of specific latent heat of fusion and of vaporisation are given in Figure 3.16.

Figure 3.16

material	specific latent heat of fusion/kJ kg^{-1}	specific latent heat of vaporisation/kJ kg^{-1}
ice/water	330	2260
ethanol	108	840
copper	205	4840
sulphur	38.1	

Example

Use the information given in Figure 3.16 to determine the heat energy required to melt 50 g of sulphur at its normal melting point.

$$heat\ energy\ required = m \times L_f$$

$$= \frac{50}{1000} \times 38.1 \times 1000$$

$$= \textbf{1900 J}$$

Now it's your turn

Where appropriate, use the information given in Figure 3.16.

(a) Calculate the heat energy required to
 i) melt 50 g of ice at 0 °C,
 ii) evaporate 50 g of water at 100 °C.
 Ans: 16.5 kJ, 113 kJ

(b) Using your answers to **(a)**, determine how many times more energy is required to evaporate a mass of water than to melt the same mass of ice.
 Ans: 6.8

Exchanges of heat energy

When a hot object and a cold object come into contact, heat energy passes from the hot object to the other so that the two objects reach the same temperature. The law of conservation of energy applies in that the heat energy gained by the cold object is equal to the heat energy lost by the hot object. This does, of course, assume that no energy is lost to the surroundings. This simplification enables temperatures to be calculated.

Example

A mass of 0.30 kg of water at 95 °C is mixed with 0.50 kg of water at 20 °C. Calculate the final temperature of the water, given that the specific heat capacity of water is 4200 J kg^{-1} K^{-1}.

Hint: always start by writing out a word equation containing all the gains and losses of heat energy.

$$\text{heat energy lost by hot water} = \text{heat energy gained by cold water}$$
$$(m \times c \times \Delta\theta_1) = (M \times c \times \Delta\theta_2)$$
$$0.30 \times 4200 \times (95 - t) = 0.50 \times 4200 \times (t - 20)$$

where t is the final temperature of the water.

$$1260 \times (95 - t) = 2100 \times (t - 20)$$
$$119\,700 - 1260t = 2100t - 42\,000$$
$$161\,700 = 3360t$$
$$t = 48\ ^\circ\text{C}$$

Example

A mass of 12 g of ice at 0 °C is placed in a drink of mass 210 g at 25 °C. Calculate the final temperature of the drink, given that the specific latent heat of fusion of ice is 334 kJ kg^{-1} and that the specific heat capacity of water and the drink is 4.2 kJ kg^{-1} K^{-1}.

$$\begin{array}{ccc} \text{energy lost} & \text{energy gained} & \text{energy gained} \\ \text{by drink} & = \text{by melting ice} & + \text{by ice-water} \end{array}$$
$$(m \times c \times \Delta\theta_1) = (M \times L_f) + (M \times c \times \Delta\theta_2)$$

$$M \times c \times \Delta\theta_2 = \frac{12}{1000} \times 4.2 \times 1000 \times (t - 0)$$

$$= 50.4t$$

$$\frac{210}{1000} \times 4.2 \times 1000 \times (25 - t) = \frac{12}{1000} \times 334 \times 1000 + 50.4t$$

where t is the final temperature of the drink. Simplifying,

$$22\,050 - 882t = 4008 + 50.4t$$
$$18\,042 = 932.4t$$
$$t = 19\,°C$$

Now it's your turn

Use the data of Figures 3.13 and 3.16 where appropriate.

1   A lump of copper of mass 120 g is heated in a gas flame. It is then transferred to a mass of 450 g of water, initially at 20 °C. The final temperature of the copper and the water is 31 °C. Calculate the temperature of the gas flame.
    Ans: 480 °C

2   Steam at 100 °C is passed into a mass of 350 g of water, initially at 15 °C. The steam condenses. Calculate the mass of steam required to raise the temperature of the water to 80 °C.
    Ans: 41 g

Section 3.3 summary

★   Specific heat capacity is numerically equal to the heat energy required to raise the temperature of unit mass of substance by one degree.

★   The heat energy ΔQ required to raise the temperature of a mass m of substance of specific heat capacity c by an amount $\Delta\theta$ is given by the expression

$$\Delta Q = mc\Delta\theta.$$

★   Specific heat capacity has the SI unit of $J\,kg^{-1}\,K^{-1}$.

★   The heat capacity of an object is numerically equal to the heat energy required to raise the temperature of the whole body by one degree.

★   The heat energy ΔQ required to raise the temperature of an object having a heat capacity C by an amount $\Delta\theta$ is given by the expression

$$\Delta Q = C\Delta\theta.$$

★   Heat capacity has the SI unit of $J\,K^{-1}$.

★ Specific latent heat of fusion is numerically equal to the quantity of heat energy required to convert unit mass of solid to liquid without any change in temperature.

★ Specific latent heat of vaporisation is numerically equal to the quantity of heat energy required to convert unit mass of liquid to vapour without any change in temperature.

★ When a substance of mass m changes its state the quantity of heat energy ΔQ required is given by

$$\Delta Q = mL,$$

where L is the appropriate specific latent heat.

★ Specific latent heat has the SI unit of J kg^{-1}.

Section 3.3 questions

In the following questions, use the data of Figures 3.13 and 3.16 as appropriate.

1   A piece of copper of mass 170 g is cooled in a freezer. It is then dropped into water at 0 °C, causing 4.0 g of water to freeze. Determine the temperature inside the freezer.

2   A liquid-in-glass thermometer consists of a mass of 62 g of glass and 3.5 g of mercury.

   (a) Calculate the thermal capacity of the thermometer.
   (b) The thermometer is used to measure the temperature of some glycerol of mass 90 g. Before the thermometer is inserted into the glycerol, the thermometer records 18 °C and the temperature of the glycerol is 42 °C. Calculate the temperature recorded on the thermometer when placed in the glycerol.
   (c) Using your answer to (b), suggest why such thermometers cannot be used to measure reliably the temperature of a small mass of substance.

3   (a) A jet of steam at 100 °C is directed into a hole in a large block of ice. After the steam has been switched off, the condensed steam and the melted ice are both at 0 °C. The mass of water collected in the hole is 206 g. Calculate the mass of steam condensed.
   (b) Suggest why a scald with steam is much more serious than one involving boiling water.

4   A mass of 450 g of frozen peas is taken from a freezer at −20 °C. The peas are immediately placed in a saucepan containing 1100 g of boiling water. The saucepan has a thermal capacity of 900 J K^{-1}. The final temperature of the saucepan, water and peas is 83 °C.

   (a) Calculate the specific heat capacity of peas.
   (b) The saucepan and its contents are then heated using a heater which provides 1200 J of thermal energy each second. Determine how long it takes to bring the water back to its boiling point.

3.4 Power

We have seen that energy is the ability to do work. Consider a family car and a grand prix racing car which both contain the same amount of fuel. They are capable of doing the same amount of work, but the racing car is able to travel much faster. This is because the engine of the racing car can convert the chemical energy of the fuel into useful energy at a much faster rate. The engine is said to be more powerful. Power is the rate of doing work. Power is given by the formula

$$\bigstar \; power = \frac{work\ done}{time\ taken}$$

The unit of power is the watt (symbol W) and is equal to a rate of working of 1 joule per second. This means that a light bulb of power 1 W will convert 1 J of electrical energy to other forms of energy (e.g. light and heat) every second.

Some typical values of power are shown in Figure 3.17.

Figure 3.17

power to operate a small calculator	50 µW
light power from a torch	4 mW
loudspeaker output	5 W
man working continuously	100 W
horse working continuously	750 W
electric fire	3 kW
motor car engine	80 kW
electric train	5 MW
electricity generating station output	2 GW

Power, like energy, is a scalar quantity.

Care must be taken when referring to power. It is common in everyday language to say that a strong person is 'powerful'. In physics, strength, or force, and power are *not* the same. Large forces may be exerted without any movement and thus no work is done and the power is zero! For example, a large rock resting on the ground is not moving, yet it is exerting a large force.

Consider a force F which moves a distance x at constant speed v in the direction of the force in time t. Referring back to section 3.1, the work done W by the force is given by

$$W = Fx$$

Dividing both sides of this equation by time t gives

$$\frac{W}{t} = F\frac{x}{t}$$

Now, $\dfrac{W}{t}$ is the rate of doing work, i.e. the power P and $\dfrac{x}{t} = v$. Hence

★ $P = Fv$

★ *power = force × speed*

Example

A small electric motor is used to lift a weight of 1.5 N through a vertical distance of 120 cm in 2.7 s. Calculate the useful power output of the motor.

$$
\begin{aligned}
work\ done &= force \times distance\ moved \\
&= 1.5 \times 1.2 \qquad \text{(the distance must be in metres)} \\
&= 1.8\ J \\
power &= work\ done/time\ taken \\
&= (1.8/2.7) \\
&= \textbf{0.67 W}
\end{aligned}
$$

Now it's your turn

1   Calculate the electrical energy converted into thermal energy when an electric fire, rated at 2.4 kW, is left switched on for a time of 3.0 minutes.
    Ans: 4.32×10^5 J

2   The output power of the electric motors of a train is 3.6 MW when the train is travelling at 30 m s^{-1}. Calculate the total force opposing the motion of the train.
    Ans: 120 kN

3   A boy of mass 60 kg runs up a flight of steps in a time of 1.8 s. There are 22 steps and each one is of height 20 cm. Calculate the useful power developed in the boy's legs. (The acceleration of free fall is 10 m s^{-2}.)
    Ans: 1.5 kW

The kilowatt-hour

Every household suffers from having to pay the 'electricity bill'. Electricity is vital in modern living and this energy does not come free of charge. It is important to realise that what is paid for is electrical energy, not electrical power. Since many electrical appliances in the home have a power of the order of kilowatts and we use them for hours, the joule, as a unit of energy, is too small. For example, an electric fire of power 3 kW, used for 2 hours would use $3000 \times 2 \times 60 \times 60 = 21600000$ joules i.e. 21.6 million joules of energy! Electrical energy is purchased in kilowatt-hours (kW h).

> One kilowatt-hour is the energy expended when work is done at the rate of 1 kilowatt for a time of 1 hour.

$$1 \text{ kW h} = 1.0 \text{ kW} \times 1 \text{ hour}$$
$$= 1000 \text{ W} \times 3600 \text{ s}$$
$$1 \text{ kW h} = 3.6 \times 10^6 \text{ J}$$

A digital electricity meter.

The kilowatt-hour is sometimes referred to as the Unit of energy. Electricity meters in the home are often shown as measuring Units, where 1 Unit = 1 kW h.

Example

Calculate the cost of using an electric fire, rated at 2.5 kW for a time of 6.0 hours, given that 1 kW h of energy costs 7.0 pence.

$$\text{energy used} = 2.5 \times 6.0$$
$$= 15 \text{ kW h}$$
$$\text{cost} = 15 \times 7.0$$
$$= \textbf{105 pence}$$

Now it's your turn

1   A television set is rated at 280 W. Calculate the cost of watching a three-hour film given that 1 kilowatt hour of electrical energy costs 8 pence.
    Ans: 6.7 p

2   An electric kettle is rated at 2.4 kW. Electrical energy costs 8 p per kW h. The kettle takes 1.0 minute to boil sufficient water for two mugs of coffee. Calculate the cost of making this amount of coffee on three separate occasions.
    Ans: 0.96 p

3   Electrical energy generating companies sometimes measure their output in gigawatt-years. Calculate the number of kilowatt-hours in 6.0 gigawatt-years.
    Ans: 5.3×10^{10}

Section 3.4 summary

★ Power is the rate of doing work.

★ *Work done = power × time taken.*

★ The unit of power is the watt (W).

★ 1 watt = 1 joule per second.

★ *Power = force × speed.*

★ Electrical energy may be measured in kilowatt-hours (kW h).

★ 1 kW h is the energy expended when work is done at the rate of 1000 watts for a time of 1 hour.

Section 3.4 questions

1 The lights in a school laboratory have a total power of 600 W and are left on permanently for 7.0 hours each day. In order to reduce fuel bills, it is decided to have the lights switched on only when there are people in the laboratory. This amounts to a total time of 4.5 hours per day. Assuming that the laboratory is used for 200 days each year, calculate the saving, given that 1 kW h of energy costs 7.0 p.

2 A car travelling at speed v along a horizontal road moves against a resistive force F given by the equation

$$F = 400 + kv^2,$$

where F is in newtons, v in m s^{-1} and k is a constant.

At speed $v = 15$ m s^{-1}, the resistive force F is 1100 N.

(a) Calculate, for this car:
   i) the power necessary to maintain the speed of 15 m s^{-1},
   ii) the total resistive force at a speed of 30 m s^{-1},
   iii) the power required to maintain the speed of 30 m s^{-1}.

(b) Determine the energy expended in travelling 1.2 km at a constant speed of:
   i) 15 m s^{-1},
   ii) 30 m s^{-1}.

(c) Using your answers to (b), suggest why, during a fuel shortage, the maximum permitted speed of cars may be reduced.

3.5 Moment of a force

When a force acts on an object, the force may cause the object to move in a straight line. It could also cause the object to spin (rotate).

Figure 3.18 *Turning effect on a metre rule.*

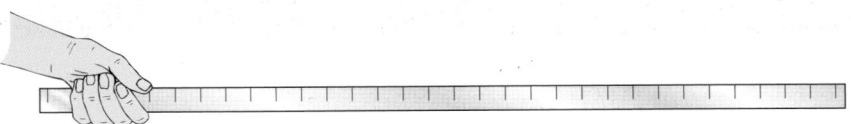

Think about a metre rule held in the hand at one end so that the rule is horizontal (Figure 3.18). If a weight is hung from the ruler we can feel a turning effect on the ruler. The turning effect increases if the weight is increased or it is moved further from the hand along the ruler. The turning effect acts at the hand where the metre rule is pivoted. Keeping the weight and its distance along the rule constant, the turning effect can be changed by holding the ruler at an angle to the horizontal. The turning effect becomes smaller as the rule approaches the vertical position.

> ★ The turning effect of a force is called the moment of the force.

The moment of a force depends on the magnitude of the force and also on the distance of the force from the pivot or fulcrum. This distance must be defined precisely. In the simple experiment above, we saw that the moment of the force depended on the angle of the ruler to the horizontal. Varying this angle meant that the line of action of the force from the pivot varied (see Figure 3.19). The distance required when finding the moment of a force is the perpendicular distance d of the line of action of the force from the pivot.

> ★ The moment of a force is defined as the product of the force and the perpendicular distance of the line of action of the force from the pivot.

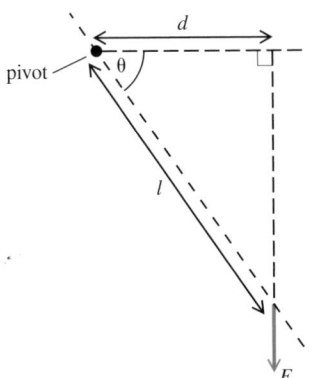

Figure 3.19 *Finding the moment of a force.*

Referring to Figure 3.19, the force has magnitude F and acts at a point distance l from the pivot. Then, when the ruler is at angle θ to the horizontal,

$$moment\ of\ force = F \times d$$
$$= F \times l \cos \theta.$$

Since force is measured in newtons and distance is measured in metres, the unit of the moment of a force is newton-metre (N m).

Example

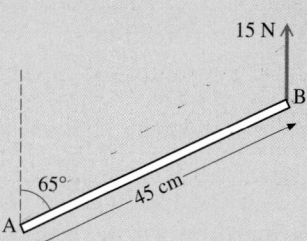

Figure 3.20

In Figure 3.20, a light rod AB of length 45 cm is held at A so that the rod makes an angle of 65° to the vertical. A vertical force of 15 N acts on the rod at B. Calculate the moment of the force about the end A.

$$moment\ of\ force = force \times perpendicular\ distance\ from\ pivot$$
$$= 15 \times 0.45 \sin 65$$
(remember that the distance must be in metres)
$$= \mathbf{6.1\ N\ m}$$

Now it's your turn

Referring to Figure 3.20, calculate the moment of the force about A for a vertical force of 25 N with the rod at an angle of 30° to the vertical.
Ans: 5.6 N m

Couples

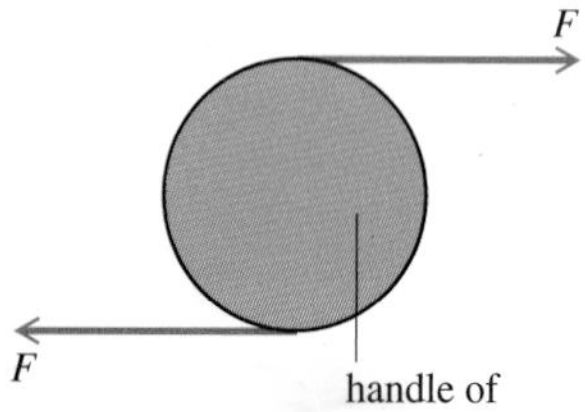

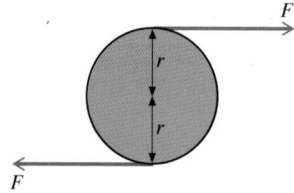

Figure 3.21 *Two forces acting as a couple.*

When a screwdriver is used, we apply a turning effect to the handle. We do not apply one force to the handle because this would mean the screwdriver would move sideways. Rather, we apply two forces of equal size but opposite direction on opposite sides of the handle (see Figure 3.21).

★ A **couple** consists of two forces, equal in magnitude but opposite in direction whose lines of action do not coincide.

Consider the two parallel forces, each of magnitude F acting as shown in Figure 3.22 on opposite ends of a diameter of a disc of radius r. Each force produces a moment about the centre of the disc of magnitude Fr in a clockwise direction. The total moment about the centre is $F \times 2r$ or $F \times perpendicular\ distance\ between\ the\ forces$.

Figure 3.22 *Torque of a couple.*

Although a turning effect is produced, this turning effect is not called a moment because it is produced by two forces, not one. Instead, this turning effect if referred to as a **torque**. The unit of torque is the same as that of the moment of a force i.e. newton-metre.

3.5 Moment of a force

> ★ The torque of a couple is the product of one of the forces and the perpendicular distance between the forces.

It is interesting to note that, in engineering, the tightness of nuts and bolts is often stated as the maximum torque to be used when screwing up the nut on the bolt. Spanners used for this purpose are called torque wrenches because they have a scale on them to indicate the torque which is being applied.

Tightening a wheel nut requires the application of a torque.

Example

Calculate the torque produced by two forces, each of magnitude 30 N, acting in opposite directions with their lines of action separated by a distance of 25 cm.

$$torque = force \times separation\ of\ forces$$
$$= 30 \times 0.25 \qquad \text{(distance in metres)}$$
$$= \textbf{7.5 N m}$$

Now it's your turn

The torque produced by a person using a screwdriver is 0.18 N m. This torque is applied to the handle of diameter 4.0 cm. Calculate the force applied to the handle.

Ans: 4.5 N

The principle of moments

A metre rule may be balanced on a pivot so that the rule is horizontal. Hanging a weight on the rule will make the rule rotate about the pivot. Moving the weight to the other side of the pivot will make the rule rotate in the opposite direction. Hanging weights on both sides of the pivot as shown in Figure 3.23 means that the ruler may rotate clockwise or anticlockwise or remain horizontal. In this horizontal position, there is no resultant turning effect and so the total turning effect of the forces in the clockwise direction equals the total turning effect in the anticlockwise direction. You can check this very easily with the apparatus of Figure 3.23!

Figure 3.23

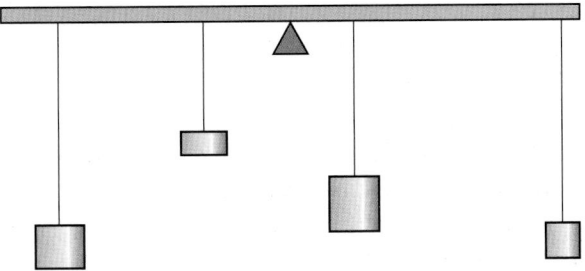

When a body has no tendency to change its speed of rotation, it is said to be in *rotational equilibrium*.

⭐ The principle of moments states that, for a body to be in rotational equilibrium, the sum of the clockwise moments about any point must equal the sum of the anticlockwise moments about that same point.

Example

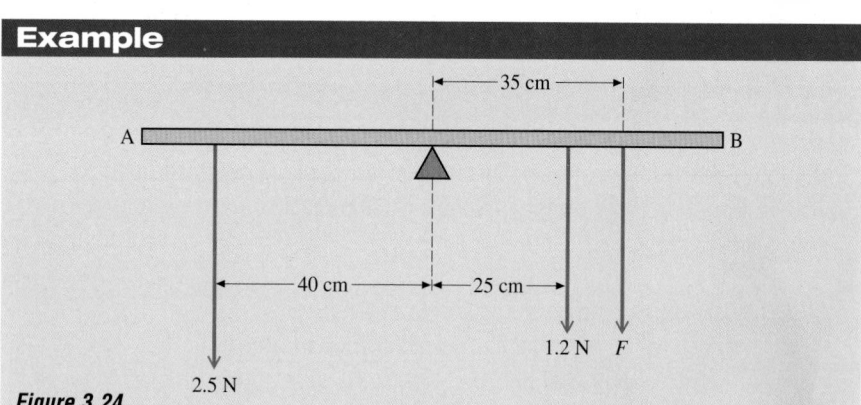

Figure 3.24

Some weights are hung from a light rod AB as shown in Figure 3.24. The rod is pivoted. Calculate the magnitude of the force *F* required to balance the rod horizontally.

Sum of clockwise moments = $(0.25 \times 1.2) + 0.35F$
Sum of anticlockwise moments = 0.40×2.5
By the principle of moments
$$(0.25 \times 1.2) + 0.35F = 0.40 \times 2.5$$
$$0.35F = 1.0 - 0.3$$
$$F = \mathbf{2.0\ N}$$

Now it's your turn

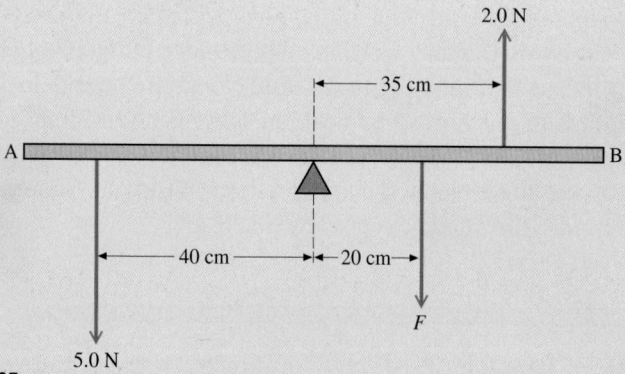

Figure 3.25

Some weights are hung from a light rod AB as shown in Figure 3.25. The rod is pivoted. Calculate the magnitude of the force *F* required to balance the rod horizontally.

Ans: 13.5 N

Centre of gravity

An object may be made to balance at a particular point. When it is balanced at this point, the object does not turn and so all the weight on one side of the pivot is balanced by the weight on the other side. Supporting the object at the pivot means that the only force which has to be applied at the pivot is one to stop the object falling – that is, a force equal to the weight of the object. Although all parts of the object have weight, the whole weight of the object appears to act at this balance point. This point is called the **centre of gravity** (C.G.) of the object.

★ The centre of gravity of an object is the point at which the whole weight of the object may be considered to act.

The weight of a body can be shown as a force acting vertically downwards at the centre of gravity. For a uniform body such as a ruler, the centre of gravity is at the geometrical centre.

Equilibrium

The principle of moments gives the condition necessary for a body to be in rotational equilibrium. However, the body could still have a resultant force acting on it which would cause it to accelerate linearly. Thus, for complete equilibrium, there cannot be any resultant force in any direction.

★ For a body to be in equilibrium,
  1 the sum of the forces in any direction must be zero,
  2 the sum of the moments of the forces about any point must be zero.

Section 3.5 summary

★ The moment of a force is a measure of the turning effect of the force.

★ The moment of a force is the product of the force and the perpendicular distance of the line of action of the force from the pivot.

★ A couple consists of two equal forces acting in opposite directions whose lines of action do not coincide.

★ The torque of a couple is a measure of the turning effect of the couple.

★ The torque of a couple is the product of one of the forces and the perpendicular distance between the lines of action of the forces.

CHAPTER 3

★ The principle of moments states that the sum of the clockwise moments about a point is equal to the sum of the anticlockwise moments about the point.

★ The centre of gravity of a body is the point at which the whole weight of the body may be considered to act.

★ For a body to be in equilibrium,

1 the sum of the forces in any direction must be zero,

2 the sum of the moments of the forces about any point must be zero.

Section 3.5 questions

1 A uniform rod has a weight of 14 N. It is pivoted at one end and held in a horizontal position by a thread tied to its other end, as shown in Figure 3.26. The thread makes an angle of 50° with the horizontal. Calculate:
(a) the moment of the weight of the rod about the pivot,
(b) the tension T in the thread required to hold the rod horizontally.

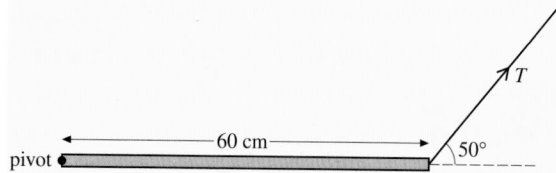

Figure 3.26

2 A ruler is pivoted at its centre of gravity and weights are hung from the ruler as shown in Figure 3.27. Calculate:
(a) the total anticlockwise moment about the pivot,
(b) the magnitude of the force F.

3 A uniform plank of weight 120 N rests on two stools as shown in Figure 3.28. A weight of 80 N is placed on the plank, midway between the stools. Calculate:
(a) the force acting on the stool at A,
(b) the force acting on the stool at B.

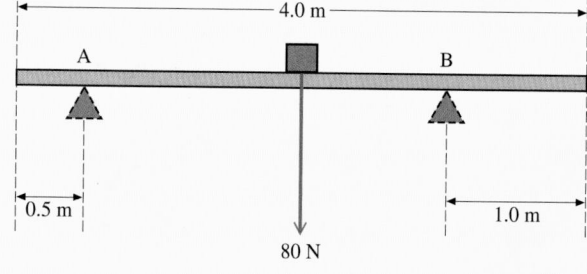

Figure 3.28

4 A nut is to be tightened to a torque of 16 N m. Calculate the force which must be applied to the end of a spanner of length 24 cm in order to produce this torque.

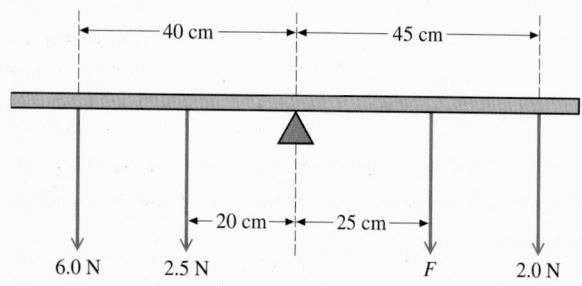

Figure 3.27

Exam Questions

1 (a) State what is meant by:
   i) work done,
   ii) power.

(b) i) Show that, when a gas expands by an amount ΔV against a constant pressure p, the work done W is given by

$$W = p\Delta V.$$

   ii) Explain why, when the gas in i) expands suddenly, it is likely to cool.

(c) The air in a car tyre has a volume of 7.8×10^3 cm^3. The tyre suddenly bursts and the air expands to 3.5 times its original volume in a time of 2.5 ms. Atmospheric pressure is 1.0×10^5 Pa. Calculate:
   i) the work done by the air in the tyre during the expansion,
   ii) the mean power dissipated in the tyre burst.

2 (a) Explain, by reference to two examples from different areas of physics, what is meant by *potential* energy.

(b) i) State the law of conservation of energy.
   ii) Explain how the law may be applied to a ball dropped from a height h on to a horizontal surface and which finally comes to rest on the surface after several bounces.

(c) A car of mass 950 kg is travelling along a straight horizontal road at a speed of 32 m s^{-1}. It is brought to rest by applying its brakes. All of the kinetic energy of the car may be assumed to be converted into thermal energy of the discs of the brakes. There are four discs, each of mass 1.9 kg and made of metal of specific heat capacity 420 J kg^{-1} K^{-1}. Calculate the temperature rise of the brakes when the car stops.

3 (a) Define:
   i) specific heat capacity,
   ii) specific latent heat of vaporisation.

(b) i) Explain what is meant by the thermal capacity of a body.
   ii) A metal has mass m and specific heat capacity c. Derive an expression relating m and c to the thermal capacity C of the metal.

   iii) In what circumstances is thermal capacity a more useful concept than specific heat capacity?

(c) An electric kettle has a power output of 2.3 kW. The kettle is used to heat a mass of 950 g of water of specific heat capacity 4200 J kg^{-1} K^{-1}.
   i) Calculate the time taken to heat the water from 15 °C to 100 °C.
   ii) The kettle is left switched on when the water temperature reaches 100 °C. Calculate the rate at which the water is boiled away. (The specific latent heat of vaporisation of water is 2.26 MJ kg^{-1}.)

(d) The kettle in (c) is used six times each day to boil water. On each occasion, approximately 200 g excess water is heated. Given that 1 kW h of energy costs 6.5 p, estimate the extra cost involved each year.

4 An electric heater is placed in some water of mass 700 g. The heater is switched on and the variation with time t of the temperature θ of the water is shown in Figure 3.29.

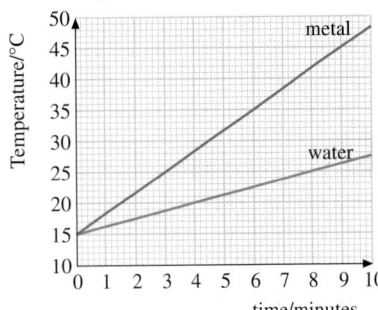

Figure 3.29

(a) i) Given that the specific heat capacity of water is 4200 J kg^{-1} K^{-1}, use Figure 3.29 to determine the power of the heater.
   ii) State two assumptions made in your calculation in i).

(b) The heater is then removed from the water and is placed in a hole in a block of metal of mass 1.2 kg. The heater is switched on and readings of temperature θ and time t are taken. These are also shown on Figure 3.29.
   i) Determine the specific capacity of the metal.
   ii) Suggest why the experiment would not be suitable for the determination of the specific heat capacity of an insulator such as concrete.

CHAPTER FOUR

Force
and collisions

The study of collisions between elementary particles, nuclei, atoms and molecules is a very important means of investigating the properties of such bodies. Automobile engineers make detailed measurements on collisions between prototype vehicles and concrete walls in order to improve the safety aspects of the design.

In Chapter 2 we described motion in terms of displacement, velocity and acceleration. Now we will try to explain why bodies move. We shall realise that a force is required to make a body accelerate. This will introduce Newton's three laws of motion, the fundamental laws which state and define what forces do in relation to motion. We shall then see how Newton's laws help us to explain what happens in collisions. The total momentum of a system of colliding bodies remains constant. This leads us to the principle of conservation of momentum, one of the important conservation laws of physics.

4.1 Force

When you push a trolley in a supermarket or pull a case behind you at an airport, you are exerting a force. When you hammer in a nail, a force is being exerted. When you drop a book and it falls to the floor, the book is falling because of the force of gravity. When you lean against a wall or sit on a chair, you are exerting a force. Forces can change the shape or dimensions of bodies. You can crush a drinks can by squeezing it and applying a force; you can stretch a rubber band by pulling it. In everyday life, we have a good understanding of what is meant by force and the situations in which forces are involved. In physics the idea of force is used to add detail to the descriptions of moving objects.

As with all physical quantities, a method of measuring force must be established. One way of doing this is to make use of the fact that forces can change the dimensions of bodies in a reproducible way. It takes the same force to stretch a spring by the same amount (provided the spring is not overstretched by applying a very large force). This principle is used in the spring balance. A scale shows how much the spring has been extended, and the scale can be calibrated in terms of force. Laboratory spring balances are often called newton balances, because the newton is the SI unit of force.

Forces are vector quantities: they have magnitude as well as direction. Forces acting on a body are often shown by means of a force diagram drawn to scale, in which the forces are represented by lines of length proportional to the magnitude of the force, and in the appropriate direction (see Chapter 1).

4.2 Force and motion

The Greek philosopher Aristotle believed that the natural state of a body was a state of rest, and that a force was necessary to make it move and to keep it moving. In this argument, the greater the force, the greater the speed of the body.

Nearly two thousand years later, Galileo questioned this idea. He suggested that motion at a constant speed could be just as natural a state as the state of rest. He introduced an understanding of the effect of friction on motion.

Imagine a heavy box being pushed along a rough floor at constant speed. This may take a considerable force. The force required can be reduced if the floor is made smooth and polished, and reduced even more if a lubricant, for example grease, is applied between the box and the floor. We can imagine a situation where, when friction is reduced to a vanishingly small value, the force required to push the box at constant speed is also vanishingly small.

Figure 4.1

Galileo realised that the force of friction was a force which opposed the pushing force. When the box is moving at constant speed, the pushing force is exactly equal to the frictional force, but in the opposite direction, so that there is a net force of zero acting on the box. In the situation of vanishingly small friction, the box will continue to move with constant speed, because there is no force to slow it down.

Isaac Newton (1642–1727) used Galileo's ideas to produce a theory of motion, expressed in his three laws of motion. The first law of motion re-states Galileo's deduction about the natural state of a body. It is:

Isaac Newton.

> ★ every body continues in its state of rest, or of uniform motion in a straight line, unless compelled to change that state by a net force.

This law tells us what a force does: it disturbs the state of rest or uniform motion of a body.

> The property of a body to stay in a state of rest or uniform motion is called **inertia**.

Newton's second law tells us what happens if a force is exerted on a body. It causes the velocity to change. A force exerted on a body at rest makes it move – it gives it a velocity. A force exerted on a moving body may make its speed increase or decrease, or change its direction of motion. A change in speed or velocity is an acceleration. Newton's second law defines the magnitude of this acceleration. It also introduces the idea of the mass of body. Mass is a measure of the inertia of a body. The bigger the mass, the more difficult it is to change its state of rest or uniform motion. A simplified form of Newton's second law is:

> ★ for a body of constant mass, its acceleration is directly proportional to the net force applied to it.

The direction of the acceleration is in the direction of the net force. In a word equation, the law is:

> ★ *force = mass × acceleration*

and in symbols,

> ★ $F = ma$

where F is the force, m is the mass and a is the acceleration. Here we have made the constant of proportionality equal to unity (that is, we use an equals sign rather than a proportionality sign) by choosing quantities with units which will give us this simple relation. In SI units, the force F is in newtons, the mass m in kilograms and the acceleration a in metres per second per second.

> ★ One newton is defined as the force which will give a mass of one kilogram an acceleration of one metre per second per second.

When a force is applied to a body, that force is always applied by another body. When you push a supermarket trolley, the trolley experiences a force. That force has been applied by another body – you. Newton understood that the body on which the force is exerted applies another force back on the body which is applying the force. The supermarket trolley exerts a force on you as well. Newton's third law relates these two forces.

Figure 4.2

> Whenever one body exerts a force on another, the second exerts an equal and opposite force on the first.

Very often this law is stated as

> 'to every action, there is an equal and opposite reaction'.

But this statement does not highlight the very important point that the action force and the reaction force act on different objects. To take the example of the supermarket trolley, the action force exerted by you on the trolley is equal and opposite to the reaction force exerted by the trolley on you.

Newton's third law has applications in every branch of everyday life. We walk because of this law. When you take a step forward, your foot presses against the ground. The ground then exerts an equal and opposite force on you. This is the force, on you, which propels you in your path. Space-rockets work because of the law. To expel the exhaust gases from the rocket, the rocket exerts a force on the gases. The gases exert an equal and opposite force on the rocket, sending it forward.

Examples

1   An object of mass 1.5 kg is to be accelerated at 2.2 m s^{-2}. What force is required?

    By Newton's second law, $F = ma = 1.5 \times 2.2 =$ **3.3 N**.

2   A car of mass 1.5 tonnes (1.5×10^3 kg), travelling at 80 km h^{-1}, is to be stopped in 11 s. What force is required?

    The acceleration of the car can be obtained from $v = u + at$ (see Chapter 2). The initial speed u is 80 km h^{-1}, or 22 m s^{-1}. The final speed v is 0. Then $a = -22/11 = -2.0$ m s^{-2}. This is negative because the car is decelerating.
    By Newton's second law, $F = ma = 1.5 \times 10^3 \times 2.0 =$ **3.0 × 10^3 N**.

Now it's your turn

1   A force of 5.0 N is applied to a body of mass 3.0 kg. What is the acceleration of the body?
    Ans: 1.7 m s^{-2}

2   In a catapult, a stone of mass 50 g is accelerated to a speed of 8.0 m s^{-1} from rest over a distance of 30 cm. What average force is applied by the rubber of the catapult?
    Ans: 5.3 N

4.3 | Weight

We saw in Chapter 2 that all objects released near the surface of the Earth fall with the same acceleration (the acceleration of free fall) if air resistance can be neglected. The force causing this acceleration is the gravitational attraction of the Earth on the object, or the force of gravity. The force of gravity which acts on an object is called the *weight* of the object. We can apply Newton's second law to the weight. For a body of mass m falling with the acceleration of free fall g, the weight W is given by

$$\bigstar \quad W = mg.$$

The SI unit of force is the newton (N). This is also the unit of weight. Note that the weight of an object is obtained from its mass in kilograms by multiplying the mass by the acceleration of free fall, 9.8 m s^{-2}. Thus a mass of one kilogram has a weight of 9.8 N. Because weight is a force and force is a vector, we ought to be aware of the direction of the weight of an object. It is towards the centre of the Earth. Because weight always has this direction, we do not need to specify direction every time we give the magnitude of the weight of objects.

How do we measure mass and weight? If you hang an object from a spring balance (a newton balance), you are measuring its weight. The unknown weight of the object is balanced by a force provided by the spring in the balance. From a previous calibration, this force is related to the extension of the spring. There is the possibility of confusion here. Laboratory newton balances may, indeed, be calibrated in newtons. But all commercial spring balances, for example the balances at vegetable counters in supermarkets, are calibrated in kilograms. Such balances are really measuring the weight of the vegetables, but the scale reading is in mass units, because there is no distinction between mass and weight in everyday life. The average shopper thinks of 5 kg of potatoes as having a weight of 5 kg. In fact, the mass of 5 kg has a weight of 49 N.

Figure 4.3 Newton balance.

The word 'balance' in the spring balance and in the laboratory top-pan balance relates to the balance of forces. In each case, the unknown force (the weight) is equalled by a force which is known through calibration.

A way of comparing masses is to use a beam balance. Here the weight of the object is balanced against the weight of some masses, which have previously been calibrated in mass units. The word 'balance' here refers to the equilibrium of the beam: when the beam is level, the moment of the weight in one pan is equal and opposite to the moment of the weight in the other. Because the beam lengths are equal on each side of the pivot, the fact that the moments are equal means that the weights are equal. Further, because weight is given by the product of mass and the acceleration of free fall, the equality of the weights means that the masses are also equal.

Figure 4.4 Beam balance.

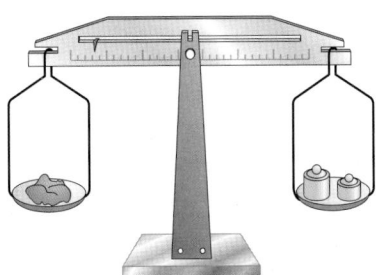

We have introduced the idea of weight by thinking about an object in free fall. But objects at rest also have weight: the gravitational attraction on a book is the same whether it is falling or whether it is resting on a table. The fact that the book is at rest tells us, by Newton's first law, that the resultant force acting on it is zero. So there must be another force acting on the book which exactly balances its weight. Here the table exerts an upwards force on the book (Newton's third law). This force is equal in magnitude to the weight but opposite in direction. It is a **normal contact force:** 'contact' because it occurs due to the contact between book and table, and 'normal' because it acts perpendicularly to the plane of contact. The weight and normal contact forces are shown in Figure 4.5.

Figure 4.5 *Book resting on a table: the normal contact force balances the weight.*

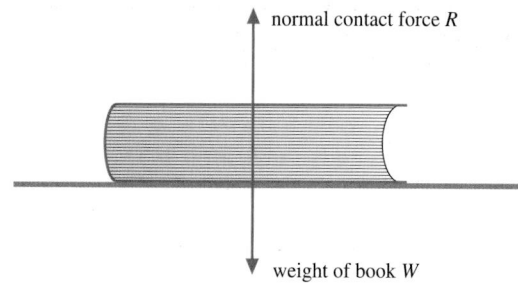

normal contact force *R*

weight of book *W*

The book remains at rest on the table because the weight *W* of the book downwards is exactly balanced by the normal contact force *R* exerted by the table on the book. The vector sum of these forces is zero, so the book is in equilibrium. A very common mistake is to state that 'By Newton's third law, *W* is equal to *R*'. But these two forces are both acting on the book, and cannot be related by the third law. Third-law forces always act on *different* bodies. To see the application of the third law, think about the normal contact force *R*. This is an upwards force exerted by the *table*. The reaction to this is a downwards force *R'* exerted by the *book*. By Newton's third law, these forces are equal and opposite. This situation is illustrated in Figure 4.6.

Figure 4.6 *Book resting on a table: action and reaction forces.*

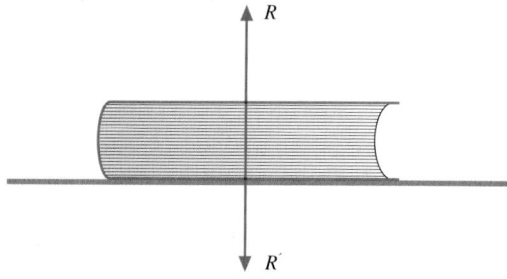

R

R'

Having dealt with the action and reaction forces between book and table, we ought to think about the reaction force to the weight of the book, considered as an action force, even when it is not on the table. This is not so easy, because there does not seem to be an obvious candidate for a second force. But remember that the weight is due to the gravitational attraction of the Earth on the book. If the Earth attracts the book, the book also attracts the Earth. This gravitational force of the book on the Earth is the reaction force. We can test whether the two forces do, indeed, act on different bodies. The action force (the weight of the book) acts on the book. The reaction force (the attraction of the Earth to the book) acts on the Earth. Thus, the condition that action and reaction forces should act on different bodies is satisfied.

4.3 Weight

4.4 Problem solving

In dealing with problems involving Newton's laws, start by drawing a general sketch of the situation. Then consider each body in your sketch. Show all the forces acting on that body, both known forces and unknown forces you may be trying to find. Here it is a real help to try to draw the arrows which represent the forces in approximately the correct direction and approximately to scale. Label each force with its magnitude, or with a symbol if you do not know the magnitude. For each force, you must know the cause of the force (gravity, friction, and so on), and you must also know *on* what object that force acts and *by* what object it is exerted. This labelled diagram is referred to as a *free-body diagram*, because it detaches the body from the others in the situation. Having established all the forces acting on the body, you can use Newton's second law to solve for these forces. This procedure is illustrated in the example which follows.

Newton's second law equates the resultant force acting on a body to the product of its mass and its acceleration. In some problems, the system of bodies is in equilibrium. They are at rest, or are moving in a straight line with uniform speed. In this case, the acceleration is zero, so the resultant force is also zero. In other cases, the resultant force is not zero and the objects in the system are accelerating.

For whichever case applies, you should remember that forces are vectors. You will probably have to resolve the forces into two components at right angles, and then apply the second law to each set of components separately. Problems can often be simplified by making a good choice of directions for resolution. You will end up with a set of equations, based on the application of Newton's second law, which must be solved to determine the unknown quantity.

Example

A box of mass 5.0 kg is pulled along a horizontal floor by a force P of 25 N, applied at an angle of 20 ° to the horizontal (Figure 4.7). A frictional force F of 20 N acts parallel to the floor. Calculate the acceleration of the box.

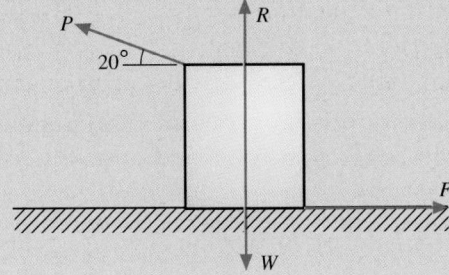

Figure 4.7

The free-body diagram is shown in Figure 4.7. Resolving the forces parallel to the floor, the component of the pulling force, acting to the left, is 25 cos 20 ° = 23.5 N. The frictional force, acting to the right, is 20 N. The resultant force to the left is thus 23.5 − 20.0 = 3.5 N.
By Newton's second law, $a = F/m = 3.5/5.0 = $ **0.70 m s^{-2}**.

Now it's your turn

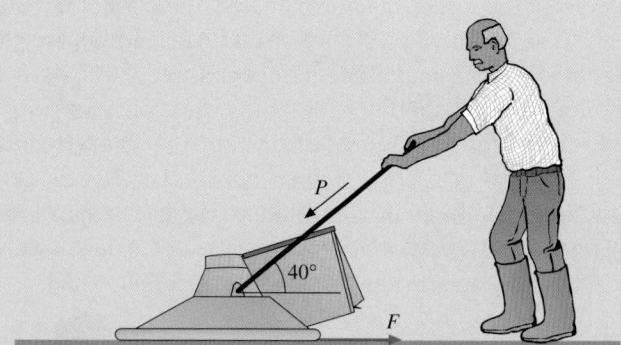

Figure 4.8

A gardener pushes a lawnmower of mass 18 kg at constant speed. To do this requires a force P of 80 N directed along the handle, which is at an angle of 40° to the horizontal (Figure 4.8). Calculate the horizontal retarding force F on the mower. If this retarding force were constant, what force, applied along the handle, would accelerate the mower from rest to 1.2 m s^{-1} in 2.0 s?
Ans: 61 N; 94 N

Sections 4.1–4.4 summary

★ Newton's laws of motion define *force*. The laws are:

First law: Every body continues in its state of rest, or of uniform motion in a straight line, unless acted upon by a force

Second law: Force is proportional to mass × acceleration or $F = ma$, where force F is in newtons, mass m is in kilograms and acceleration a is in metres per second per second

Third law: Whenever one body exerts a force on another, the second exerts an equal and opposite force on the first

★ Weight W and mass m are related by $W = mg$, where g is the acceleration of free fall.

Sections 4.1–4.4 questions

1 A net force of 95 N accelerates an object at 1.9 m s^{-2}. Calculate the mass of the object.

2 A parachute trainee jumps from a platform 3.0 m high. When he reaches the ground, he bends his knees to cushion the fall. His torso decelerates over a distance of 0.65 m. Calculate:

   (a) the speed of the trainee just before he reaches the ground,

   (b) the deceleration of his torso,

   (c) the average force exerted on his torso (of mass 45 kg) by his legs during the deceleration.

3 If the acceleration of a body is zero, does this mean that no forces act on it?

4 A railway engine pulls two carriages of equal mass with uniform acceleration. The tension in the coupling between the engine and the first carriage is T. Deduce the tension in the coupling between the first and second carriages.

5 What is your mass? What is your weight?

4.5 | Momentum

We shall now introduce a quantity called **momentum**, and see how Newton's laws may be related to it.

> ★ The momentum of a particle is defined as the product of its mass and its velocity.

in words:

$$\bigstar \quad momentum = mass \times velocity$$

and in symbols:

$$\bigstar \quad p = mv$$

The unit of momentum is the unit of mass times the unit of velocity, that is kg m s^{-1}. Al alternative unit is the newton second (N s). Momentum, like velocity, is a vector quantity.

Newton's first law states that every body continues in a state of rest, or of uniform motion in a straight line, unless acted upon by a force. We can express this law in terms of momentum. If a body maintains its uniform motion in a straight line, its momentum is unchanged. If a body remains at rest, again its momentum (zero) does not change. Thus, an alternative statement of the first law is that the momentum of a particle remains constant unless an external force acts on the particle. As an equation,

$$\bigstar \quad p = \text{constant}$$

This is a special case, for a single particle, of a very important conservation law: the principle of conservation of momentum. The word 'conservation' here means than that the quantity remains constant.

Newton's second law can also be expressed in terms of momentum. We already have it in a form which relates the force acting on a body to the product of the mass and the acceleration of the body. Remember that the acceleration of a body is the rate of change of its velocity. The product of mass and acceleration then is just the mass times the rate of change of velocity. For a body of constant mass, this is just the same as the rate of change of mass times velocity. But mass times velocity is momentum, so the product of mass and acceleration is identical with the rate of change of momentum. Thus, the second law may be stated as:

> $\bigstar$ the force acting on a body is equal to the rate of change of its momentum.

This is the formal definition of the second law. Expressed in terms of symbols,

$$\bigstar \quad F = ma = m(\Delta v/\Delta t) = \Delta(mv)/\Delta t = \Delta p/\Delta t$$

Continuing with the idea of force being equal to rate of change of momentum, the third law relating to action and reaction forces becomes: the rate of change of momentum due to the action force on one body is equal and opposite to the rate of change of momentum due to the reaction force on the other body. Simplifying this by removing the idea of rates of change, this becomes: when two bodies exert action and reaction forces on each other, their changes of momentum are equal and opposite.

4.6 The principle of conservation of momentum

Figure 4.9 System of two particles.

We have already seen that Newton's first law states that the momentum of a single particle is constant, if no external force acts on the particle. Now think about a system of two particles (Figure 4.9). We allow these particles to exert some sort of force on each other: it could be gravitational attraction or, if the particles were charged, it could be electrostatic attraction or repulsion.

These two particles are isolated from the rest of the universe, and experience no outside forces at all. If the first particle exerts a force F on the second, Newton's third law tells us that the second exerts a force $-F$ on the first. The minus sign indicates that the forces are in opposite directions. As we saw in the last section, we can express this law in terms of change of momentum. The change of momentum of the second particle as a result of the force exerted on it by the first is equal and opposite to the change of momentum of the first particle as a result of the force exerted on it by the second. Thus, the changes of momentum of the individual particles cancel out, and the momentum of the system of two particles remains constant. The particles have merely exchanged some momentum. The situation is expressed by the equation

$$\bigstar \quad p = p_1 + p_2 = \text{constant}$$

where p is the total momentum, and p_1 and p_2 are the individual momenta.

We could extend this idea to a system of three, four, or finally any number n of particles.

> $\bigstar$ If no external force acts on a system, the total momentum of the system remains constant, or is conserved.

A system on which no external force acts is often called an *isolated system*. The fact that the total momentum of an isolated system is constant is the *principle of conservation of momentum*. It is a direct consequence of Newton's third law of motion.

4.7 Collisions

We now use the principle of conservation of momentum to analyse a system consisting of two colliding particles. (If you want a real example to think about, try billiard balls.)

Consider two particles A and B making a direct, head-on collision. Particle A has mass m_1 and is moving with velocity u_1 in the direction from left to right; B has mass m_2 and has velocity u_2 in the direction from right to left (Figure 4.10). As velocity is a vector quantity, this is the same as saying that the velocity is $-u_2$ from left to right. The particles collide. After the collision they have velocities $-v_1$ and v_2 respectively in the direction from left to right. That is, both particles are moving back along their directions of approach.

According to the principle of conservation of momentum, the total momentum of this isolated system remains constant, whatever happens as a result of the interaction of the particles. Thus, the total momentum before the collision must be equal to the total momentum after the collision. The

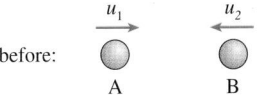

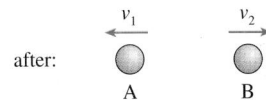

Figure 4.10 *Collision between two particles.*

momentum before the collision is

$$m_1u_1 - m_2u_2$$

and the momentum after is

$$-m_1v_1 + m_2v_2.$$

Because total momentum is conserved,

$$m_1u_1 - m_2u_2 = -m_1v_1 + m_2v_2.$$

Knowing the masses of the particles and the velocities before collision, this equation would allow us to calculate the relation between the velocities after the collision.

The way to attack collision problems is as follows. First, draw a labelled diagram showing the colliding bodies before collision. Draw a separate diagram showing the situation after the collision. Take care to define the directions of all the velocities. Then obtain an expression for the total momentum before the collision, bearing in mind that momentum is a vector quantity. Similarly, find the total momentum after the collision, taking the same reference direction. Then equate the momentum before the collision to the momentum afterwards.

Examples

1   What is the magnitude of the momentum of an α-particle of mass 6.6×10^{-27} kg travelling with a speed of 2.0×10^7 m s^{-1}?

$p = mv = 6.6 \times 10^{-27} \times 2.0 \times 10^7 =$ **1.3×10^{-19} kg m s^{-1}**.

2   A cannon of mass 1.5 tonnes (1.5×10^3 kg) fires a cannon-ball of mass 5.0 kg. The speed with which the ball leaves the cannon is 70 m s^{-1} relative to the Earth (Figure 4.11). What is the initial speed of recoil of the cannon?

Figure 4.11

The system under consideration is the cannon and the cannon-ball. The total momentum of the system before firing is zero. Because the total momentum of an isolated system is constant, the total momentum after firing must also be zero. That is, the momentum of the cannon-ball, which is $5.0 \times 70 = 350$ kg m s^{-1} to the right, must be exactly balanced by the momentum of the cannon. If the initial speed of recoil is v, the momentum of the cannon is $1500v$ to the left. Thus, $1500v = 350$ and $v = \mathbf{0.23}$ **m s^{-1}**.

Now it's your turn

1 What is the magnitude of the momentum of an electron of mass 9.1×10^{-31} kg travelling with a speed of 7.5×10^6 m s^{-1}?
Ans: 6.8×10^{-24} kg m s^{-1}

2 An ice-skater of mass 80 kg, initially at rest, pushes his partner, of mass 65 kg, away from him so that she moves with an initial speed of 1.5 m s^{-1}. What is the initial speed of the skater after this manoeuvre?
Ans: 1.2 m s^{-1}

4.8 The momentum-impulse equation

We now introduce a quantity called **impulse** and relate it to a change in momentum.

★ If a constant force F acts on a body for a time Δt, the **impulse** of the force is given by $F\Delta t$.

The unit of impulse is given by the unit of force, the newton, multiplied by the unit of time, the second: it is the newton second (N s).

We know from Newton's second law that the force acting on a body is equal to the rate of change of momentum of the body. We have already expressed this as the equation

$$F = \Delta p/\Delta t.$$

If we multiply both sides of this equation by Δt, we obtain

★ $F\Delta t = \Delta p$

We have already defined $F\Delta t$ as the impulse of the force. The right-hand side of the equation (Δp) is the change in the momentum of the body. So, Newton's second law tells us that the impulse of a force is equal to the change in momentum. This is the **impulse-momentum equation**. It is often used in dealing with forces that act over a short interval of time, as in a collision. The forces between colliding bodies are seldom constant throughout the collision, but the equation can be applied to obtain information about the average force acting.

Note that the idea of impulse explains why there is an alternative unit for momentum. In section 4.5 we introduced the kg m s^{-1} and the N s as possible units for momentum. The kg m s^{-1} is the logical unit, the one you arrive at if you take momentum as being the product of mass and velocity. The N s comes from the impulse-momentum equation: it is the unit of impulse, and because impulse is equal to change of momentum, it is also a unit for momentum.

Example

Some tennis players can serve the ball at a speed of 55 m s^{-1} (about 120 miles per hour). The tennis ball has a mass of 60 g. In an experiment it is determined that the ball is in contact with the racket for 25 ms during the serve. Calculate the average force exerted by the racket on the ball.

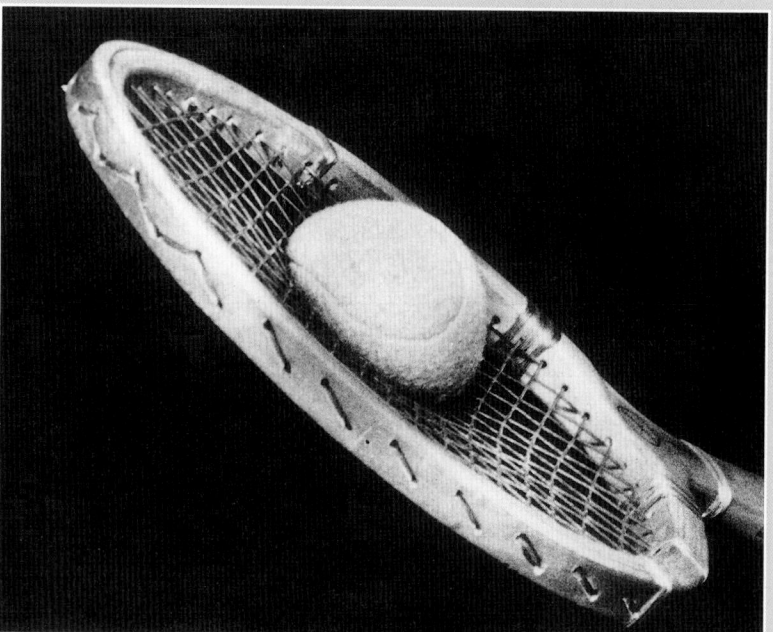

The change in momentum of the ball as a result of the serve is $0.060 \times 55 = 3.3$ kg m s^{-1}. By the impulse-momentum equation, the change in momentum is equal to the impulse of the force. Since impulse is the product of force and time, $Ft = 3.3$ N s.

Here t is 0.025 s; thus $F = 3.3/0.025 =$ **132 N**.

Now it's your turn

A golfer hits a ball of mass 45 g at a speed of 40 m s^{-1}. The golf club is in contact with the ball for 3.0 ms. Calculate the average force exerted by the club on the ball.

Ans: 600 N

4.9 Elastic and inelastic collisions

In some collisions, kinetic energy is conserved as well as momentum. By the **conservation of kinetic energy**, we mean that the total kinetic energy of the colliding bodies before collision is the same as the total kinetic energy afterwards. This means that no energy is lost in the permanent deformation of the colliding bodies, or as heat and sound. There is a transformation of energy during the collision: while the colliding bodies are in contact, some of the kinetic energy is transformed into elastic potential energy, but as the bodies separate, it is transformed into kinetic energy again.

Using the same notation for the masses and speed of the colliding particles as in Section 4.7 (see Figure 4.12), the total kinetic energy of the particles before collision is:

$$\tfrac{1}{2}m_1u_1^2 + \tfrac{1}{2}m_2u_2^2.$$

The total kinetic energy afterwards is:

$$\tfrac{1}{2}m_1v_1^2 + \tfrac{1}{2}m_2v_2^2.$$

before: $\quad u_1 \quad\quad u_2$

A $\quad$ B

after: $\quad v_1 \quad\quad v_2$

A $\quad$ B

Figure 4.12 *Collision between two particles.*

★ If the collision is elastic, the kinetic energy before collision is equal to the kinetic energy after collision

$$\tfrac{1}{2}m_1u_1^2 + \tfrac{1}{2}m_2u_2^2 = \tfrac{1}{2}m_1v_1^2 + \tfrac{1}{2}m_2v_2^2.$$

Note that because energy is a scalar, the directions of motion of the particles are not indicated by the signs of the various terms.

In solving problems about elastic collisions, this equation is useful because it gives another relation between masses and velocities, in addition to that obtained from the principle of conservation of momentum.

Elastic collisions occur in the collisions of atoms and molecules. We shall see in Chapter 5 that one of the most important assumptions in the kinetic theory of gases is that the collisions of the gas molecules with the walls of the container are perfectly elastic. However, in larger scale collisions, such as those of billiard balls, collisions cannot be perfectly elastic. (The 'click' of billiard balls on impact indicates that a very small fraction of the total energy of the system has been transformed into sound.) Nevertheless, we often make the assumption that such a collision is perfectly elastic.

★ Collisions in which the total kinetic energy is not the same before and after the event are called **inelastic**.

Total energy must, of course, be conserved. But in an inelastic collision the kinetic energy that does not reappear in the same form is transformed into heat, sound and other forms of energy. In an extreme case, all the kinetic energy may be lost. A lump of plasticine dropped on to the floor does not bounce. All the kinetic energy it possessed just before hitting the floor has been transformed into the work done in flattening the lump, and (a much smaller amount) into the sound energy emitted as a 'squelch'.

★ Although kinetic energy may or may not be conserved in a collision, momentum is always conserved, and so is total energy.

The truth of this statement may not be entirely obvious, especially when considering examples such as the lump of plasticine which was dropped on to the floor. Surely the plasticine had momentum just before the collision with the floor, but had no momentum afterwards? True! But for the system of the lump of plasticine alone, external forces (the attraction of the Earth on the plasticine, and the force exerted by the floor on the plasticine on impact) were acting. When external forces act, the conservation principle does not apply. We need to consider a system in which no external forces act. Such a system is the lump of plasticine and the Earth. While the plasticine falls towards the floor, gravitational attraction will also pull the Earth towards the plasticine. Conservation of momentum then applies in that the total momentum of plasticine and Earth remains constant throughout the process:

before the collision, and after it. The effects of the transfer of the plasticine's momentum to the Earth are not noticeable due to the difference in mass of the two objects.

Example

A billiard ball A moves with speed u_A directly towards a similar ball B which is at rest (Figure 4.13). The collision is perfectly elastic. What are the speeds v_A and v_B after the collision?

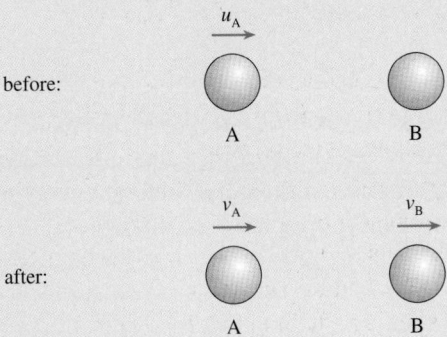

Figure 4.13

It is convenient to take the direction from left to right as the direction of positive momentum. If the mass of a billiard ball is m, the total momentum of the system before the collision is mu_A. By the principle of conservation of momentum, the total momentum after collision is the same as that before, or

$$mu_A = mv_A + mv_B.$$

The collision is perfectly elastic, so the total kinetic energy before the collision is the same as that afterwards, or

$$\tfrac{1}{2}mu_A^2 = \tfrac{1}{2}mv_A^2 + \tfrac{1}{2}mv_B^2.$$

Solving these equations gives $v_A = 0$ and $v_B = u_A$. That is, ball A comes to a complete standstill, and ball B moves off with the same speed as which ball A struck it. (Another solution is possible algebraically: $v_A = u_A$ and $v_B = 0$. This corresponds to a non-collision. Ball A is still moving with its initial speed, and ball B is still at rest. In cases where algebra gives us two possible solutions, we need to decide which is the one which is physically appropriate.)

Now it's your turn

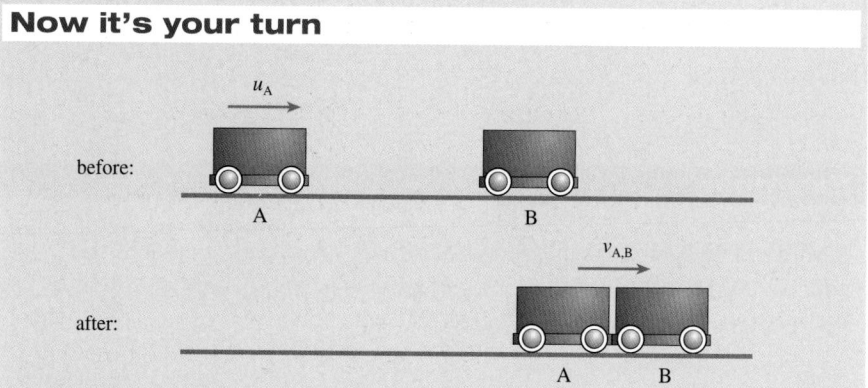

Figure 4.14

A trolley A moves with speed u_A towards a trolley B of equal mass which is at rest (Figure 4.14). The trolleys stick together and move off as one with speed $v_{A,B}$. Find $v_{A,B}$. What fraction of the initial kinetic energy of trolley A is converted into other forms in this inelastic collision?

Ans: 0.5; $u_A/2$

Sections 4.5–4.9 summary

★ The momentum of p of a body is the product of its mass m and its velocity v, described as $p = mv$. Momentum has units kg m s^{-1} or N s. It is a vector quantity.

★ Newton's laws of motion can be stated in terms of momentum.

First law: The momentum of a particle remains constant unless an external force acts on the particle; p = constant

Second law: The force acting on a body is equal to the rate of change of its momentum; $F = \Delta p/\Delta t$

Third law: When two bodies exert action and reaction forces on each other, their changes of momentum are equal and opposite

★ The principle of conservation of momentum states that the total momentum of an isolated system is constant. An isolated system is one on which no external force acts.

★ In collisions between bodies, application of the conservation of momentum principle shows that the total momentum of the system before the collision is equal to the total momentum after the collision.

★ An elastic collision is one in which the total kinetic energy remains the same.

★ An inelastic collision is one in which the total kinetic energy is not the same before and after the event.

★ Although kinetic energy may or may not be conserved in a collision, momentum is always conserved, and so is total energy.

★ The impulse of a force F is the product of the force and the time Δt for which it acts: impulse $= F\Delta t$. The unit of impulse is the N s.

★ The impulse-momentum equation states that the impulse of a force acting on a body is equal to the change of momentum of the body: $F\Delta t = \Delta p$.

Sections 4.5–4.9 questions

1 Calculate the magnitude of the momentum of a car of mass 1.5 tonnes (1.5×10^3 kg) travelling at a speed of 22 m s^{-1}.

2 When a certain space rocket is taking off, the propellant gases are expelled at a rate of 900 kg s^{-1} and a speed of 40 km s^{-1}. Calculate the thrust on the rocket.

3 An insect of mass 4.5 mg, flying with a speed of 0.12 m s^{-1}, encounters a spider's web, which brings it to rest in 2.0 ms. Calculate the force exerted by the insect on the web.

4 An atomic nucleus at rest emits an α-particle of mass 4 u. The speed of the α-particle is found to be 5.6×10^6 m s^{-1}. Calculate the speed with which the daughter nucleus, of mass 218 u, recoils.

5 A heavy particle of mass m_1, moving with speed u, makes a head-on collision with a light particle of mass m_2, which is initially at rest. The collision is perfectly elastic, and m_2 is very much less than m_1. Describe the motion of the particles after the collision.

6 A light body and a heavy body have the same momentum. Which has the greater kinetic energy?

Exam Questions

1 A bullet of mass 12 g is fired horizontally from a gun with a velocity of 180 m s^{-1}. It hits, and becomes embedded in, a block of wood of mass 2000 g, which is freely suspended by long strings, as shown in Figure 4.15.

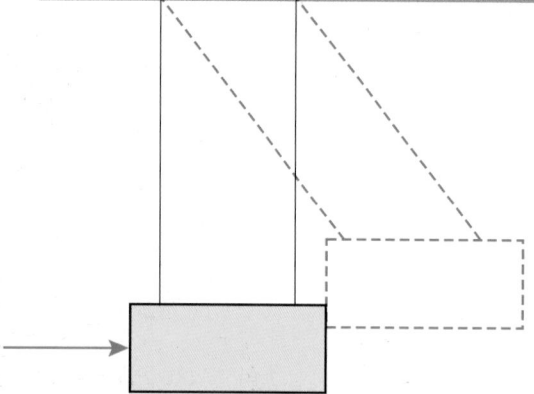

Figure 4.15

Calculate:

(a) i) the magnitude of the momentum of the bullet just before it enters the block,

ii) the magnitude of the initial velocity of the block and bullet after impact,

iii) the kinetic energy of the block and embedded bullet immediately after the impact.

(b) Deduce the maximum height above the equilibrium position to which the block, and embedded bullet, rises after impact.

2 A nucleus A of mass 222 u is moving at a speed of 350 m s^{-1}. While moving, it emits an α-particle of mass 4 u. After the emission, it is determined that the daughter nucleus, of mass 218 u, is moving with speed 300 m s^{-1}. Calculate the speed of the α-particle.

3 A safety feature of modern cars is the air-bag, which, in the event of a collision, inflates and is intended to decrease the risk of serious injury. Use the concept of impulse to explain why an air-bag might have this effect.

4 Two frictionless trolleys A and B, of mass m and $3m$ respectively, are on a horizontal track (Figure 4.16). Initially they are clipped together by a device which incorporates a spring, compressed between the trolleys. At time $t = 0$ the clip is released. The velocity of trolley B is then u to the right.

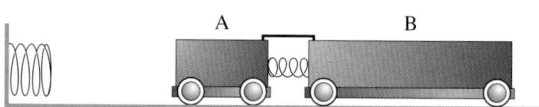

Figure 4.16

(a) Calculate the velocity of trolley A as the trolleys move apart.

(b) At time $t = t_1$, trolley A collides elastically with a fixed spring and rebounds. (Compression and expansion of the spring take a negligibly short time.) Trolley A catches up with trolley B at time $t = t_2$.

i) Calculate the velocity of trolley A between $t = t_1$ and $t = t_2$.

ii) Find an expression for t_2 in terms of t_1.

(c) When trolley A catches up with trolley B at time t_2 the clip operates so as to link them again, this time without the spring between them, so that they move together with velocity v. Calculate the common velocity v in terms of u.

(d) Initially the trolleys were at rest and the total momentum of the system was zero. However, the answer to (c) shows that the total momentum after $t = t_2$ is not zero. Discuss this result with reference to the principle of conservation of momentum.

5 A ball of mass m makes a perfectly elastic head-on collision with a second ball, of mass M, initially at rest. The second ball moves off with half the original speed of the first ball.

(a) Express M in terms of m.

(b) Determine the fraction of the original kinetic energy retained by the ball of mass m after the collision.

CHAPTER FIVE
Thermal physics

The aim of this chapter is to describe a number of phenomena associated with temperature and thermal energy. We shall start by looking at the physical basis of thermometers, which are instruments to measure temperature, and then see how scales of temperature are defined. The fundamental scale of temperature, the thermodynamic scale, is based on the properties of an ideal gas. To develop the idea of the temperature of a gas, we shall need to think of the gas as a collection of very small particles in continuous random motion. This is the basis of the kinetic theory of gases. Finally, we shall consider the energy of the particles of a gas and link this internal energy to the quantities of work and heat through the first law of thermodynamics.

Engraving of Montgolfier balloon beginning its flight. The Montgolfier brothers were the first to fly by hot-air balloon. Lighting a fire below the balloon caused the air in it to expand, forcing some out and reducing the overall density of the balloon. When the upthrust on the balloon became greater than its weight, it rose above the ground.

5.1 Temperature

Our everyday idea of temperature is based on our sense of touch. Putting your hand into a bowl containing ice immediately gives a sense of cold; putting the other hand into a bowl of warm water gives the sensation of heat.

Intuitively, we would say that the water is at a higher temperature than the ice.

Figure 5.1

In physics we look for ways of defining and measuring quantities. In the case of temperature, we will first look at ways of measuring this quantity, and then think about the definition.

Many physical properties change with temperature. Most materials (solids, liquids and gases) expand as their temperature is increased. The electrical resistance of a metal wire increases as the temperature of the wire is increased. If two wires of different metals are twisted together at one end, and the other ends are connected to a voltmeter, the reading on the voltmeter depends on the temperature of the junction of the wires. All these properties may be made use of in different types of thermometer. A thermometer is an instrument for measuring temperature. The physical property on which a particular thermometer is based is called the thermometric property, and the working material of the thermometer, the property of which varies with temperature, is called the thermometric substance. Thus, in the familiar mercury-in-glass thermometer, the thermometric substance is mercury and the thermometric property is the length of the mercury thread in the capillary tube of the thermometer.

Remember that temperature measures the degree of 'hotness' of a body. It does not measure the amount of heat.

Temperature scales

Each type of thermometer can be used to establish its own temperature scale. To do this, the fact that substances change state (from solid to liquid, or from liquid to gas) at fixed temperatures is used to define reference temperatures, which are called fixed points. By taking the value of the thermometric property at two fixed points, and dividing the range of values into a number of equal steps (or degrees), we can set up what is called an empirical scale of temperature for that thermometer. ('Empirical' means 'derived by experiment'.) If the fixed points are the melting point of ice and the boiling point of water, and if we choose to have one hundred equal degrees between the temperatures corresponding to these fixed points, taken as 0 degrees and 100 degrees respectively, we arrive at the empirical centigrade scale of temperature for that thermometer. If the values of the thermometric property P are P_i and P_s at the ice- and steam-points respectively, and if the property has the value P_θ at an unknown temperature θ, the unknown temperature is given by

$$\bigstar \quad \theta = \frac{100(P_\theta - P_i)}{(P_s - P_i)}$$

on the empirical centigrade scale of this particular thermometer. This equation is illustrated in graphical form in Figure 5.2. It is important to realise that the choice of a different thermometric substance and thermometric property would lead to a different centigrade scale. Agreement between scales occurs only at the two fixed points. This happens because the property may not vary linearly with temperature.

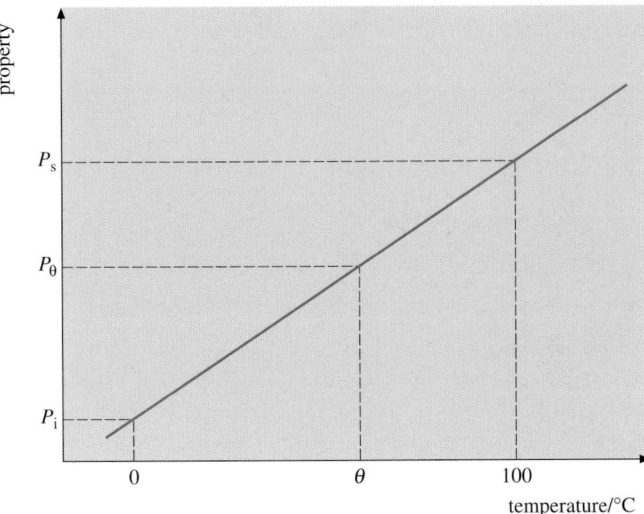

Figure 5.2 *Empirical centigrade scale.*

This situation, with temperature values depending on the type of thermometer on which they are measured, is clearly unsatisfactory for scientific purposes. It is found that the differences between empirical scales are small in the case of thermometers based on gases as thermometric substances. In the constant-volume gas thermometer (Figure 5.3) the pressure of a fixed volume of gas is used as the thermometric property. The differences between the scales of different gas thermometers become even less as the pressures used are reduced. This is because the lower the pressure of a real gas, the more linear is the variation of pressure or volume. If we set up an empirical centigrade scale for a real gas in a constant-volume thermometer by obtaining the pressures of the gas at the ice-point (0 °C) and the steam-point (100 °C) we can extrapolate the graph of pressure p against the centigrade temperature θ to find the temperature at which the pressure of the gas would become zero. This is shown in Figure 5.4. The extrapolated temperature will be found to be close to -273 degrees on the empirical centigrade constant-volume gas thermometer scale. If the experiment is repeated with lower and lower pressures of gas in the thermometer, the extrapolated temperature tends to a value of -273.15 degrees. This temperature is known as the **absolute zero** of temperature.

Figure 5.3 *Constant-volume gas thermometer.*

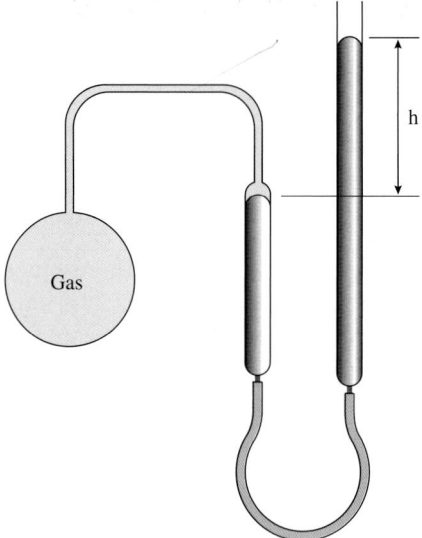

Figure 5.4 *Graph of p against θ for constant-volume gas thermometer.*

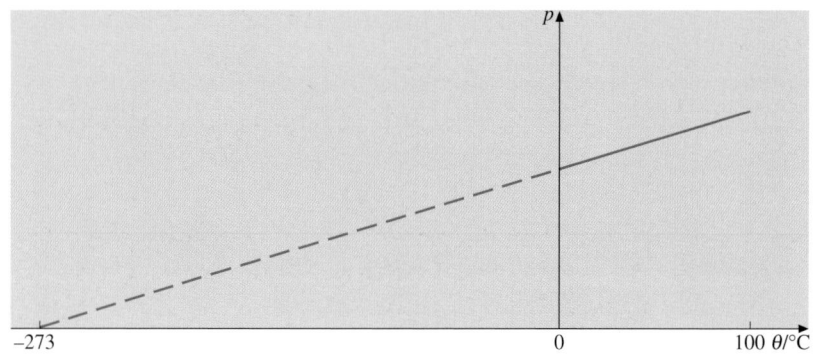

Example

The length of the mercury column in a mercury-in-glass thermometer is 25 mm at the ice-point and 180 mm at the steam-point. What is the temperature when the length of the column is 55 mm?

Using the equation $\theta = 100(P_\theta - P_i)/(P_s - P_i)$, the thermometric property P is the length of the mercury column. Thus,
$\theta = 100(55 - 25)/(180 - 25) = \textbf{19.4 °C}$ on the centigrade scale of this mercury-in-glass thermometer.

Now it's your turn

A resistance thermometer has resistance 95.2 Ω at the ice-point and 138.6 Ω at the steam-point. What resistance would be obtained at a temperature of 19.4 °C on the centigrade scale of this resistance thermometer?

Ans: 103.6 Ω

Thermodynamic temperature

As mentioned above, decreasing the pressure of a real gas makes it behave more and more like an ideal gas. For an ideal gas, the relation between pressure p, volume V and temperature T is

$$\bigstar \quad \frac{pV}{T} = \text{constant.}$$

Here T is the **thermodynamic temperature**. We now use the property of an ideal gas to define the thermodynamic scale of temperature. We have already seen from the idea of extrapolating the p, θ graph for a constant-volume gas thermometer that there seems to be a natural zero of temperature, absolute zero. This is used as one of the fixed points of the thermodynamic scale. The upper fixed point is taken as the triple point of water, the temperature at which ice, water and water vapour are in equilibrium. This is found to be less dependent on environmental conditions, such as pressure, than the ice-point. The thermodynamic temperature of the triple point of water is taken as 273.16 units, by international agreement. This defines the kelvin (symbol K), the unit of thermodynamic temperature.

> ★ One kelvin is the fraction 1/273.16 of the thermodynamic temperature of the triple point of water.

Thus, if a constant-volume gas thermometer gives a pressure reading of p_{tr} at the triple point, and a pressure reading of p at an unknown temperature T, the unknown temperature (in K) is given by

$$\bigstar \quad T = 273.16(p/p_{tr}).$$

The Celsius scale

Why choose 273.16 as the number of units between the two fixed points of this scale? The reason is that this number will give 100 K between the ice- and steam-points, allowing agreement between the thermodynamic temperature scale and a centigrade scale based on the pressure of an ideal gas. The ideal-gas centigrade scale is based on experiments with real gases at decreasing pressures. This agreement is based on a slightly awkward linking up of the theoretical thermodynamic scale and the empirical constant-volume gas thermometer scale. To avoid this complication, a new scale, the Celsius scale, was defined by international agreement.

> ★ The unit of temperature on the Celsius scale is the degree Celsius (°C), which is exactly equal to the kelvin.

The equation linking temperature t on the Celsius scale and thermodynamic temperature T is

$$\bigstar \quad t = T - 273.15$$

In this equation, t is measured in °C and T in K. Note that the degree sign ° always appears with the Celsius symbol, but it is never used with the kelvin symbol K.

Don't be confused by the numbers 273.15 and 273.16. Absolute zero on the ideal gas constant-volume scale is −273.15 degrees. The fact that 273.16 occurs in the definition of the kelvin means that the temperature of the triple point of water is 0.01 °C. For many purposes, in particular in calculations involving conversions of temperatures from the Celsius to the thermodynamic scale or the other way, it is sufficient to work with the number 273.

Example

The pressure reading of a constant-volume gas thermometer is 1.50×10^4 Pa at the triple point of water. When the bulb is placed in a certain liquid, the pressure is 4.28×10^3 Pa. Find the temperature of the liquid.

We use the equation $T = 273.16(p/p_{tr})$. Because the pressure readings are given to three significant figures only, we can approximate the 273.16 to 273. Thus, $T = 273 \times (4.28 \times 10^3/1.50 \times 10^4) = $ **77.9 K**. (The liquid is liquid nitrogen at its boiling-point.)

Now it's your turn

The pressure reading of a constant-volume gas thermometer is 1.367×10^3 Pa at the steam-point. What is the pressure reading at the triple point of water?

Ans: 1.001×10^3 Pa

5.2 Thermometers

We have already mentioned some of the many different types of thermometer. In choosing a thermometer for a particular application, we need to consider a number of aspects. These include accuracy, sensitivity (the distance between divisions on its scale), and the range of temperatures it is able to measure. In some cases the thermometer has to measure rapidly-varying temperatures, in which case its speed of response will be important, and whether it can be directly read, or requires time-consuming adjustments by the operator. For other applications it is important that the sensitive part of the thermometer is small and does not absorb much heat, so that it does not change the temperature of the object during measurement.

Gas thermometers (Figure 5.3) are important because they provide a link with the thermodynamic scale. However, they are bulky and inconvenient. Because of the adjustments necessary, they are certainly not suitable for the measurement of rapidly-varying temperatures. The bulb containing the gas may have a volume of as much as a litre (1000 cm³), so a gas thermometer could not be used for measuring the temperature of a small body.

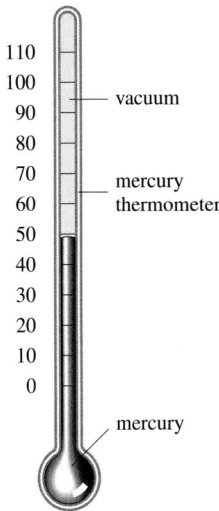

Figure 5.5 *Mercury-in-glass thermometer.*

The most familiar type of thermometer is probably the liquid-in-glass type (Figure 5.5), based on the expansion of the liquid. Such thermometers are convenient, sensitive and moderately quick-acting. The range of the thermometer is limited by the liquid used and the glass containing it. Mercury-in-glass thermometers cover the range from about −40 °C to 350 °C; ethanol-in-glass (the ethanol is normally coloured with a red dye) from about −120 °C to 80 °C. An advantage of liquid-in-glass thermometers is, that over their temperature range, their empirical centigrade scale is very close to the thermodynamic scale.

An important class of thermometers is based on the temperature variation of electrical resistance. The resistance of a metal wire increases with increasing temperature. The range of metal resistance thermometers is very wide, from about −260 °C to 1700 °C. The temperature sensor is a coil of fine wire, often platinum. The variation with temperature of the resistance of a metal wire is not exactly linear with thermodynamic temperature, and for accurate work the resistance thermometer needs to be calibrated at a number of standard temperatures. However, over a small range of temperature, the variation of resistance is linear, as shown in Figure 5.6. Semiconducting materials are also used as resistance thermometers: these are called **thermistors** (Figure 5.7). The electrical resistance of such devices decreases very rapidly with increasing temperature. (Note that this is in the opposite sense to the behaviour of a metal wire.) Because of the rapid change of resistance, thermistors are very sensitive devices, but their scales are very non-linear compared with thermodynamic temperature, and calibration is essential. By selection of suitable semiconducting materials, a wide range of temperatures can be covered. Their very small size (some smaller than a pin-head) can be used to measure the temperature of small objects, and also varying temperatures. A simple circuit allows the resistance of the thermistor, or a quantity related to the resistance, to be displayed on a meter, which can then be calibrated in terms of temperature. A common use of a thermistor thermometer is as a temperature sensor in car radiators.

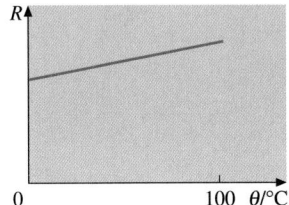

Figure 5.6 *Resistance of a metal wire over a small range of temperatures.*

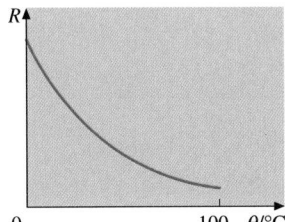

Figure 5.7 *Resistance of a thermistor over a small range of temperatures.*

Another type of electrical thermometer makes use of the **thermoelectric effect**. When the junctions of two different conductors (Figure 5.8) are at different temperatures, an e.m.f. is developed. The device is known as a thermocouple. The relation between the e.m.f. and the temperature difference is not linear with thermodynamic temperature (see Figure 5.9), and calibration is essential. The e.m.f. generated is rather small (for a

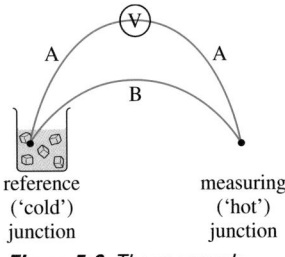

Figure 5.8 *Thermocouple.*

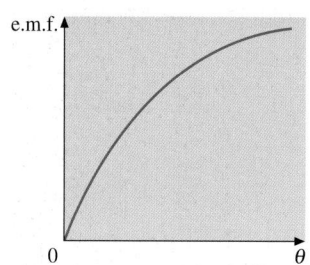

Figure 5.9 *Graph of thermocouple e.m.f. against temperature.*

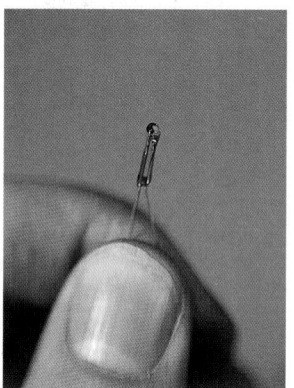

Thermistors are useful temperature measuring devices.

thermocouple made of a copper wire and a constantan wire, it is about 5 mV for a temperature difference of 100 °C), but it is easy to arrange a circuit to amplify this e.m.f. and provide a direct reading, which can then be calibrated. Thermocouples cover a large temperature range. Because the sensing part (the junction between the two wires) is very small the thermocouple can be used to measure rapidly-varying temperatures, and the temperatures of small objects.

Sections 5.1–5.2 summary

★ Empirical centigrade scale of temperature:
$\theta = 100(P_\theta - P_i)/(P_s - P_i)$, where θ is the centigrade temperature on the scale of a thermometer based on a property P. P_s, P_i and P_θ are the values of the property at the steam-point, ice-point and at temperature θ respectively.

★ The kelvin (K), unit of thermodynamic temperature, is the fraction 1/273.16 of the thermodynamic temperature of the triple point of water.

★ Thermodynamic scale of temperature: for a constant-volume gas thermometer, $T = 273.16(p/p_{tr})$, where T is the thermodynamic temperature, p is the pressure reading at temperature T and p_{tr} is the reading at the triple point of water.

★ Celsius scale: $t = T - 273.15$, where t is the Celsius temperature (in °C) and T is the thermodynamic temperature (in K).

★ The thermometric properties of common thermometers include the length of a column of liquid (in liquid-in-glass thermometers), the pressure of a gas (in constant-volume gas thermometers), the resistance of a coil of wire or of a sample of semiconductor (in electrical resistance thermometers), and the thermoelectric e.m.f. (in thermocouples).

Sections 5.1–5.2 questions

1 The e.m.f. of a certain thermocouple is 5.60 mV when one junction is placed in melting ice and the other in steam. With one junction in melting ice and the other in a boiling liquid, the e.m.f. is −2.46 mV. Calculate the boiling point of the liquid on the centigrade scale of this thermocouple.

2 The gain of an operational amplifier is found to depend on temperature. At 5 °C, the gain is 110; at 18 °C, it is 125. Assuming that the gain depends linearly on Celsius temperature, estimate the gain when the op-amp is used on a hot day, at a temperature of 25 °C.

5.3 The gas laws

Experiments in the seventeenth and eighteenth centuries showed that the volume, pressure and temperature of a given sample of gas are all related. The precise relation between these quantities and the mass of gas in the sample is called the **equation of state** of the gas.

For a given mass of gas, it was found that

> ★ the volume V of the gas is inversely proportional to its pressure p, provided that the temperature is held constant.

Robert Boyle.

Expressed mathematically, this is

$$★ \quad V \propto 1/p,$$

or

$$★ \quad pV = \text{constant.}$$

Another way of writing this equation is

$$★ \quad p_1V_1 = p_2V_2,$$

where p_1 and V_1 are the initial pressure and volume of the gas, and p_2 and V_2 are the final values after a change of pressure and volume carried out at constant temperature.

A graph of V against $1/p$ is a straight line through the origin (Figure 5.10). This relation is known as **Boyle's law**, after Robert Boyle (1627–91), who first stated this relation as a result of his own experiments.

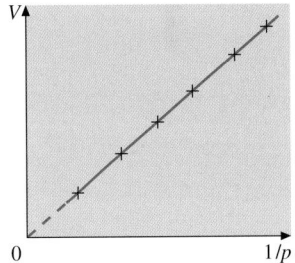

Figure 5.10 *Graph of V against 1/p.*

The effect of temperature on the volume of a gas was investigated by the French scientist Jacques Charles (1746–1823). Charles found that the graph of volume V against temperature θ is a straight line (Figure 5.11). Because gases liquefy when the temperature is reduced, experimental points could not be obtained below the liquefaction temperature. But if the graph was projected backwards, it was found that it cut the temperature axis at about $-273\ °C$.

In the description of the constant-volume gas thermometer, we have already met the effect of temperature on the pressure of a gas. This effect was investigated by another Frenchman, Joseph Gay-Lussac (1778–1850).

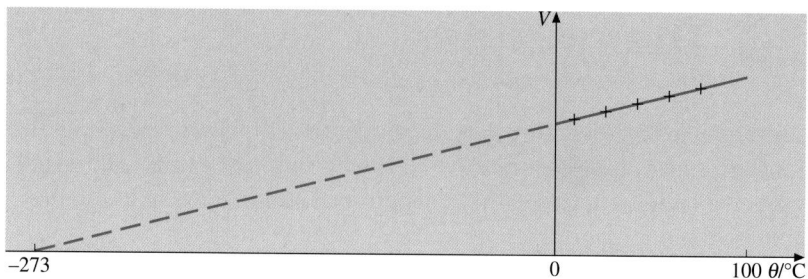

Remember that the graph of pressure p against temperature θ is a straight line, which, if projected like the volume-temperature graph, also meets the temperature axis at about $-273\,°C$ (Figure 5.4). We used this fact to introduce the thermodynamic scale of temperature and the idea of the absolute zero of temperature.

If Celsius temperatures are converted to thermodynamic temperatures, Charles' results are expressed as:

 for a given mass of gas maintained at constant pressure, the volume V of the gas is directly proportional to its thermodynamic temperature T.

Expressed mathematically, this is

$$\bigstar\ V \propto T.$$

This is **Charles' law**. Another way of writing the equation is

$$\bigstar\ \frac{V_1}{T_1} = \frac{V_2}{T_2},$$

where V_1 and T_1 are the initial volume and temperature and V_2 and T_2 are the final values.

The corresponding relation between pressure and temperature is:

 for a given mass of gas maintained at constant volume, the pressure p of the gas is directly proportional to its thermodynamic temperature T.

Mathematically,

$$\bigstar\ p \propto T.$$

This is **Gay-Lussac's law**, or the **law of pressures**. The equation can also be written as

$$\bigstar \quad \frac{p_1}{T_1} = \frac{p_2}{T_2},$$

where the subscripts 1 and 2 again refer to initial and final values.

We can combine the three gas laws into a single relation between pressure p, and volume V and temperature T for a fixed mass of gas. This relation is

$$\bigstar \quad pV \propto T$$

or

$$\bigstar \quad \frac{p_1 V_1}{T_1} = \frac{p_2 V_2}{T_2}.$$

All the laws relate to a fixed mass of gas. Another series of experiments could be carried out to find out how the volume of a gas, held at constant pressure and temperature, depends on the mass of gas present. It would be found that the volume is proportional to the mass. This would give the combined relation

$$pV \propto mT,$$

where m is the mass of gas, or

$$pV = AmT$$

where A is a constant of proportionality. However, this is not a very useful way of expressing the relation, as the constant A has different values for different gases. We need to find a way of including the quantity of gas. The way to do this is to express the fixed mass of gas in Boyle's, Charles' and Gay-Lussac's laws in terms of the number of moles of gas present.

We have met the mole in Chapter 1, as one of the base units of the SI system. It is defined as the amount of substance which contains as many elementary entities (atoms or molecules, in the case of a gas) as there are atoms in 0.012 kg of carbon-12. For a simpler way of thinking about the mole, see the note below.

> $\bigstar$ The **mole** (abbreviated mol) is the amount of substance which contains as many elementary entities as there are atoms in 0.012 kg of carbon-12.

> ★ The **Avogadro constant** N_A is the number of atoms in 0.012 kg of carbon-12. That is, it is the number of atoms in a mole of atoms. It has the value 6.02×10^{23} mol^{-1}.

The **relative atomic mass** A_r is the ratio of the mass of an atom to one-twelfth of the mass of an atom of carbon-12. Since the definitions of the mole and the Avogadro constant relate to 0.012 kg (or 12 g) of carbon-12, the relative atomic mass is numerically equal to the mass in grams of a mole of atoms.

Similarly, the **relative molecular mass** M_r is the ratio of the mass of a molecule to one-twelfth of the mass of an atom of carbon-12, and is numerically equal to the mass in grams of a mole of molecules.

> ★ The **atomic mass unit** u is one-twelfth of the mass of the mass of an atom of carbon-12, and has the value 1.66×10^{-27} kg. (This quantity is also called the unified atomic mass constant m_θ.)

> ★ Thus, a simpler way of finding the mass of one mole of a particular element is to take its nucleon number (see Chapter 12) expressed in grams. For example, the nucleon number of argon (^{40}Ar) is 40. One mole of argon then has mass 40 g.

Returning to the combination of the three gas laws and the number n of moles of the gas, we have

$$pV \propto nT,$$

or, putting in a new constant of proportionality R,

> ★ $pV = nRT.$

R is called the **molar gas constant** (sometimes the **universal gas constant**, because it has the same value for all gases). It has the value 8.3 J mol^{-1} K^{-1}.

Sometimes the equation $pV = nRT$ is expressed in the form

> ★ $pV_m = RT,$

where the molar volume V_m is the volume occupied by one mole of the gas. Another version is

$$\bigstar \quad pV = NkT,$$

where N is the number of molecules in the gas and k is a constant called the **Boltzmann constant**. It has the value 1.38×10^{-23} J K^{-1}. The molar gas constant R and the Boltzmann constant k are connected through the Avogadro constant N_A:

$$\bigstar \quad R = kN_A.$$

Strictly speaking, the laws of Boyle, Charles and Gay-Lussac are not really laws, as their validity is restricted. They are accurate for real gases only if the pressure of the sample is not too great, and if the gas is well above its liquefaction temperature. But they can be used to define an ideal gas.

> $\bigstar$ An ideal gas is one which obeys the equation
> $$pV = nRT,$$
> at all pressures, volumes and temperatures.

We have already used the ideal gas equation to introduce the idea of thermodynamic temperature. For approximate calculations, the ideal gas equation can be used with real gases.

Examples

1 Find the volume occupied by 1 mole of air at standard temperature and pressure (273 K and 1.01×10^5 Pa), taking R as 8.3 J K^{-1} mol^{-1} for air.

Since $pV = nRT$, $V = nRT/p$. Substituting the values $n = 1$ mol, $R = 8.3$ J K^{-1} mol^{-1}, $T = 273$ K and $p = 1.01 \times 10^5$ Pa, $V = (1 \times 8.3 \times 273)/1.01 \times 10^5 = \mathbf{2.24 \times 10^{-2}}$ **m**^{3}.

2 Find the number of molecules per cubic metre of air at standard temperature and pressure.

We have just shown that the volume occupied by one mole of air at standard temperature and pressure is 2.24×10^{-2} m^3. One mole of air contains N_A molecules, where N_A is the Avogadro constant (6.02×10^{23} mol^{-1}). Thus the number of molecules per cubic metre of air is $6.02 \times 10^{23}/2.24 \times 10^{-2} = \mathbf{2.69 \times 10^{25}}$ **m**$^{-3}$.

It is useful to remember these two quantities, the *molar volume* of a gas and the *number density* of molecules in it. They give an idea of the relatively small volume occupied by a mole of gas at standard temperature and pressure (a cube of side about 28 cm), and the enormous number of molecules in every cubic metre of a gas under these conditions.

3 A syringe contains $25 \times 10^{-6} \, \text{m}^3$ of helium gas at a temperature of 20 °C and a pressure of 5.0×10^4 Pa. The temperature is increased to 400 °C and the pressure on the syringe is increased to 2.4×10^5 Pa. Find the new volume of gas in the syringe.

Boyle's law, pV = constant, and Charles' law, V/T = constant, may be combined to give pV/T = constant. This is written in the form $p_1V_1/T_1 = p_2V_2/T_2$ and re-arranged as $V_2 = p_1V_1T_2/p_2T_1$. Substituting the values $p_1 = 5.0 \times 10^4$ Pa, $V_1 = 25 \times 10^{-6} \, \text{m}^3$, $T_2 = 673$ K, $p_2 = 2.4 \times 10^5$ Pa, $T_1 = 293$ K (again note that temperatures are converted from °C to K), $V_2 = 5.0 \times 10^4 \times 25 \times 10^{-6} \times 673/2.4 \times 10^5 \times 293 = \mathbf{12 \times 10^{-6} \, m^3}$.

Now it's your turn

1 The number of molecules per cubic metre of air at standard temperature and pressure is about $2.7 \times 10^{25} \, \text{m}^{-3}$. What is the average separation of these molecules?
Ans: 3.3×10^{-9} m

2 The mean mass of one mole of air (which is made up mainly of nitrogen, oxygen and argon) is 0.029 kg. What is the density of air at standard temperature and pressure?
Ans: $1.29 \, \text{kg m}^{-3}$

3 The volume of a sample of gas is $3.2 \times 10^{-2} \, \text{m}^3$ when the pressure is 8.6×10^4 Pa and the temperature is 27 °C. How many moles of gas are there in the sample? How many molecules? What is the number of molecules per cubic metre?
Ans: 1.1 mol; 6.7×10^{23}; $2.1 \times 10^{25} \, \text{m}^{-3}$

4 A sample of air has volume $15 \times 10^{-6} \, \text{m}^3$ when the pressure is 5.0×10^5 Pa. The pressure is reduced to 3.0×10^5 Pa, without changing the temperature. What is the new volume?
Ans: $2.5 \times 10^{-5} \, \text{m}^3$

5 A sample of gas, originally at standard temperature and pressure, has volume $3.0 \times 10^{-5} \, \text{m}^3$ under these conditions. When the pressure is increased to 4.0×10^5 Pa, the temperature rises to 40 °C. What is the new volume?
Ans: $8.6 \times 10^{-6} \, \text{m}^3$

Section 5.3 summary

★ Boyle's law states that for a fixed mass of gas at constant temperature, the product of the pressure of a gas and its volume is a constant: $pV =$ constant or $p_1V_1 = p_2V_2$.

★ Charles' law states that if the pressure of a fixed mass of gas is kept constant, the ratio of volume to thermodynamic temperature is a constant: $V/T =$ constant or $V_1/T_1 = V_2/T_2$.

★ Gay-Lussac's law (the law of pressures) states that if the volume of a fixed mass of gas is kept constant, the ratio of pressure to thermodynamic temperature is a constant: $p/T =$ constant or $p_1/T_1 = p_2/T_2$.

★ The equation of state for an ideal gas relates the pressure p, volume V and thermodynamic temperature T of n moles of gas: $pV = nRT$ where $R = 8.3$ J K^{-1} mol^{-1}, the molar gas constant.

★ For N molecules of gas: $pV = NkT$ where $k = 1.38 \times 10^{-23}$ J K^{-1}, the Boltzmann constant.

★ The relation between R and k is $R = N_A k$ where $N_A = 6.02 \times 10^{23}$ mol^{-1}, the Avogadro constant.

★ Another expression of the ideal gas equation (or of Boyle's law and Charles' law combined) is $pV/T =$ constant or $p_1V_1/T_1 = p_2V_2/T_2$.

★ *Standard temperature and pressure* means a temperature of 273 K and a pressure of 1.01×10^5 Pa.

Section 5.3 questions

1 On a day when the atmospheric pressure is 102 kPa and the temperature is 8 °C, the pressure in a car tyre is 190 kPa above atmospheric pressure. After a long journey the temperature of the air in the tyre rises to 29 °C. Calculate the pressure above atmospheric of the air in the tyre at 29 °C. Assume that the volume of the tyre remains constant.

2 A helium-filled balloon is released at ground level, where the temperature is 17 °C and the pressure is 1.0 atmosphere. The balloon

rises to a height of 2.5 km, where the pressure is 0.75 atmospheres and the temperature is 5 °C. Calculate the ratio of the volume of the balloon at 2.5 km to that at ground level.

3 A metal box in the form of a cube of side 30 cm is filled with air at atmospheric pressure (1.01×10^5 Pa) at 15 °C. The box is sealed and heated in an oven to 200 °C. Calculate the net force on each side of the box.

5.4 A microscopic model of a gas

One of the aims of physics is to describe and explain the behaviour of various systems. For mechanical systems, this involves calculating the motion of the parts of the system in detail. For example, we have already seen how to predict the motion of a stone thrown in a uniform gravitational field. Using the equations of uniformly-accelerated motion, it is not too difficult to calculate the position and velocity of the stone at any time. However, there are some cases in which it is quite impossible to describe what happens to each component of the system.

This sort of problem arises if we try to describe the properties of a gas in terms of the motion of each of its molecules. The difficulty is that the numbers are so vast. One cubic metre of atmospheric air contains about 3×10^{25} molecules. There is no practical method of determining the position and velocity of every single molecule at a given time. Even the most advanced computer would be unable to handle the calculation of the motions of such a very large number of molecules.

In some ways, the fact that the gas is made up of such an enormous number of molecules is an advantage. It means that we can give a large-scale description of the gas in terms of only a few variables. These variables are quantities such as pressure p, volume V and temperature T. They tell us about average conditions in the gas, instead of describing the behaviour of each molecule. We have already met the experimental laws relating to these quantities. Our aim now is to relate the ideal gas equation, which deals with the large-scale (macroscopic) quantities p, V and T, to the small-scale (microscopic) behaviour of the particles of the gas. We shall do this by taking averages over the very large numbers of molecules involved. We shall find that we can derive the equation for Boyle's law when we make very simple assumptions about the atoms or molecules which make up the gas. This is the **kinetic theory of an ideal gas**. We shall also see that temperature can be related to the kinetic energy of the molecules of the gas.

The kinetic theory

An explanation of how a gas exerts a pressure was developed by Robert Boyle in the seventeenth century, and in greater detail by Daniel Bernoulli in the eighteenth. The basic idea is that the gas consists of atoms or molecules moving about at great speed. The gas exerts a pressure on the walls of its container because of the continued impacts of the molecules with the walls.

We shall make some simplifying assumptions about the molecules of the gas. The assumptions of the kinetic theory of an ideal gas are:

 1 All molecules behave as identical, hard, perfectly elastic spheres.
2 The volume of the molecules is negligible compared with the volume of the containing vessel.
3 There are no forces of attraction or repulsion between molecules.
4 The motion of the molecules is completely random.

In the addition, the number of molecules must be very large, so that average behaviour can be considered.

Suppose that the container is a cube of side L (Figure 5.12). The motion of each molecule can be resolved into x-, y- and z-components. For convenience we shall take the x-, y- and z-directions to be parallel to the edges of the cube.

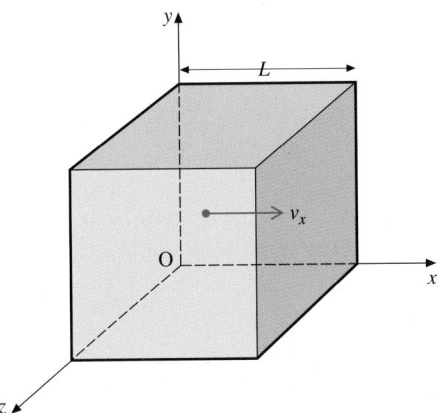

Figure 5.12 *Gas molecule in a cubical container.*

Consider one molecule. Let the x-component of its velocity be v_x. When this molecule collides with the wall of the container perpendicular to the x-axis, the x-component of velocity will be exactly reversed, because the collision of the molecule with the wall is perfectly elastic. The time taken for the molecule to move between the two walls perpendicular to the x-axis is L/v_x. To make the round trip from one wall to the opposite one and back again takes $2L/v_x$. This is the time between one collision of the molecule with a wall and its next collision with the same wall.

When the molecule strikes a wall, the component of velocity is reversed in direction, from v_x to $-v_x$. Thus, during each collision with a wall, the x-component of momentum changes by

$$\Delta p_x = 2mv_x,$$

where m is the mass of the molecule. The rate at which this molecule transfers momentum to the wall is

change of momentum/time between collisions $= 2mv_x/(2L/v_x) = mv_x^2/L.$

From Newton's Second Law, this rate of change of momentum is the average force exerted by this particular molecule on the wall through its collisions with the wall. If there are N molecules in the container, the total force is Nmv_x^2/L. Pressure is force divided by area, and the area of the wall is L^2, so the pressure exerted by all N molecules on this wall is Nmv_x^2/L^3. The volume V of the container is L^3, giving us

$$p = Nmv_x^2/V$$

as an expression for the pressure.

This expression relates only to the x-component of velocity of the molecules. For a molecule moving with velocity v, an extension of Pythagoras' theorem to three dimensions gives the relation between v and the three components of velocity v_x, v_y and v_z: $v^2 = v_x^2 + v_y^2 + v_z^2$. Because we are dealing with a large number of molecules in random motion, the average value of the components in the components in the x-direction will be the same as those in the y-direction or the z-direction. Therefore, taking the averages, $\langle v_x^2 \rangle = \langle v_y^2 \rangle = \langle v_z^2 \rangle$, and $\langle v_x^2 \rangle = \frac{1}{3}\langle v^2 \rangle$. The notation $\langle v_x^2 \rangle$ means 'the average value of v_x^2'. Our expression for the pressure can now be written

$$pV = \tfrac{1}{3}Nm\langle v^2 \rangle.$$

Since the average kinetic energy of a molecule is

$$\langle E_k \rangle = \tfrac{1}{2}m\langle v^2 \rangle,$$

there seems to be a link between our kinetic theory equation for pV and energy. We can find this relation by re-writing the pV equation as

$$pV = \tfrac{2}{3}N(\tfrac{1}{2}m\langle v^2 \rangle) = \tfrac{2}{3}N\langle E_k \rangle.$$

We have already stated the ideal gas law in the forms $pV = nRT = NkT$. Indeed, even real gases obey this law reasonably well under normal conditions. If our kinetic theory model and the subsequent theoretical derivation of the equation are correct, we can bring together the two equations for pV. This will allow us to relate the temperature of a gas to the average kinetic energy of its molecules.

$$pV = \tfrac{2}{3}N\langle E_k \rangle = NkT$$

and

$$\bigstar \quad \langle E_k \rangle = \tfrac{3}{2}kT.$$

This is an important result. We have obtained a relation between the average kinetic energy of a molecule in a gas and the temperature of the gas. This will allow us to obtain an idea of the average speed of the molecules.

Since

$$\langle E_k \rangle = \tfrac{1}{2}m\langle v^2 \rangle = \tfrac{3}{2}kT,$$

$$\langle v^2 \rangle = 3kT/m$$

and

$$\bigstar \quad \sqrt{\langle v^2 \rangle} = \sqrt{3kT/m}.$$

The quantity $\sqrt{\langle v^2 \rangle}$ is called the **root-mean-square speed**, or the **r.m.s. speed v_{rms}**, of the molecules. It is not exactly equal to the average speed of the molecules, but is often taken as being so. The average speed is about 0.92 of the root-mean-square speed. The difference between the r.m.s. speed and the average speed is highlighted in Example 1 below.

Note that the r.m.s. speed is proportional to the square root of the $\langle v \rangle$ thermodynamic temperature of the gas, and inversely proportional to the square root of the mass of the molecule. Thus, at a given temperature, lighter molecules move faster, on average, than heavier molecules. For a given gas, the higher the temperature, the faster the molecules move.

Examples

1 The speeds of seven molecules in a gas are numerically equal to 2, 4, 6, 8, 10, 12 and 14 units. Find the numerical values of **(a)** the mean speed $\langle v \rangle$, **(b)** the mean speed squared $\langle v \rangle^2$, **(c)** the mean square speed $\langle v^2 \rangle$, **(d)** the r.m.s. speed.

(a) $\langle v \rangle = (2 + 4 + 6 + 8 + 10 + 12 + 14)/7 =$ **8 units**.
(b) $\langle v \rangle^2 = 8^2 =$ **64 units²**.
(c) $\langle v^2 \rangle = (4 + 16 + 36 + 64 + 100 + 144 + 196)/7 =$ **80 units²**.
(d) r.m.s. speed $= (79.7)^{1/2} =$ **8.94 units**.

2 Find the total kinetic energy of the molecules in one mole of an ideal gas at standard temperature.

We know that the average kinetic energy of one molecule is $\tfrac{3}{2}kT$. For one mole of molecules, that is N_A molecules, the energy is $\tfrac{3}{2}N_A kT$ or $\tfrac{3}{2}RT$. Substituting the values $R = 8.3$ J K^{-1} mol^{-1} and $T = 273$ K, we have $E_k = \tfrac{3}{2} \times 8.3 \times 273 =$ **3400 J mol^{-1}**.

3 Find the root-mean-square speed of the molecules in nitrogen gas at 27 °C. The mass of a nitrogen molecule is 4.6×10^{-26} kg.

Since $v_{rms} = \sqrt{3kT/m}$, we have $v_{rms} = (3 \times 1.38 \times 10^{-23} \times 300/4.6 \times 10^{-26})^{1/2} =$ **520 m s^{-1}**.

Now it's your turn

1 Find the average kinetic energy of a molecule in an ideal gas at a temperature of 500 K.
Ans: 1.0×10^{-20} J

2 Find the root-mean-square speed of the molecules in hydrogen gas at 500 °C. The relative molecular mass of hydrogen is 2.
Ans: 3.1×10^4 m s^{-1}

Section 5.4 summary

★ Assumptions of the kinetic theory of gases:

1 Molecules behave as identical, hard, perfectly elastic spheres
2 Volume of the molecules is negligible compared with the volume of the containing vessel
3 There are no forces of attraction or repulsion between atoms
4 There are many molecules, all moving randomly

★ Kinetic theory pV equation: $pV = \frac{1}{3}Nm\langle v^2 \rangle$

★ Average kinetic energy of a molecule: $\langle E_k \rangle = \frac{1}{2}m\langle v^2 \rangle = \frac{3}{2}kT$

★ Root-mean square speed of molecules: $v_{rms} = \sqrt{3kT/m}$

Section 5.4 questions

1 Estimate the root-mean-square speed of helium atoms near the surface of the Sun, where the temperature is about 6000 K. (Mass of helium atom = 6.6×10^{-27} kg.)

2 In the kinetic theory of the pressure of a gas, the impulse delivered by a molecule to the wall of the container is proportional to the velocity of the molecule. Why then does the kinetic theory equation for pV include the square of the velocity?

5.5 | Internal energy

We have seen that the molecules of an ideal gas possess kinetic energy, and that this kinetic energy is proportional to the thermodynamic temperature of the gas. The sum of the kinetic energies of all the molecules is called the **internal energy** of the ideal gas. Not all molecules have the same kinetic energy, because they are moving with different speeds, but the sum of all the kinetic energies will be a constant if the gas is kept at a constant temperature.

For a real gas, the situation is a little more complicated. Because the molecules of a real gas exert forces on each other, at any instant there will be a certain potential energy associated with the positions the molecules occupy in space. Because the molecules are moving randomly, the potential energy of a given molecule will also vary randomly. But at a given temperature the total potential energy of all the molecules will remain constant. If the temperature changes, the total potential energy will also change. Furthermore, the molecules of a real gas collide with each other, and will interchange kinetic energy during the collision. For this real gas, the internal energy is given by the sum of the potential energies and the kinetic energies of all the molecules. It is important to realise that looking at a single molecule will give us very little information. Its kinetic energy will be changing all the time as it collides with other gas molecules, and its potential energy is also changing as its position relative to the other molecules in the gas changes. This single molecule has a kinetic energy which is part of the very wide range of kinetic energies of the molecules of the gas. We say that there is a *distribution* of molecular kinetic energies. The distribution is illustrated in Figure 5.13. Similarly, there is a distribution of molecular potential energies. But by adding up the kinetic and potential energies of all the molecules in the gas, the random nature of the kinetic and potential energies of the single molecule is removed.

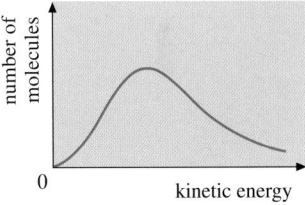

Figure 5.13 Distribution of molecular kinetic energies.

The idea of internal energy can be extended to all states of matter. In a liquid, intermolecular forces are stronger as the molecules are closer together, so the potential energy contribution to internal energy becomes more significant. The kinetic energy contribution is still due to the random motion of the molecules in the liquid. We can think of a solid as being made up of atoms or molecules which oscillate (vibrate) about equilibrium positions. Here, the potential energy contribution is caused by the strong binding forces between atoms, and the kinetic energy contribution due to the motion of the vibrating atoms.

The concept of internal energy is particularly useful as it helps us to distinguish between temperature and heat. Using an ideal gas as an example, temperature is a measure of the *average* kinetic energy of the molecules. It therefore does not depend on how many molecules are present in the gas. Internal energy, however, is the *total* kinetic energy of the molecules, and clearly does depend on how many molecules there are. In general heat refers to a *transfer* of energy from one substance to another, often as a result of a temperature difference. If two objects at different temperatures are placed in contact, there will be a flow of heat from the object at the higher temperature to the one at the lower. The direction of heat flow is determined by the difference in temperature, not by any difference in internal energies. If 10 g of a liquid at 30 °C is placed in contact, or mixed into, 100 g of the same liquid at 20 °C, the direction of heat flow is from the liquid at 30 °C to the liquid at 20 °C, even though the liquid at 20 °C has a greater internal energy than the smaller mass of liquid at 30 °C.

5.6 The first law of thermodynamics

We have already met the law of conservation of energy (Chapter 3). There, it was stated in the following form: Energy can neither be created nor destroyed, it can only be transformed from one form to another. In this section we shall see how this conservation law may be re-stated in relation to terms like work, heat and internal energy. This will lead to an understanding of the **first law of thermodynamics**.

Thermodynamics is the study of processes involving the transfer of heat and the doing of work. In thermodynamics it is necessary to define the **system** under consideration. For example, the system may be an ideal gas in a cylinder fitted with a piston, or an electric heating coil in a container of liquid.

In Chapter 3 we established the scientific meaning of work. Work is done when energy is transferred by mechanical means. We have just seen that heat is a transfer of energy due to a difference in temperature. Work and heat both involve a transfer of energy, but by different means.

We also know that the internal energy of a system is the total energy, kinetic and potential, of the various parts of the system. For a system consisting of an ideal gas, the internal energy is simply the total kinetic energy of all the atoms or molecules of the gas. For such a system, we would expect the internal energy to increase if heat were added to the gas, or if work were done on it. In both cases we are adding energy to the system. By the law of conservation of energy, this energy cannot just disappear; it must be transformed to another type of energy. It appears as an increase in the internal energy of the gas, that is the total kinetic energy of the molecules is increased. But if the *total* kinetic energy of the molecules increases, their *average* kinetic energy is also increased. Because the average kinetic energy is a measure of temperature, the addition of energy to the ideal gas shows up as an increase in its temperature.

We can express this transformation of energy as an equation.

> ★ The change in internal energy of a system is equal to the sum of the heat added to the system and the work done on it.

The change in internal energy is given the symbol ΔU, heat added is represented by Q and work done on the system by W. The equation is then

$$\bigstar \quad \Delta U = Q + W.$$

⭐ Note the sign convention which has been adopted. A positive value of Q means that heat has been *added to the system*. A positive value of W means that work is done *on the system*.

If the system does work, then we show this by writing $-W$. If heat leaves the system, we show this by writing $-Q$. Take care! There is another sign convention which takes the work done by the system as a positive quantity. To avoid confusion, write down the sign convention you are using every time you quote the first law.

Let's see how the first law of thermodynamics applies to some simple processes. First think about a change in the pressure and volume of a gas in a cylinder fitted with a piston. The cylinder and piston are thoroughly insulated, so that no heat can enter or leave the gas. (The thermodynamic name for such a change is an **adiabatic** change. That is, no thermal energy is allowed to enter or leave the system.) If no heat enters or leaves the gas, Q in the first law equation is zero. Thus

$$\Delta U = 0 + W.$$

If work is done on the gas by pushing the piston in, W is positive (remember the sign convention!) and ΔU will also be positive. That is, the internal energy increases, and because temperature is proportional to internal energy, the temperature of the gas rises. An adiabatic change can be achieved even if the cylinder and piston are not well-insulated. Moving the piston rapidly, so that the heat has no time to flow in or out, is just as effective. You will have noticed that a cycle pump gets hot as a result of brisk pumping: this is because the gas in the pump is being compressed adiabatically. Work is being done on the gas, the pump strokes are too rapid for the heat to escape, and the internal energy, and hence the temperature, increases. Another example is the diesel engine, where air in the cylinder is compressed so rapidly that the temperature rises to a point that, when fuel is injected into the cylinder, it is above its ignition temperature.

Now think about an electric kettle containing water. Here the element provides heat to the system. The quantity Q in the first law equation is positive (the sign convention is heat added, Q positive). No mechanical work is done on or by the water, so W in the first law equation is zero. Thus

$$\Delta U = W + 0.$$

The fact that W is positive means that ΔU is also positive. Internal energy, and hence temperature, increases.

When a substance changes from solid to liquid, intermolecular bonds are broken, thus increasing the potential energy component of the internal energy. During the melting process, the temperature does not change, and therefore the kinetic energy of the molecules does not change. Most substances expand on melting, and thus external work is done. By the first law, thermal energy must be supplied to the system, and this thermal energy is the latent heat.

Volume changes associated with evaporation are much greater than those associated with melting. The external work done is much greater during vaporisation, and thus latent heat of vaporisation is much greater than latent heat of fusion.

Example

200 J of heat is added to a system, which does 150 J of work. Find the change in internal energy of the system.

We use the first law of thermodynamics in the form $\Delta U = Q + W$ with the sign convention that Q is positive if heat is supplied to the system and W is positive if work is done on the system. Here $Q = 200$ J and $W = -150$ J (the system is doing the work, hence the minus sign!). Hence $\Delta U = 200 - 150 = \textbf{50 J}$. This change in internal energy is an increase.

Now it's your turn

An **isothermal** change is one which takes place at constant temperature. Explain why, in any isothermal change, the change in internal energy is zero. In such a change, 200 J of heat is added to a system. How much work is done on or by the system?

Ans: 200 J of work is done *by* the system

Sections 5.5–5.6 summary

★ The internal energy of a system is the sum of the kinetic and potential energies of the various parts of the system. For an ideal gas, the internal energy is the total kinetic energy of the molecules. Internal energy is a measure of the temperature of the system.

★ The first law of thermodynamics expresses the law of conservation of energy. The change in internal energy ΔU of a system is equal to the sum of the heat Q added to the system and the work W done on it,

$$\Delta U = Q + W.$$

(Sign convention: positive Q, heat is added to the system; positive W, work is done on the system.)

★ An adiabatic change is one in which no heat enters or leaves the system, i.e. $Q = 0$.

★ An isothermal change is one in which the temperature, and hence the internal energy, does not change, i.e. $\Delta U = 0$.

Sections 5.5–5.6 questions

1 An ideal gas expands isothermally, doing 250 J of work. What is the change in internal energy? How much heat is absorbed in the process?

2 50 J of heat energy is supplied to a fixed mass of gas in a cylinder. The gas expands, doing 20 J of work. Calculate the change in internal energy of the gas.

Exam Questions

1 Over the range of temperature between about 250 K and 450 K the resistance R of a certain platinum wire is given by

$$R = a + bT + cT^2,$$

where T is the thermodynamic temperature and $a = -1.32\ \Omega$, $b = +4.31 \times 10^{-2}\ \Omega\ K^{-1}$, $c = -6.04 \times 10^{-6}\ \Omega\ K^{-2}$.

(a) A certain solid melts at 150.0 °C (degrees Celsius). Calculate the resistance of the wire at this temperature.

(b) If the wire were used as a resistance thermometer, calculate the melting point of the solid on the empirical centigrade scale of this thermometer.

2 Consider one mole of gas at standard temperature and pressure (273 K and 1.01×10^5 Pa).

(a) Calculate:
  i) the number of molecules there are per cubic metre of this gas (this is known as the Loschmidt number N_L),
  ii) the average distance apart of the molecules.

(b) If the diameter of a molecule of this gas is about 2.5×10^{-10} m, estimate the fraction of each cubic metre occupied by matter.

3 In the ideal gas equation $pV = nRT$, could one use values of Celsius temperature instead of thermodynamic temperature by using an appropriate value R^* of the molar gas constant?

4 A volume of 1.50×10^3 m³ of hydrogen, at a pressure of 2.00×10^5 Pa and a temperature of 30 °C, is heated until both volume and pressure are doubled. Calculate:

(a) the final temperature,

(b) the mass of hydrogen. (Mass of one mole of hydrogen = 2.00×10^{-3} kg.)

5 Two containers, each of equal, constant volume, are connected by a tube of negligible volume. When the temperature of both containers is 27 °C, the pressure of the gas in the containers is 2.1×10^5 Pa. One container is heated to a temperature of 127 °C, whilst the other is maintained at 27 °C. Calculate the final pressure in the containers.

6 The kinetic theory leads to the expression $\langle E \rangle = \frac{3}{2}kT$ for the average kinetic energy of a molecule of a gas. The constant k does not depend on the type of molecule. Can this result really be true both for hydrogen and chlorine? (The mass of a chlorine molecule is about 35 times that of hydrogen molecule.) Explain why.

7 The mass of an oxygen molecule is 16 times the mass of a hydrogen molecule. Equal masses of hydrogen and oxygen gases are mixed, and are allowed to come to equilibrium. Calculate:

(a) the ratio of numbers of molecules,
(b) the ratio of the average kinetic energy per molecule,
(c) the ratio of pressures exerted on the walls of the container.

8 An ideal gas is compressed adiabatically to one-third of its volume. The work done on the gas is 500 J.

(a) State how much heat enters or leaves the gas.
(b) State the change in internal energy of the gas.
(c) What happens to the temperature of the gas?

CHAPTER SIX

Electricity

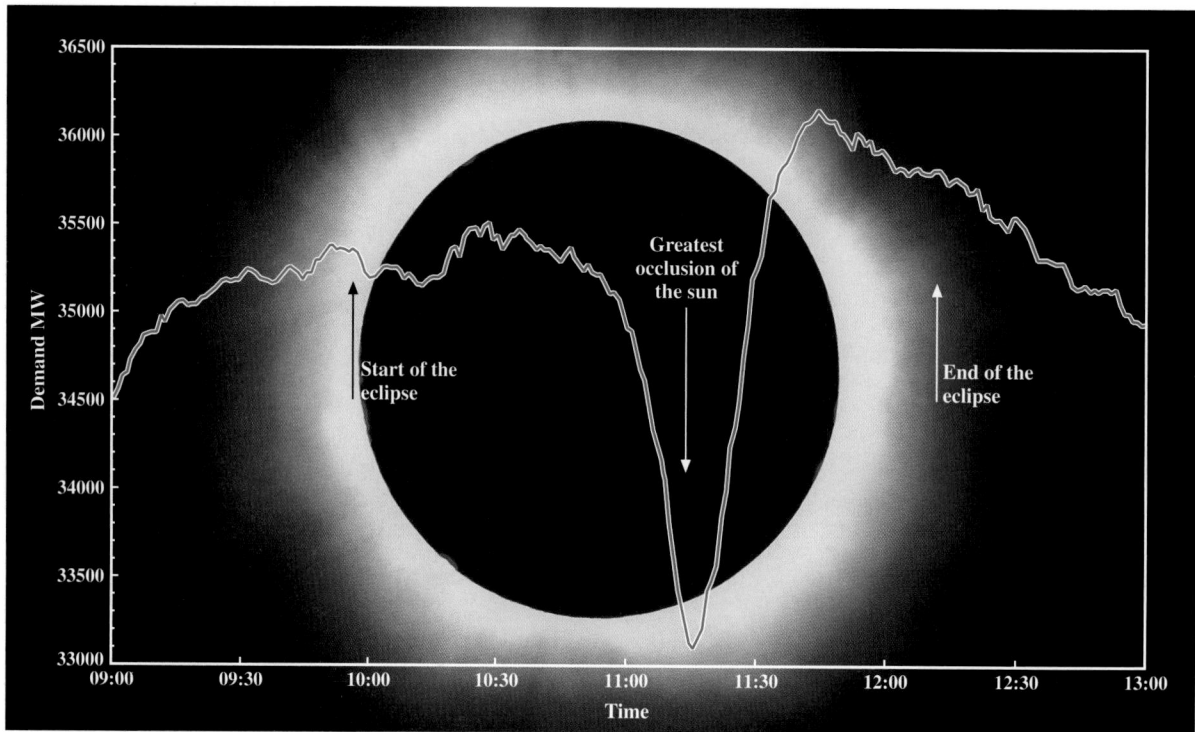

Electricity is so vital to modern life that physicists and engineers need to understand it fully in order that we can enjoy all the benefits it brings. The aim of this chapter is to consider fundamental ideas about electric current and electric circuits in order to provide a basis for further study of electricity. We start by thinking about current as the flow of charge. We link the idea of potential difference to energy changes in a circuit. Resistance comes in as a measure of opposition to the flow of charge. The idea of resistance introduces Ohm's law. Kirchhoff's first and second laws are seen as the laws of conservation of charge and of energy, applied to simple electric circuits. Moving on to circuit theory, we shall derive the rules for combining resistors in series and in parallel, and also analyse complete circuits.

Electricity is a convenient way of carrying energy into homes and offices to do useful jobs. Since electricity cannot be stored, the electricity companies have to anticipate the demand. This involves keeping an eye on the weather and the TV programme listings. Cold weather obviously causes an increase in demand, as does the end of a popular programme when people decide to switch on kettles. The graph shows the electricity demand in the UK on the day of the Solar Eclipse, 11 August 1999. The lowest point on the graph is at eclipse; the resulting surge in demand as people resumed their activities is the largest ever seen in the UK.

6.1 Charge and current

All matter is made up of tiny particles called atoms, and each atom consists of a positively-charged nucleus with negatively-charged electrons moving around it.

Charge is measured in units called **coulombs** (symbol C). The charge on an electron is -1.6×10^{-19} C. Normally atoms have equal numbers of positive and negative charges, so that their overall charge is zero. But for some atoms it is relatively easy to remove an electron, leaving an atom with an unbalanced number of positive charges. This is called a **positive ion**.

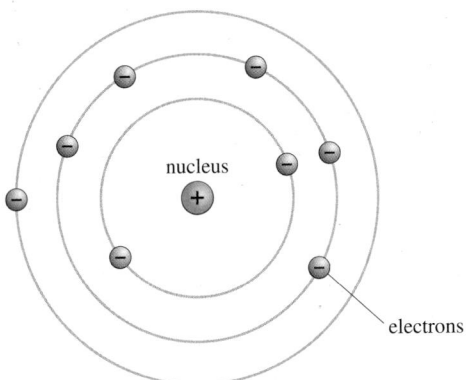

Figure 6.1 Atoms consist of a positively-charged nucleus with negative electrons outside.

Atoms in metals have one or more electrons which are not held tightly to the nucleus. These **free** (or mobile) **electrons** wander at random throughout the metal. But when a battery is connected across the ends of the metal, the free electrons drift towards the positive terminal of the battery, producing an **electric current**.

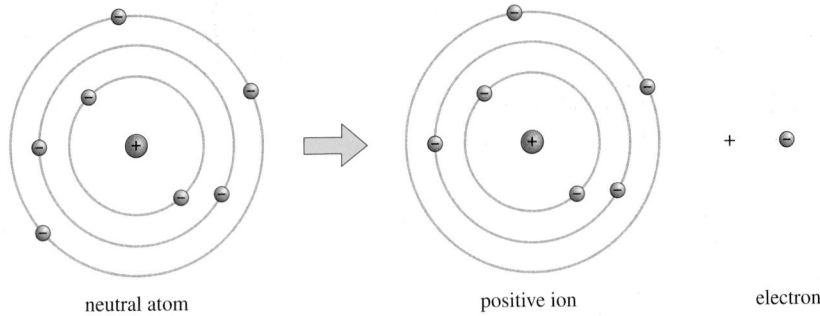

Figure 6.2 An atom with one or more electrons missing is a positive ion.

The size of the electric current is given by the rate of flow of charge, and is measured in units called amperes (or amps for short), with symbol A. A current of 3 amperes means that 3 coulombs pass a point in the circuit every second. In 5 seconds, a total charge of 15 coulombs will have passed the point. So,

$$\bigstar \quad charge = current \times time$$

or

$$\bigstar \quad \Delta q = I\Delta t,$$

where the charge q is in coulombs when the current I is in amperes and the time t is in seconds.

> ★ The coulomb is that charge passing a point in a circuit when there is a current of one ampere for one second.

Example

The current in the filament of a torch bulb is 0.03 A. How much charge flows through the bulb in 1 minute?

Using $\Delta q = I\Delta t$, $\Delta q = 0.03 \times 60$ (remember the time must be in seconds!), so $\Delta q = \mathbf{1.8\ C}$.

Now it's your turn

1   Calculate the current when a charge of 240 C passes a point in a circuit in a time of 2 minutes.
    Ans: 2.0 A

2   In a silver-plating experiment, 9.65×10^4 C of charge is needed to deposit a certain mass of silver. Calculate the time taken to deposit this mass of silver when the current is 0.20 A.
    Ans: 4.83×10^5 s

3   The current in a wire is 200 mA. Calculate **(a)** the charge which passes a point in the wire in 5 minutes, **(b)** the number of electrons needed to carry this charge. (Electron charge $e = -1.6 \times 10^{-19}$ C.)
    Ans: (a) 60 C; (b) 3.75×10^{20}

Conventional current

Early studies of the effects of electricity led scientists to believe that an electric current was the flow of 'something'. In order to develop a further understanding of electricity, they needed to know the direction of flow. It was decided that this flow in the circuit should be from the positive terminal of the battery to the negative. This current is called the **conventional current**, and is in the direction of flow of positive charge. We now know, in a metal, that the electric current is the flow of electrons in the opposite direction, from the negative terminal to the positive terminal. However, laws of electricity had become so firmly fixed in the minds of people that the idea of conventional current has persisted. But be warned! Occasionally we need to take into account the fact that electron flow is in the opposite direction to conventional current.

Potential difference

A cell makes one end of the circuit positive and the other negative. The cell is said to set up a **potential difference** across the circuit. Potential difference (p.d. for short) is measured in volts (symbol V), and is often called the voltage. You should never talk about the potential difference or voltage *through* a device, because it is in fact a difference *across* the ends of the device. The potential difference provides the energy to move charge through the device.

Resistance

Connecting wires in circuits are often made from copper, because copper offers little opposition to the movement of electrons. The copper wire is said to have a low electrical resistance. In other words, copper is a good conductor.

Some materials, such as plastics, are poor conductors. These materials are said to be insulators, because under normal circumstances they conduct little or no current.

★ The resistance R of a wire is defined as the ratio of the potential difference V across the wire to the current I in it,

or

$$\bigstar \; R = \frac{V}{I}$$

where the resistance is in ohms when the potential difference is in volts and the current in amperes. The symbol for ohms is the Greek capital letter omega, Ω. We have defined resistance in terms of a wire, but many devices have resistance. The general term for such a device is a **resistor**. (Note that the resistance of a resistor is measured in ohms, just as the volume of a tank is measured in m^3. We do not refer to the 'm^3' of a tank, nor to the 'ohms' of a resistor.)

The relation between resistance, potential difference and current means that, for a given potential difference, a high resistance means a small current, while a low resistance means a large current.

Example

The current in an electrical immersion heater in a school experiment is 6.3 A when the p.d. across it is 12 V. Calculate the resistance of the heater.

Since $R = V/I$, the resistance $R = 12/6.3 = $ **1.9 Ω**.

Now it's your turn

The current in a light-emitting diode is 20 mA when it has a potential difference of 2.0 V across it. Calculate its resistance.
Ans: 100 Ω

Potential difference and energy

The potential difference between two points in a circuit measures the electrical energy transferred, or the work done, by each coulomb of charge as it moves from one point to the other. We already know that the unit of potential difference is the volt (V). Energy W is measured in joules, and charge q in coulombs.

$$\star \quad potential\ difference = \frac{energy\ transferred\ (or\ work\ done)}{charge}$$

or

$$\star \quad V = \frac{W}{q}.$$

We can turn this relation round to get an expression for the electrical energy transferred or converted when a charge q is moved through a potential difference V

$$\star \quad Energy\ transferred\ (work\ done) = potential\ difference \times charge$$

$$\star \quad W = Vq$$

In the photograph, one lamp is connected to the 240 V mains supply and the other to a 12 V car battery. Both lamps have the same current yet the 240 V lamp glows more brightly. This is because the energy supplied to each coulomb of charge in the 240 V lamp is twenty times greater than for the 12 V lamp.

Example

Electrons in a particular TV tube are accelerated by a potential difference of 20 kV between the filament and the screen. The charge of the electron is -1.6×10^{-19} C. Calculate the gain in kinetic energy of each electron.

Since $V = W/q$, then $W = Vq$. The electrical energy transferred to the electron shows itself as the kinetic energy of the electron. Thus, kinetic energy $= Vq = 20 \times 10^3 \times 1.6 \times 10^{-19} = \mathbf{3.2 \times 10^{-15}}$ **J**. (Don't forget to turn the 20 kV into volts!)

Now it's your turn

1 An electron in a particle accelerator is said to have 1 MeV (1 mega electron volt) of energy when it has been accelerated through a potential difference of 1 million volts. Calculate the energy, in joules, gained by the electron.
   Ans: 1.6×10^{-13} J

2 A torch bulb is rated 2.2 V, 0.25 A. Calculate **(a)** the charge passing through the bulb in one second, **(b)** the energy transferred by the passage of each coulomb of charge.
   Ans: (a) 0.25 C; (b) 0.55 J

Electrical power

Remember that power P is the rate of doing work, or of transferring energy. Remember also that $V = W/q$. Divide each term on the right-hand side of this equation by time t, so that $V = (W/t)/(q/t)$. W/t is power P, and q/t is current I, so

$$\star \quad potential\ difference = \frac{power}{current}$$

or

$$\star \quad V = \frac{P}{I}$$

(W/q is a quotient, so dividing W by t and dividing q by t at the same time does not affect the final value because the t terms cancel each other out.)

The power is measured in watts (W) when the potential difference is in volts (V) and the current is in amperes (A). A voltmeter can measure the p.d. across a device and an ammeter the current through it; the equation above can then be used to calculate the power in the device.

Electrical heating

When an electric current passes through a resistor, it gets hot. This heating effect is sometimes called **Joule heating**. The electrical power produced (dissipated) is given by $V = P/I$, which can be rearranged to give $P = VI$. We can obtain alternative expressions for power in terms of the resistance R of the resistor.

Since $V = IR$, then

$$\star \quad P = I^2R$$

and

$$\star \quad P = \frac{V^2}{R}$$

For a given resistor, the power dissipated depends on the square of the current. This means that if the current is doubled, the power will be four times as great. Similarly, a doubling of voltage will increase the power by a factor of four.

Example

An electric immersion heater used in a school experiment has a current of 6.3 A when the p.d. across it is 12 V. Calculate the power of the heater.

Since $P = VI$, power = $12 \times 6.3 =$ **76 W**.

Now it's your turn

1   Show that a 100 W lamp connected to the 240 V mains supply will draw the same current as a 5 W car lamp connected to a 12 V battery. (See page 134)

2   An electric kettle has a power of 2.2 kW at 240 V. Calculate **(a)** the current in the kettle, **(b)** the resistance of the kettle element.
Ans: (a) 9.2 A; (b) 26 Ω

Section 6.1 summary

★   Electric current is a flow of charge: $I = \Delta q/\Delta t$

★   Conventional current is a flow of positive charge from positive to negative. In metals, current is carried by electrons, which travel from negative to positive

★   Resistance R of a resistor is defined as $R = V/I$

★   Potential difference (or voltage) measures the electrical energy transferred by each coulomb of charge: $V = W/q$

★   Electrical power $P = VI = I^2R = V^2/R$

Section 6.1 questions

1   A 240 V heater takes a current of 4.2 A. Calculate:

   (a)   the charge that passes through the heater in 3 minutes,

   (b)   the heat energy produced by the heater,

   (c)   the resistance of the heater.

2   A small torch has a 3.0 V battery connected to a bulb of resistance 15 Ω.

   (a)   Calculate:
      i)   the current in the bulb,
      ii)   the power delivered to the bulb.

   (b)   The battery supplies a constant current to the bulb for 2.5 hours. Calculate the total energy delivered to the bulb.

3   The capacity of storage batteries is rated in ampere-hours (A h). An 80 A h battery can supply a current of 80 A for 1 hour, or 40 A for 2 hours, and so on. Calculate the total energy, in J, stored in a 12 V, 80 A h car battery.

4   An electric kettle is rated at 2.2 kW, 240 V. The supply voltage is reduced to 230 V. Calculate the new power of the kettle.

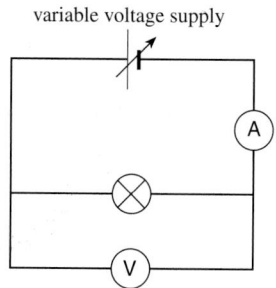

Figure 6.3 *Circuit for plotting graphs of voltage against current.*

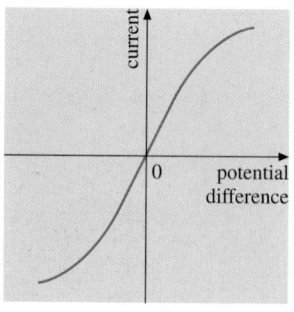

Figure 6.4 *Voltage against current graph for a light bulb.*

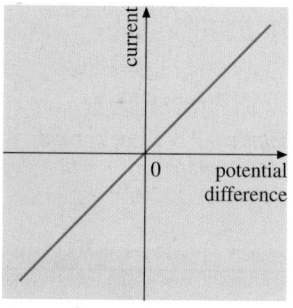

Figure 6.5 *Voltage against current graph for a constantan wire.*

Portrait of Ohm.

6.2 More about resistance

Ohm's law

The relation between the potential difference across an electrical component and the current through it may be investigated using the circuit of Figure 6.3. For example, if a light bulb is to be investigated, adjust the voltage across the bulb and measure the corresponding currents and voltages. The variation of current with potential difference is shown in Figure 6.4.

The resistance R of the bulb can be calculated from $R = V/I$. At first the resistance is constant (where the graph is a straight line), but then the resistance increases with current (where the graph curves).

If the light bulb is replaced by a length of constantan wire, the graph of the results is as shown in Figure 6.5. It is a straight line through the origin. This shows that for constantan wire, the current is proportional to the voltage. The resistance of the wire is found to stay the same as the current increases. The difference between Figures 6.4 and 6.5 is that the temperature of the constantan wire was constant for all currents used in the experiment, whereas the temperature of the filament of the light bulb increased to about 1500 °C.

Graphs like Figure 6.5 would be obtained for wires of any metal, provided that the temperature of the wires did not change during the experiment. The graph illustrates a law discovered by the German scientist Georg Simon Ohm. (Ohm's name is now used as the name of the unit of resistance.)

> ★ Ohm's law states that, for a conductor at constant temperature, the current in the conductor is proportional to the potential difference across it.

Conductors whose voltage against current graphs are a straight line through the origin, like that in Figure 6.5, are said to obey Ohm's law. It is found that Ohm's law applies to metal wires, provided that the current is not too large. What does 'too large' mean here? It means that the current must not be so great that there is a pronounced heating effect, causing an increase in temperature of the wire.

A light bulb filament consists of a thin metal wire. Why does it not obey Ohm's law? (Figure 6.4 shows that the voltage against current graph is not a straight line.) This is because, as stated previously, the temperature of the filament does not remain constant. The increase in current causes the temperature to increase so much that the filament glows.

Voltage-current graphs

When other devices are tested in the same way, voltage-current graphs like those in Figures 6.6 to 6.8 are obtained.

We have already met the thermistor as a type of electrical resistance thermometer (Chapter 5). Thermistors are made from semiconducting material. If a thermistor is held at constant temperature in a water bath, its voltage-current graph is a straight line. At a higher temperature, the gradient is steeper, showing that the resistance is lower. Figure 6.6 illustrates this. The resistance of a sample of semiconductor decreases as the temperature increases. This is the opposite behaviour to a metal: the resistance of a metal wire increases as the temperature increases. Thermistors are used as temperature sensors, making use of the fact that their resistance changes considerably with temperature.

Light-dependent resistors (LDRs) are also made from semiconducting material. As their name suggests, their resistance can be altered by a change in light intensity: the brighter the light, the lower the resistance. LDRs are used in light sensors and camera exposure meters.

Diodes are also made from semiconducting material. The diode conducts when the current is in the direction of the arrowhead on the symbol. This condition is called **forward bias**. When the voltage is reversed, there is negligible bias. This is called **reverse bias**. Figure 6.8 shows this important difference in the voltage-current graph when the voltage is reversed. Diodes are used to change alternating current into direct current in devices called **rectifiers**.

Resistance and temperature

All solids are made up of atoms that constantly vibrate about their equilibrium positions. The higher the temperature, the greater the amplitude of vibration.

Electric current is the flow of free electrons through the material. As the electrons move, they collide with the vibrating atoms, so their movement is impeded. The more the atoms vibrate, the greater is the chance of collision. This means that the current is less and the resistance is greater.

A temperature rise can cause an increase in the number of free electrons.

If there are more electrons free to move, this may outweigh the effect due to the vibrating atoms, and thus the flow of electrons, or the current, will increase. The resistance is therefore reduced. This is the case in semiconductors. Insulators, too, show a reduction in resistance with temperature rise.

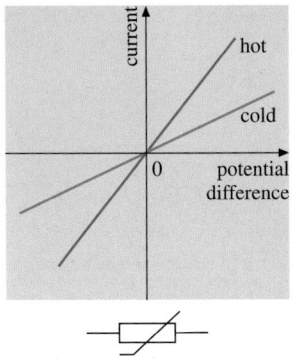

circuit symbol for a thermistor

Figure 6.6 *Voltage against current graphs for a thermistor.*

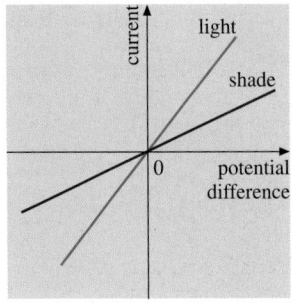

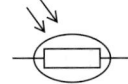

circuit symbol for a light dependent resistor

Figure 6.7 *Voltage against current graphs for a light-dependent resistor.*

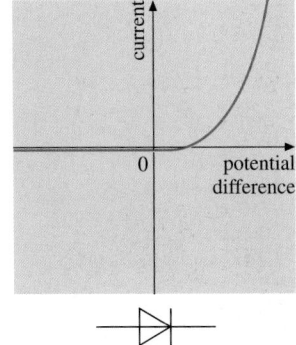

circuit symbol for a diode

Figure 6.8 *Voltage against current graph for a diode.*

For metals there is no increase in the number of free electrons. The increased amplitude of vibration of the atoms makes the resistance of metals increase with temperature.

Resistivity

All materials have some resistance to a flow of charge. A potential difference across the material causes free charges inside to accelerate. As the charges move through the material, they collide with the atoms of the material which get in their way. They transfer some or all of their kinetic energy, and then accelerate again. It is this transfer of energy on collision that causes electrical heating.

As you might guess, the longer a wire, the greater its resistance. This is because the charges have further to go through the material; there is more chance of collisions with the atoms of the material. In fact the resistance is proportional to the length of the wire, or $R \propto l$.

Also, the thicker a wire is, the smaller its resistance will be. This is because there is a bigger area for the charges to travel through, with less chance of collision. In fact the resistance is inversely proportional to the cross-sectional area of the wire, or $R \propto 1/A$. These relations are illustrated in Figures 6.9 and 10.

Figure 6.9 *The longer the room, the greater the resistance the waiters meet.*

Figure 6.10 *The wider the room, the easier it is for the waiters to pass through.*

Finally, the resistance depends on the type of material. As previously stated, copper is a good conductor, whereas plastics are good insulators.

6.2 More about resistance

Putting all of this together gives

$$\bigstar \quad R = \frac{\rho l}{A},$$

where ρ is a constant for a particular material at a particular temperature. ρ is called the **resistivity** of the material at that temperature.

> $\bigstar$ The resistivity ρ of a material is numerically equal to the resistance between opposite faces of a cube of the material, of unit length and unit cross-sectional area.

So if $\rho = RA/l$ and if the resistance is in ohms, the cross-sectional area in metres squared and the length in metres, then the resistivity is in ohm metres (Ω m). Remember that A is the cross-sectional area through which the current is passing, not the surface area.

We have already seen that the resistance of a wire depends on temperature. Thus, resistivity also depends on temperature. The resistivities of metals increase with increasing temperature, and the resistivities of semiconductors decrease very rapidly with increasing temperature.

The values of the resistivity of some materials at room temperature are given in Figure 6.11.

material	resistivity/Ω m
metals:	
copper	1.7×10^{-8}
gold	2.4×10^{-8}
aluminium	2.6×10^{-8}
semiconductors:	
germanium (pure)	0.6
silicon (pure)	2.3×10^{3}
insulators:	
glass	about 10^{12}
perspex	about 10^{13}
polyethylene	about 10^{14}
sulphur	about 10^{15}

Figure 6.11 *Resistivities at room temperature.*

Note the enormous range of resistivities spanned by the materials in this list – a range of 23 orders of magnitude, from 10^{-8} Ω m to 10^{15} Ω m.

Note, too, that the resistivity is a property of a material, while the resistance is a property of a particular wire or device.

Example

Calculate the resistance per metre at room temperature of a constantan wire of diameter 1.25 mm. The resistivity of constantan at room temperature is 5.0×10^{-7} Ω m.

The cross-sectional area of the wire is calculated using πr^2. Area = $\pi(1.25 \times 10^{-3}/2)^2$ (don't forget to change the units from mm to m!). The resistance per metre is given by R/l, and $R/l = \rho/A$. So resistance per metre = $5.0 \times 10^{-7}/\pi(1.25 \times 10^{-3}/2)^2 =$ **0.41 Ω m^{-1}**.

Now it's your turn

1   Find the length of copper wire, of diameter 0.63 mm, which has a resistance of 1.00 Ω. The resistivity of copper at room temperature is 1.7×10^{-8} Ω m.

    Ans: 18 m

2   Find the diameter of a copper wire which has the same resistance as an aluminium wire of equal length and diameter 1.20 mm. The resistivities of copper and aluminium at room temperature are 1.7×10^{-8} Ω m and 2.6×10^{-8} Ω m respectively.

    Ans: 0.97 mm

A microscopic picture of current

Let's follow up the idea of electric current as the flow of free electrons through a conductor. We have already thought of resistance as arising from the collision of the free electrons with the atoms of the conductor. We now think about the motion of the free electrons in a wire. When a potential difference is applied across the ends of the wire, the electrons feel a force attracting them to the end of the wire connected to the positive pole of the battery. Because of this force, the electrons accelerate. As they move along the wire, they will collide with atoms, sometimes coming to rest and starting again. The effect of the collisions is that, on average, the electrons move with a steady speed from one end of the wire to the other. This speed is called the **drift speed v_d** of the charge carriers. (v_d is sometimes called the drift velocity. Is v_d a vector? There is direction involved: the electrons move towards the positive end of the wire.)

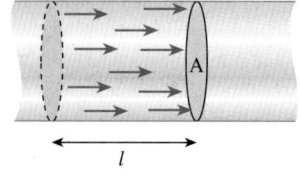

Figure 6.12 Drift speed of free electrons.

We shall now relate the drift speed v_d to the current I in the wire. Figure 6.12 shows a wire of cross-sectional area A. From the definition of average speed, in a time Δt the electrons will travel a distance l equal to $v_d \Delta t$ along the wire. In this time, all the electrons in a volume V given by Al will pass through the cross-section A. If there are n electrons per unit volume of the wire, the total number of electrons passing through this cross-section is nAl. Each electron has charge e, so the total charge passing through the cross-section is $nAle$. But we have shown that l is equal to $v_d\Delta t$. Thus, the total charge Δq passing through the cross-section is given by

$$\Delta q = nAv_d e\Delta t.$$

But the definition of current is the rate of flow of charge, or

$$I = \Delta q/\Delta t.$$

Thus we have

$$\bigstar \; I = nAv_d e.$$

If we know the number n of free electrons per unit volume of the material of the conductor, we can calculate the drift speed corresponding to a current I in a wire of known cross-sectional area. n is called the **number density** of the free electrons. It has the unit m^{-3}.

Example

A copper wire of cross-sectional area 2.0 mm² carries a current of 3.0 A. There are 8.4×10^{28} free electrons in every cubic metre of copper. Calculate the drift speed of the free electrons in the wire. (Electron charge $e = -1.6 \times 10^{-19}$ C.)

From $I = nAv_d e$, we have $v_d = I/nAe = 3.0/(8.4 \times 10^{28} \times 2.0 \times 10^{-6} \times 1.6 \times 10^{-19}) = \mathbf{1.1 \times 10^{-4}\ m\ s^{-1}}$.

Now it's your turn

An aluminium conductor has cross-sectional area 1.5 mm² and carries a current of 50 mA. There are 6.0×10^{28} free electrons in every cubic metre of aluminium. Calculate the drift speed of the free electrons in the conductor. (Electron charge $e = -1.6 \times 10^{-19}$ C.)
Ans: $3.5 \times 10^{-6}\ m\ s^{-1}$

The answers we have just obtained for typical drift speeds are perhaps surprising. Clearly, electrons carrying a current in a conductor move very slowly. In the worked example, an electron would take two and a half hours to move one metre along the copper wire. This sort of time seems out of place when you realise that the electric light comes on immediately once you switch it on. You certainly do not have to wait hours before the electrons reach the bulb and it lights up! What you have to realise is that there are free electrons at all parts of the wire. As soon as there is any movement of charge carriers anywhere in the wire, the same current is observed at all parts of the circuit.

Section 6.2 summary

* Ohm's law: for a conductor at constant temperature, the current in the conductor is proportional to the potential difference across it

* The resistance of a metallic conductor increases with increasing temperature; the resistance of a semiconductor decreases with increasing temperature

* The resistance of a thermistor decreases with temperature

* The resistance of an LDR decreases as the light intensity increases

* A diode has a low resistance when connected in forward bias, and a very high resistance in reverse bias

* Resistivity of a material is the resistance between opposite faces of a unit cube of the material

* The drift speed of electrons in a conductor is the average speed with which they move when a potential difference is applied to the conductor: $I = nAv_de$, where n is the number density of free electrons in the material of the conductor

Section 6.2 questions

1 The element of an electric kettle has resistance 26 Ω at room temperature. The element is made of nichrome wire of diameter 0.60 mm and resistivity 1.1×10^{-6} Ω m at room temperature. Calculate the length of the wire.

2 Figure 6.13 gives values of the current I through an electrical component for different potential differences V across it.

V/V	0	0.19	0.48	1.47	2.92	4.56	6.56	8.70
I/A	0	0.20	0.40	0.60	0.80	1.00	1.20	1.40

(a) Draw a diagram of the circuit that could be used to obtain these values.
(b) Calculate the resistance of the component at each value of current.
(c) Plot a graph to show the variation with current of resistance of the component.
(d) Suggest what the component is likely to be, giving a reason for your answer.

3 A 2.50 m length of wire of diameter 1.5 mm carries a current of 0.65 mA when the potential difference of 0.40 V is applied between its ends. The drift speed of the free electrons is measured in a separate experiment as 2.5×10^{-5} m s^{-1}. Calculate:

(a) the resistance of the wire,
(b) the resistivity of the material of the wire,
(c) the number density of the free electrons in the wire.

Electrical circuits

Series and parallel circuits: Kirchhoff's first law

A **series circuit** is one in which the components are connected one after another, forming one complete loop. You have probably connected an ammeter at different points in a series circuit to show that it reads the same current at each point (see Figure 6.14).

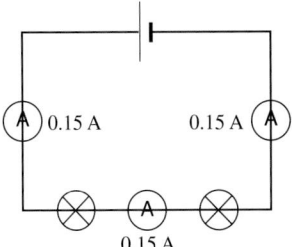

Figure 6.14 *The current at each point in a series circuit is the same.*

A **parallel circuit** is one where the current can take alternative routes in different loops. In a parallel circuit, the current divides at a junction, but the current entering the junction is the same as the current leaving it (see Figure 6.15). The fact that the current does not get 'used up' at a junction is because current is the rate of flow of charge, and charges cannot accumulate or get 'used up' at a junction. The consequence of this conservation of electric charge is known as **Kirchhoff's first law**. This law is usually stated as:

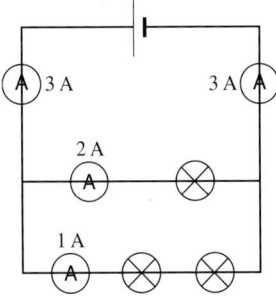

Figure 6.15 *The current divides in a parallel circuit.*

> ★ the sum of the currents entering a junction in a circuit is always equal to the sum of the currents leaving it.

In the junction shown in Figure 6.16,

$$I = I_1 + I_2 + I_3.$$

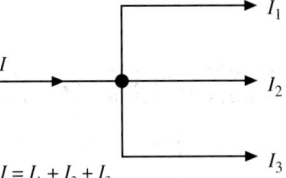

Figure 6.16 *The sum of the currents entering a junction is equal to the sum of the currents leaving it.*

Example

For the circuit of Figure 6.17, state the readings of the ammeters A_1, A_2 and A_3.

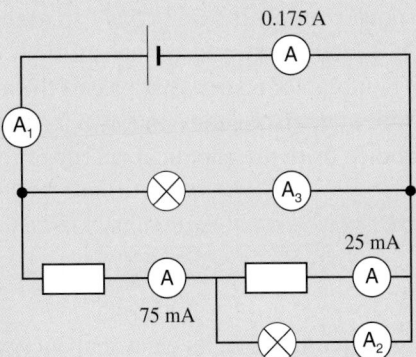

Figure 6.17

A_1 would read **175 mA**, as the current entering the power supply must be the same as the current leaving it. A_2 would read $75 - 25 = $ **50 mA**, as the total current entering a junction is the same as the total current leaving it. A_3 would read $175 - 75 = $ **100 mA**.

Now it's your turn

1 The lamps in Figure 6.18 are identical. There is a current of 0.50 A through the battery. What is the current in each lamp?

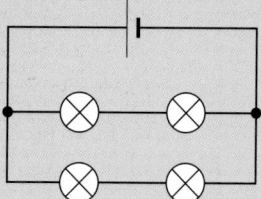

Figure 6.18

Ans: 0.25 A

2 Figure 6.19 shows one junction in a circuit. Calculate the ammeter reading.

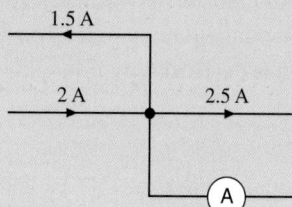

Figure 6.19

Ans: 2.0 A, with the current towards the junction

Electromotive force and potential difference

When charge passes through a power supply such as a battery, it **gains** electrical energy. The power supply is said to have an **electromotive force** (or e.m.f. for short). The electromotive force measures, in volts, the electrical energy gained by each coulomb of charge that passes through the power supply. Note that, in spite of its name, the e.m.f. is *not* a force. The energy gained by the charge comes from the chemical energy of the battery.

$$\bigstar\ e.m.f. = \frac{(energy\ converted\ from\ other\ forms\ to\ electrical)}{charge}$$

When charge passes through a resistor, its electrical energy is converted to heat energy in the resistor. The resistor has a **potential difference** across it. The potential difference measures, in volts, the electrical energy lost by each coulomb of charge that passes through it.

$$\bigstar\ p.d. = \frac{(energy\ converted\ from\ electrical\ to\ other\ forms)}{charge}$$

Conservation of energy: Kirchhoff's second law

Charge flowing round a circuit gains electrical energy on passing through the battery and loses electrical energy on passing through the rest of the circuit. From the law of conservation of energy, we know that the total energy must remain the same. The consequence of this conservation of energy is known as **Kirchhoff's second law**. This law may be stated as:

> ★ the sum of the electromotive forces in a closed circuit is equal to the sum of the potential differences.

Figure 6.20 shows a circuit containing a battery, lamp and resistor in series. Applying Kirchhoff's second law, the electromotive force in the circuit is the e.m.f. E of the battery. The sum of the potential differences is the p.d. V_1 across the lamp plus the p.d. V_2 across the resistor. Thus, $E = V_1 + V_2$. If the current in the circuit is I and the resistances of the lamp and resistor are R_1 and R_2 respectively, the p.d.s can be written as $V_1 = IR_1$ and $V_2 = IR_2$, so $E = IR_1 + IR_2$.

Figure 6.20

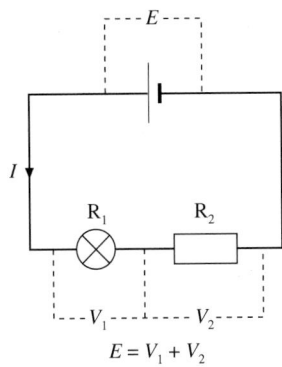

$$E = V_1 + V_2$$

It should be remembered that both electromotive force and potential difference have direction. This must be considered when working out the equation for Kirchhoff's second law. For example, in the circuit of Figure 6.21 two cells have been connected in opposition. Here the total electromotive force in the circuit is $E_1 - E_2$, and by Kirchhoff's second law $E_1 - E_2 = V_1 + V_2 = IR_1 + IR_2$.

Figure 6.21

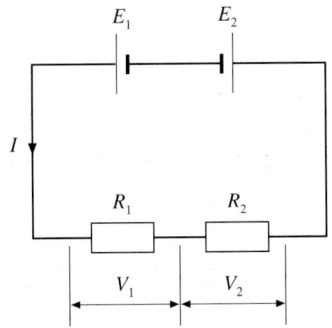

Internal resistance

When a car engine is started while the headlights are switched on, the headlights sometimes dim. This is because the car battery has resistance.

All power supplies have some resistance between their terminals, called **internal resistance**. This causes the charge circulating in the circuit to dissipate some electrical energy in the power supply itself. The power supply becomes warm when it delivers a current.

Figure 6.22

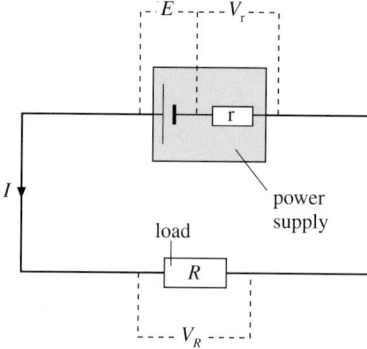

Figure 6.22 shows a power supply which has e.m.f. E and internal resistance r. It delivers a current I when connected to an external resistor of resistance R (called the load). V_R is the potential difference across the load, and V_r is the potential difference across the internal resistance. By Kirchhoff's second law,

$$E = V_R + V_r$$

The potential difference V_R across the load is thus given by

$$V_R = E - V_r.$$

V_R is called the **terminal potential difference**. That is,

> ★ the terminal potential difference is the p.d. between the terminals of a cell when a current is being delivered.

The terminal potential difference is always less than the electromotive force when the power supply delivers a current. This is because of the potential difference across the internal resistance. The potential difference across the internal resistance is sometimes called the **lost voltage**.

> ★ *lost volts = e.m.f. − terminal p.d.*

In contrast, the electromotive force is the terminal potential difference when the cell is in an open circuit (when it is delivering no current). This e.m.f. may be measured by connecting a very high resistance voltmeter across the terminals of the cell.

You can use the circuit in Figure 6.23 to show that the greater the current delivered by the power supply, the lower its terminal potential difference. As more lamps are connected in parallel to the power supply, the current increases and the lost volts, given by

lost volts = current × internal resistance,

will increase. Thus the terminal potential difference decreases.

To return to the example of starting a car with its headlights switched on, a large current (perhaps 100 A) is supplied to the starter motor by the battery. There will then be a large potential difference across the internal resistance, that is the lost voltage will be large. The terminal potential difference will drop and the lights will dim.

Returning to Figure 6.22, $V_R = IR$ and $V_r = Ir$, so $E = V_R + V_r$ becomes $E = IR + Ir$, or $E = I(R + r)$.

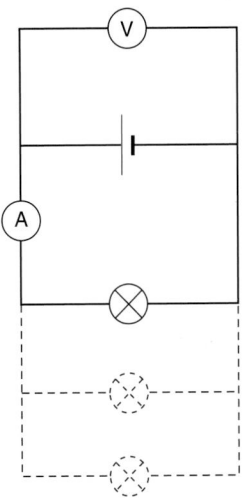

Figure 6.23 *Effect of circuit current on terminal potential difference.*

The maximum current that a power supply can deliver will be when its terminals are short-circuited by a wire of negligible resistance, so that $R = 0$. In this case, the potential difference across the internal resistance will equal the e.m.f. of the cell. The terminal p.d. is then zero. **Warning: do not try out this experiment, as the wire gets very hot; there is also a danger of the battery exploding.**

Quite often, in problems, the internal resistance of a supply is assumed to be negligible, so that the potential difference V_R across the load is equal to the e.m.f. of the power supply.

Effect of internal resistance on power from a battery

The power delivered by a battery to a variable external load resistance can be investigated using the circuit of Figure 6.24. Readings of current I and potential difference V_R across the load are taken for different values of the variable load resistor. The product $V_R I$ gives the power dissipated in the load, and the quotient V_R/I gives the load resistance R.

Figure 6.24 *Circuit for investigating power transfer to an external load.*

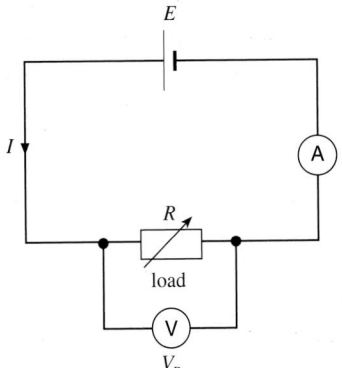

Figure 6.25 *Graph of power delivered to external load against load resistance.*

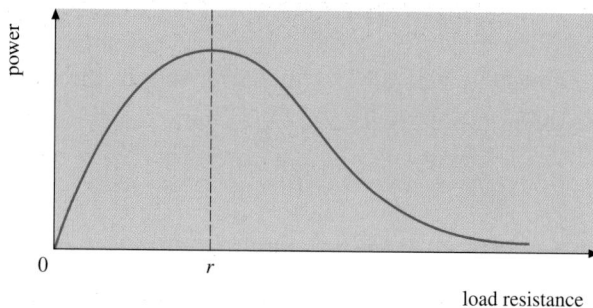

Figure 6.25 shows the variation with load resistance R of the power $V_R I$ dissipated. The graph indicates that there is a maximum power delivered by the battery at one value of the external resistance. This value is equal to the internal resistance r of the battery.

> ★ A battery delivers maximum power to a circuit when the load resistance of the circuit is equal to the internal resistance of the battery.

Example

A high-resistance voltmeter reads 13.0 V when it is connected across the terminals of a battery. The voltmeter reading drops to 12.0 V when the battery delivers a current of 3.0 A to a lamp. State the e.m.f. of the battery. Calculate the potential difference across the internal resistance (the lost volts) when the battery is connected to the lamp. Calculate the internal resistance of the battery.

The e.m.f. is **13.0 V**, since this is the voltmeter reading when the battery is delivering negligible current.

Using $V_r = E - V_R$, lost volts $= V_r = 13.0 - 12.0 = $ **1.0 V**.

Using $V_r = Ir$, $r = 1.0/3.0 = $ **0.33 Ω**.

Now it's your turn

1   Three cells, each of e.m.f. 1.5 V, are connected in series to a 15 Ω light bulb. The current in the circuit is 0.27 A. Calculate the internal resistance of each cell.
    Ans: 0.56 Ω

2   A cell of e.m.f. 1.5 V has an internal resistance of 0.50 Ω. Calculate the maximum current it can deliver. Under what circumstances does it deliver this maximum current? Calculate also the maximum power it can deliver to an external load. Under what circumstances does it deliver this maximum power?
    Ans: 3.0 A when short-circuited; 1.1 W when load resistance equals internal resistance

6.3 Electrical circuits

Section 6.3 summary

★ At any junction in a circuit, the total current entering the junction is equal to the current leaving it. This is Kirchhoff's first law, and is a consequence of the law of conservation of charge.

★ The electromotive force of a supply measures the electrical energy gained per unit of charge passing through the supply.

★ The potential difference across a resistor measures the electrical energy lost per unit of charge passing through the resistor.

★ In any closed loop of a circuit, the sum of the electromotive forces is equal to the sum of the potential differences. This is Kirchhoff's second law, and is a consequence of the law of conservation of energy.

★ The voltage across the terminals of a supply (the terminal p.d.) is always less than the e.m.f. of the supply when the supply is delivering a current, because of the lost volts across the internal resistance.

★ For a supply of e.m.f. E which has internal resistance r, $E = I(R + r)$ where R is the external circuit resistance and I is the current in the supply.

★ A supply delivers maximum power to a load when the load resistance is equal to the internal resistance of the supply.

Section 6.3 questions

1 The internal resistance of a dry cell increases gradually with age, even if the cell is not being used. However, the e.m.f. remains approximately constant. If you don't trust the 'best before' date on a pack of cells, you can check the age of a cell by connecting a low-resistance ammeter across the cell and measuring the current. For a new 1.5 V cell of a certain type, the short-circuit current should be about 30 A.

   (a) Calculate the internal resistance of a new cell.
   (b) A student carries out this test on an older cell, and finds the short-circuit current to be only 5 A. Calculate the internal resistance of this cell.

2 A torch bulb has a power supply of two 1.5 V cells connected in series. The potential difference across the bulb is 2.2 V, and it dissipates energy at the rate of 550 mW. Calculate:

   (a) the current through the bulb,
   (b) the internal resistance of each cell,
   (c) the heat energy dissipated in each cell in two minutes.

3 Two identical light bulbs are connected first in series, and then in parallel, across the same battery (assumed to have negligible internal resistance). Use Kirchhoff's laws to decide which of these connections will give the greater intensity of light.

6.4 | More on series and parallel

Resistors in series

Figure 6.26 *Resistors in series.*

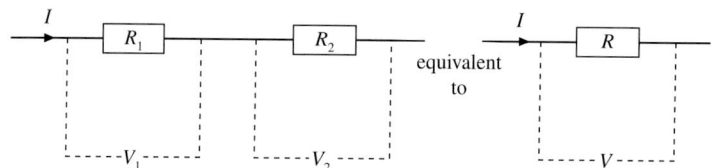

Figure 6.26 shows two resistors of resistance R_1 and R_2 connected in series, and a single resistor of resistance R equivalent to them. The current I in the resistors, and in their equivalent single resistor, is the same. The total potential difference V across the two resistors must be the same as that across the single resistor. If V_1 and V_2 are the potential differences across each resistor,

$$V = V_1 + V_2$$

But since potential difference is given by multiplying the current by the resistance,

$$IR = IR_1 + IR_2$$

Dividing by the current I,

$$R = R_1 + R_2$$

This equation can be extended so that the equivalent resistance R of several resistors connected in series is given by the expression

$$\bigstar \ \ R = R_1 + R_2 + R_3 + \ldots.$$

Thus,

> ★ the combined resistance of resistors in series is just the sum of all the individual resistances.

Resistors in parallel

Now consider two resistors of resistance R_1 and R_2 connected in parallel, as shown in Figure 6.27. The current through each will be different, but they will each have the same potential difference. The equivalent single resistor of resistance R will have the same potential difference across it, but the current will be the total current through the separate resistors.

Figure 6.27 Resistors in parallel.

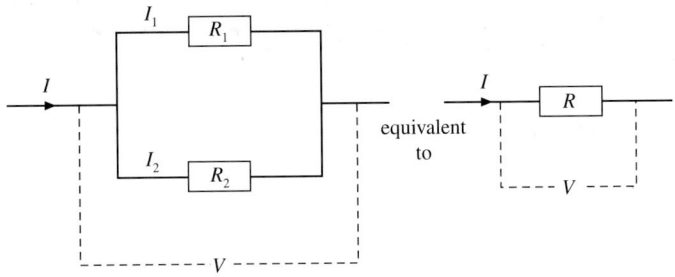

By Kirchhoff's first law,

$$I = I_1 + I_2$$

and, using Ohm's law, $I = V/R$, so

$$V/R = V/R_1 + V/R_2$$

Dividing by the potential difference V,

$$1/R = 1/R_1 + 1/R_2$$

This equation can be extended so that the equivalent resistance R of several resistors connected in parallel is given by

$$\star \quad \frac{1}{R} = \frac{1}{R_1} + \frac{1}{R_2} + \frac{1}{R_3} + \dots$$

Thus,

★ the reciprocal of the combined resistance of resistors in parallel is the sum of the reciprocals of all the individual resistances.

Note that:

1 for two identical resistors in parallel, the combined resistance is equal to half of the value of each one,

2 for resistors in parallel, the combined resistance is always less than the value of the smallest individual resistance.

Example

Calculate the equivalent resistance of the arrangement of resistors in Figure 6.28.

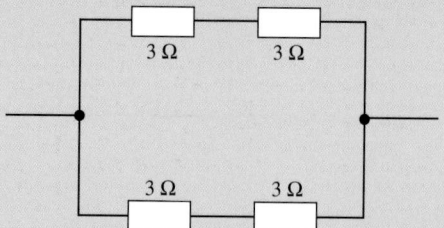

Figure 6.28

The arrangement is equivalent to two 6 Ω resistors in parallel, so the combined resistance R is given by $1/R = 1/6 + 1/6 = 2/6$. (Don't forget to find the reciprocal of this value!).

Thus $R = 3\ \Omega$.

Now it's your turn

1 Calculate the equivalent resistance of the arrangement of resistors in Figure 6.29. (*Hint:* First find the resistance of the parallel combination.)

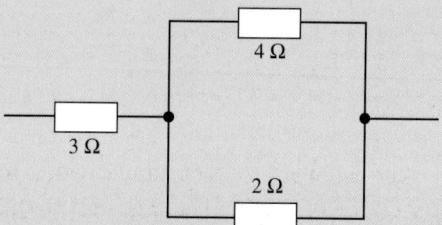

Figure 6.29

Ans: 4.3 Ω

2 Calculate the effective resistance between the points A and B in the network in Figure 6.30.

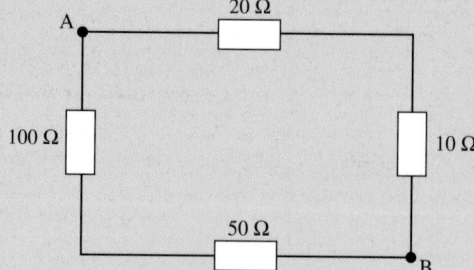

Figure 6.30

Ans: 25 Ω

Potential dividers

Two resistors connected in series with a cell each have a potential difference. They may be used to divide the e.m.f. of the cell. This is illustrated in Figure 6.31.

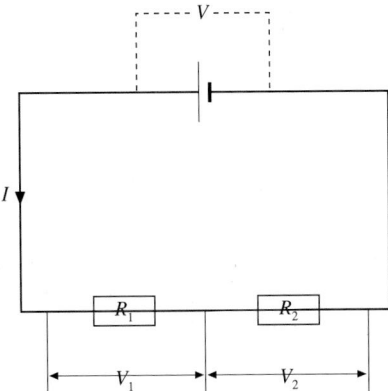

Figure 6.31 *The potential divider.*

The current in each resistor is the same, because they are in series. Thus $V_1 = IR_1$ and $V_2 = IR_2$. Dividing the first equation by the second gives $V_1/V_2 = R_1/R_2$. The ratio of the voltages across the two resistors is the same as the ratio of their resistances. If the potential difference across the combination were 12 V and R_1 were equal to R_2, then each resistor would have 6 V across it. If R_1 were twice the magnitude of R_2, then V_1 would be 8 V and V_2 would be 4 V.

Potentiometers

A potentiometer is a continuously-variable potential divider. In Figure 6.3 a variable voltage supply was used to vary the voltage across different circuit components. A variable resistor, or rheostat, may be used to produce a continuously-variable voltage.

Such a variable resistor is shown in Figure 6.32. All three connections to the variable resistor are used. The fixed ends are connected across the battery so that there is the full battery voltage across the ends AB of the resistor. As with the potential divider, the ratio of the voltages across AC and CB will be the same as the ratio of the resistances of AC and CB. When the sliding contact C is at the end B, the output voltage V_{out} will be 12 V. When the sliding contact is at end A, then the output voltage will be zero. So, as the sliding contact is moved from A to B, the output voltage varies continuously from zero up to the battery voltage. In terms of the terminal p.d. V of the cell, the output V_{out} of the potential divider is given by

$$V_{out} = \frac{VR_1}{(R_1 + R_2)}$$

A variable resistor connected in this way is called a **potentiometer**. See Figure 6.32.

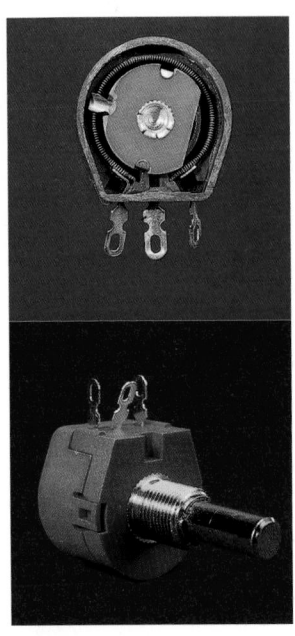

Photographs of potentiometers.

Figure 6.32 *Potentiometer circuit.*

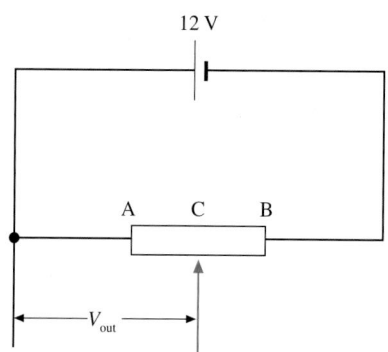

Example

A light-emitting diode (LED) is connected in series with a resistor to a 5.0 V supply. **(a)** Calculate the resistance of the series resistor required to give a current in the LED of 12 mA, with a voltage across it of 2.0 V **(b)** Calculate the potential difference across the LED when the series resistor has resistance 500 Ω.

(a) If the supply voltage is 5.0 V and the p.d. across the LED is 2.0 V, the p.d. across the resistor must be $5.0 - 2.0 = 3.0$ V. The current through the resistor is 12 mA as it is in series with the LED. Using $R = V/I$, the resistance of the resistor is $3.0/12 \times 10^{-3} = \textbf{250 Ω}$.

(b) The resistance of the LED is given by $R = V/I = 2.0/12 \times 10^{-3} = 167$ Ω. If this resistance is in series with a 500 Ω resistor and a 5.0 V supply, the p.d. across the LED is $5.0 \times 167/(167 + 500) = \textbf{1.25 V}$.

Now it's your turn

1   Figure 6.33 shows a light-dependent resistor (LDR) connected in series with a 10 kΩ resistor and a 12 V supply. Calculate:
   (a) the p.d. V_L across the LDR when it is in the dark and has resistance 8.0 MΩ,
   (b) the p.d. V_L across the LDR when it is in bright light and has resistance 500 Ω,
   (c) the resistance of the LDR in lighting conditions which make V_L equal to 4.0 V.

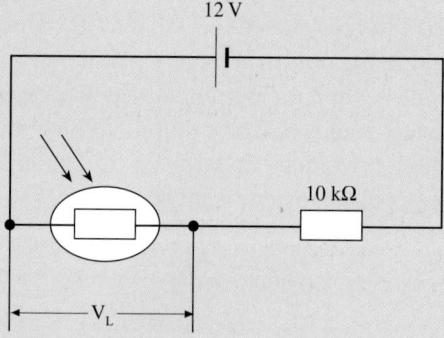

Figure 6.33

   Ans: (a) 12 V; (b) 0.6 V; (c) 5 kΩ

2 The thermistor in the potential divider circuit of Figure 6.34 has a resistance which varies between 100 Ω at 100 °C and 6.0 kΩ at 0 °C. Calculate the potential difference across the thermistor at **(a)** 100 °C, **(b)** 0 °C.

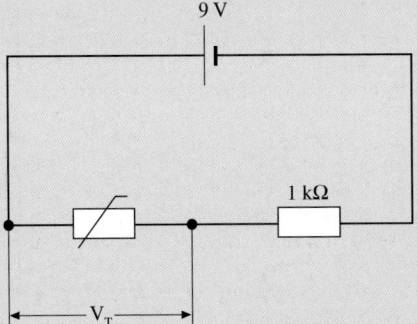

9 V

1 kΩ

V_T

Figure 6.34

Ans: (a) 0.8 V; (b) 7.7 V

Section 6.4 summary

★ The equivalent resistance R of resistors connected in series is given by $R = R_1 + R_2 + R_3 + \ldots$

★ The equivalent resistance R of resistors connected in parallel is given by $1/R = 1/R_1 + 1/R_2 + 1/R_3 + \ldots$

★ Two resistors in series act as a potential divider, where $V_1/V_2 = R_1/R_2$. If V is the supply voltage, $V_{out} = VR_1/(R_1 + R_2)$

★ A potentiometer is a variable resistor connected as a potential divider to give a continuously-variable output voltage

Section 6.4 questions

1 You are given three resistors of resistance 22 Ω, 47 Ω and 100 Ω. Calculate:
   (a) the maximum possible resistance,
   (b) the minimum possible resistance,
   that can be obtained by combining any or all of these resistors.

2 In the circuit of Figure 6.35, the currents I_1 and I_2 are equal. Calculate:
   (a) the resistance R of the unknown resistor,
   (b) the total current I_3 drawn from the battery.

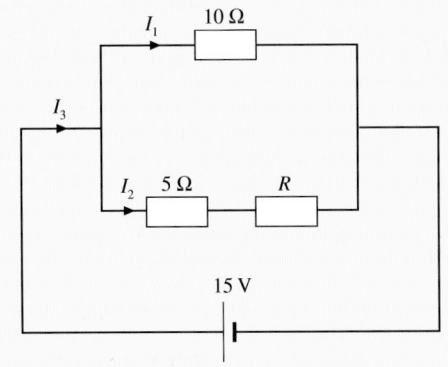

I_1 10 Ω

I_3

I_2 5 Ω R

15 V

Figure 6.35

3 Figure 6.36 shows a potential divider circuit, designed to provide p.d.s of 1.0 V and 4.0 V from a battery of e.m.f. 9.0 V and negligible internal resistance.

(a) Calculate the value of resistance R.

(b) State and explain what happens to the voltage at terminal A when an additional 1.0 Ω resistor is connected between terminals B and C in parallel with the 5.0 Ω resistor. No calculations are required.

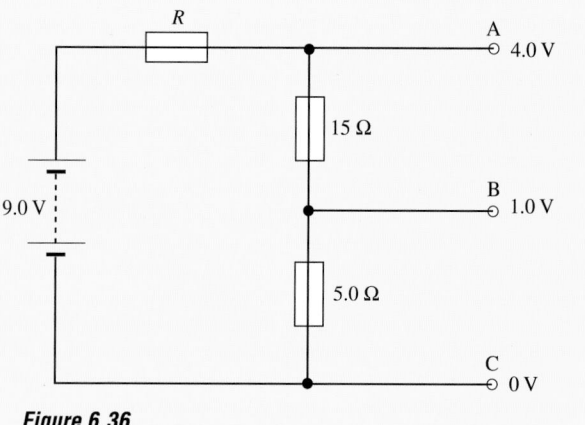

Figure 6.36

Exam Questions

1 A student designs an electrical method to monitor the position of a steel sphere rolling on two parallel rails. Each rail is made from bare wire of length 30 cm and resistance 20 Ω. The position-sensing circuit is shown in Figure 6.37. The resistance of the steel sphere and the internal resistance of the battery are negligible.

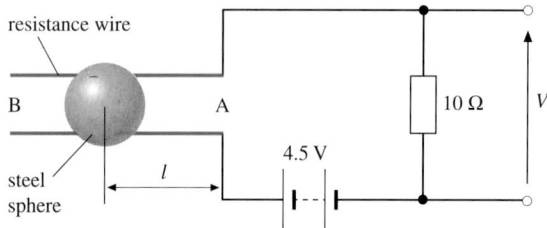

Figure 6.37

(a) State the voltage across the 10 Ω resistor when the sphere is at A, where $l = 0$.

(b) With the sphere at the end of B of the rails, calculate:

   i) the total resistance of the circuit,

   ii) the current in the 10 Ω resistor,

   iii) the output voltage V.

2 Two equations for the power P dissipated in a resistor are $P = I^2R$ and $P = V^2/R$. The first suggests that the greater the resistance R of the resistor, the more power is dissipated. The second suggests the opposite: the greater the resistance, the less the power. Resolve this apparent contradiction.

3 State the minimum number of resistors, each of the same resistance and power rating of 0.5 W, which must be used to produce an equivalent 1.2 kΩ, 5 W resistor. Calculate the resistance of each, and state how they should be connected.

4 In the circuit shown in Figure 6.38 the current leaving the battery is 1.5 A. The battery has internal resistance 1.0 Ω. Calculate:

(a) the combined resistance of the resistors that are connected in parallel in the circuit of Figure 6.38,

(b) the total resistance of the circuit,

(c) the resistance of resistor Y,

(d) the current through the 6 Ω resistor.

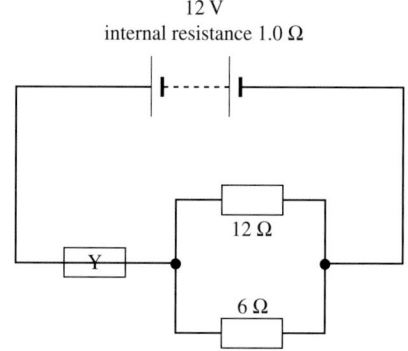

Figure 6.38

5   The current in the starter motor of a car is 160 A when starting the engine. The connecting cable has total length 1.3 m, and consists of fifteen strands of wire, each of diameter 1.2 mm. The resistivity of the metal of the strands is 1.4×10^{-8} Ω m.

(a) Calculate:
   i)   the resistance of each strand,
   ii) the total resistance of the cable,
   iii) the power loss in the cable.

(b) When the starter motor is used to start the car, 700 C of charge pass through a given cross-section of the cable.
   i)   Assuming that the current is constant at 160 A, calculate for how long the charge flows.
   ii) Calculate the number of electrons which pass a given cross-section of the cable in this time. The electron charge e is -1.6×10^{-19} C.

(c) The e.m.f. of the battery is 13.6 V and its internal resistance is 0.012 Ω. Calculate
   i)   the potential difference across the battery terminals when the current in the battery is 160 A,
   ii) the rate of production of heat energy in the battery.

6   A copper wire of length 16 m has a resistance of 0.85 Ω. The wire is connected across the terminals of a battery of e.m.f. 1.5 V and internal resistance 0.40 Ω.

(a) Calculate the potential difference across the wire and the power dissipated in it.

(b) In an experiment, the length of this wire connected across the terminals of the battery is gradually reduced.
   i)   Sketch a graph to show how the power dissipated in the wire varies with the connected length.
   ii) Calculate the length of the wire when the power dissipated in the wire is a maximum.
   iii) Calculate the maximum power dissipated in the wire.

7   In the National Grid, electrical energy is transmitted over long distances at 400 kV. An alternative might be to transmit at the domestic supply voltage of 230 V. Explain how the high-voltage option reduces power losses in the cables.

Waves

This wave is called a 'tsunami'. It is a tidal wave which has been created by an earthquake beneath the ocean bed. This kind of wave can travel at speeds of 600 km h^{-1} and can be many metres high when it reaches the shore. The energy carried by a tsunami is so large that it may demolish towns and villages which lie in its path.

The aim of this chapter is to introduce some general properties of waves. We shall meet two broad classifications of waves, transverse and longitudinal, based on the direction in which the particles move relative to the direction in which the wave transmits energy. We shall define terms such as amplitude, wavelength and frequency for a wave, and derive the relationship between speed, frequency and wavelength. We shall look at demonstrations of some properties of waves, such as reflection, refraction, diffraction and interference. In connection with interference, we shall use the principle of superposition, which tells us how waves which come together at the same place interact.

7.1 Transverse and longitudinal waves

Wave motion is a means of moving energy from place to place. For example, electromagnetic waves from the Sun carry the energy plants and trees need to survive and grow. The energy carried by sound waves causes our ear drums to vibrate. The energy carried by seismic waves (earthquakes) can devastate vast areas, causing the land to move and buildings to collapse. Waves which move energy from place to place are called **progressive waves**.

Vibrating objects act as sources of waves. For example, a vibrating tuning fork sets the air close to it into oscillation, and a sound wave spreads out from the fork. For a radio wave the vibrating objects are electrons.

There are two main groups of waves. These are **transverse waves** and **longitudinal waves**.

> ★ A transverse wave is one in which the vibrations of the particles in the wave are at right angles to the direction in which the energy of the wave is travelling.

Figure 7.1 shows a transverse wave moving along a rope. The particles of the rope vibrate up and down, whilst the energy travels at right angles to this, from A to B. There is no transfer of matter from A to B. Examples of transverse waves include light waves, surface water waves and secondary seismic waves (S-waves).

> ★ A longitudinal wave is one in which the direction of the vibrations is along the direction in which the energy of the wave is travelling.

Figure 7.2 shows a longitudinal wave moving along a stretched spring (a 'slinky'). The coils of the spring vibrate along the length of the spring, whilst the energy travels along the same line, from A to B. Note that the spring itself does not move from A to B. Examples of longitudinal waves include sound waves and primary seismic waves (P-waves).

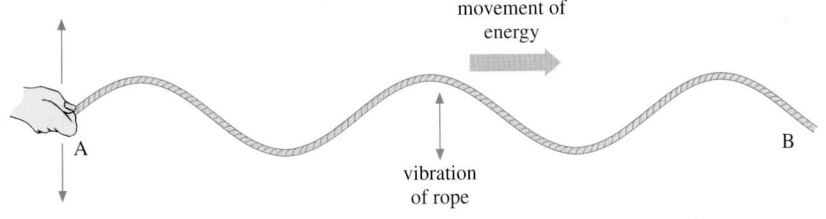

movement of energy

vibration of rope

A

B

Figure 7.1 Transverse wave in a rope.

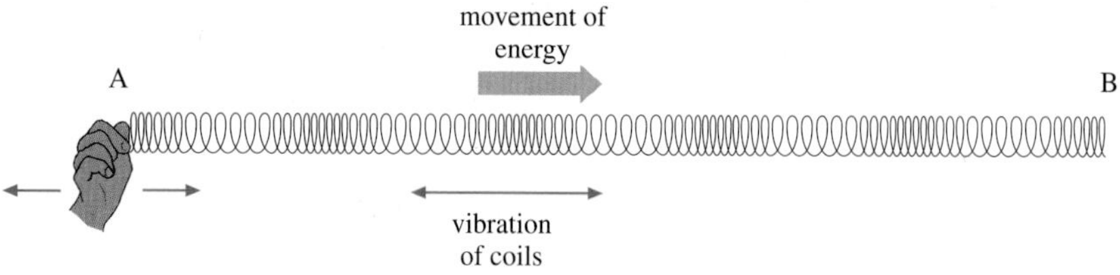

movement of
energy

A

B

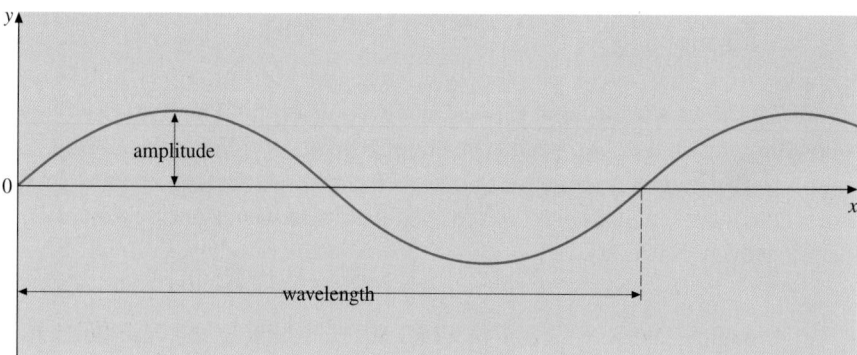

Figure 7.2 *Longitudinal wave in a slinky spring.*

vibration
of coils

Graphical representation of waves

★ The displacement of a particle on a wave is its distance from the rest position.

Displacement is a vector quantity, and can be positive or negative. A transverse wave may be represented by plotting displacement y on the y-axis against distance x along the wave, in the direction of energy travel, on the x-axis. This is shown in Figure 7.3. It can be seen that the graph is a snapshot of what is actually observed to be a transverse wave.

Figure 7.3 *Displacement-distance graph for a transverse wave.*

For a longitudinal wave, the displacement of the particles is along the direction of energy travel. However, if these displacements are plotted on the y-axis of a graph of displacement against distance, the graph has exactly the same shape as for a transverse wave (see Figure 7.3). This is very useful, in that one graph can represent both types of wave. Using this graph, wave properties may be treated without reference to the type of wave.

Referring to Figure 7.3,

★ the **amplitude** of the wave motion is defined as the maximum displacement.

Also, it can be seen that the wave repeats itself. That is, the wave can be constructed by repeating a section of the wave. The length of the smallest repetition unit is called the **wavelength**.

7.1 Transverse and longitudinal waves

⭐ One wavelength is the distance between two neighbouring peaks or two neighbouring troughs, or two neighbouring points which are vibrating together in exactly the same way (in phase).

The usual symbol for wavelength is λ, the Greek letter lambda.

Another way to represent both waves is to plot a graph of displacement y against time t. This is shown in Figure 7.4. Again, the wave repeats itself after a certain interval of time. The time for one complete vibration or to produce one complete wave is called the **period** of the wave T.

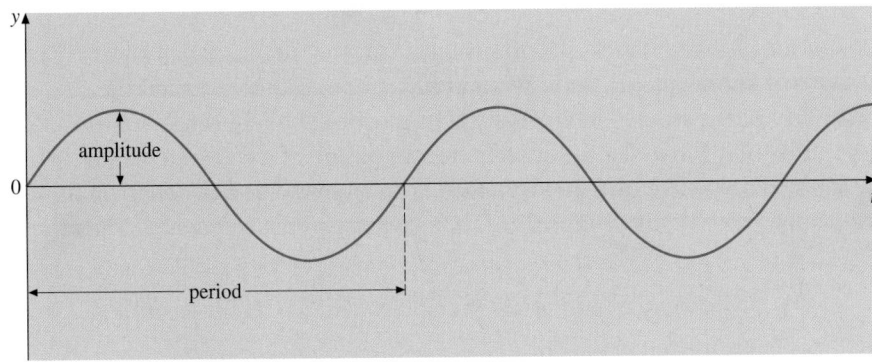

Figure 7.4 *Displacement-time graph for a wave.*

⭐ The period of the wave is the time for a particle in the wave to complete one vibration, or one cycle.

⭐ The number of complete vibrations (cycles) per unit time is called the **frequency** of the wave f.

For waves in ropes and springs, displacement and amplitude are measured in mm, m or other units of length. Period is measured in seconds (s). Frequency has the unit per second (s^{-1}) or hertz (Hz).

A term used to describe the relative positions of the crests or troughs of two waves of the same frequency is **phase**. When the crests and troughs are aligned, the waves are said to be **in phase**. When a crest is aligned with a trough, the waves are **out of phase**. When used as a quantitative measure, phase has the unit of angle (radians or degrees). Thus, when waves are out of phase, one wave is half a cycle behind the other. Since one cycle is equivalent to 2π radians or $360°$, the phase difference between waves which are exactly out of phase is π radians or $180°$.

Consider Figure 7.5, in which there are two waves of the same frequency and amplitude, but with a phase difference between them. The period T corresponds to a phase angle of 2π rad or $360°$. The two waves are out of step by a time t. Thus, phase difference is equal to $2\pi(t/T)$ rad $= 360(t/T)°$. A similar argument may be used for waves of wavelength λ which are out of step by a distance x. In this case the phase difference is $2\pi(x/\lambda)$ rad $= 360(x/\lambda)°$.

Figure 7.5 *Phase difference.*

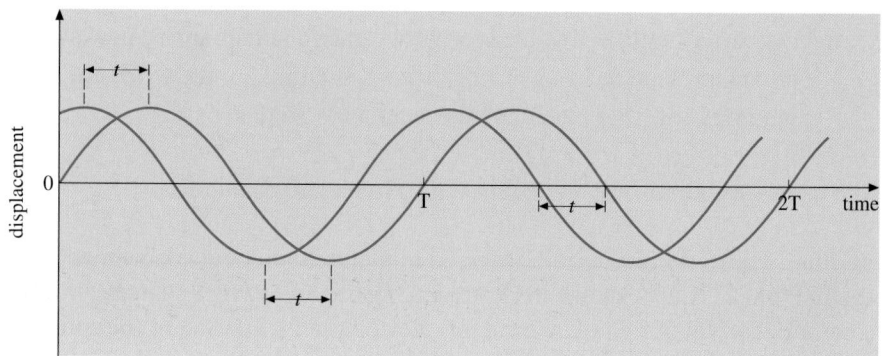

One of the characteristics of a progressive wave is that it carries energy. The amount of energy passing through unit area per unit time is called the **intensity** of the wave. The intensity is proportional to the square of the amplitude of a wave. Thus, doubling the amplitude of a wave increases the intensity of the wave by a factor of four. The intensity also depends on the frequency: intensity is proportional to the square of the frequency. Thus,

⭐ for a wave of amplitude A and frequency f, the intensity I is proportional to $A^2 f^2$.

If the waves from a point source spread out equally in all directions, we have what is called a **spherical wave**. As the wave travels further from the source, the energy it carries passes through an increasingly large area. Since the surface area of a sphere is $4\pi r^2$, the intensity is $W/4\pi r^2$, where W is the power of the source. The intensity of the wave thus decreases with increasing distance from the source.

⭐ The intensity I is proportional to $1/r^2$, where r is the distance from the source.

This relationship assumes that there is no absorption of wave energy.

Wave equation

From the definition of wavelength λ, in one cycle of the source the wave energy moves a distance λ. In f cycles, the wave moves a distance $f\lambda$. If f is the frequency of the wave, then f cycles are produced in unit time. Therefore $f\lambda$ is the distance moved in unit time. Referring to Chapter 2, speed c is the distance moved per unit time. Therefore

$$\bigstar \; c = f\lambda,$$

or

7.1 **Transverse and longitudinal waves**

⭐ *speed = frequency × wavelength.*

This is an important relationship between the speed of a wave and its frequency and wavelength.

Examples

1   A tuning fork of frequency 170 Hz produces sound waves of wavelength 2.0 m. Calculate the speed of sound.

Using $c = f\lambda$, we have $c = 170 \times 2.0 = \textbf{340 m s}^{-1}$.

2   The amplitude of a wave in a rope is 15 mm. If the amplitude were changed to 20 mm, keeping the frequency the same, by what factor would the energy carried by the rope change?

Intensity is proportional to the square of the amplitude. Here the amplitude has been increased by a factor of 20/15, so the energy carried by the wave increases by a factor of $(20/15)^2 = \textbf{1.8}$.

Now it's your turn

1   Water waves of wavelength 0.080 m have a frequency 5.0 Hz. Calculate the speed of these water waves.
Ans: 0.40 m s^{-1}

2   The speed of sound is 340 m s^{-1}. Calculate the wavelength of the sound wave produced by a violin when a note of frequency 500 Hz is played.
Ans: 0.68 m

3   The speed of light is 3.0×10^8 m s^{-1}. Calculate the frequency of red light of wavelength 6.5×10^{-7} m.
Ans: 4.6×10^{14} Hz

4   A beam of red light has twice the intensity as a second beam of the same colour. Calculate the ratio of the amplitudes of waves.
Ans: 1.4

Properties of wave motions

Although there are many different types of waves (light waves, sound waves, electrical waves, mechanical waves etc.) there are a number of basic properties which all have in common. All waves can be reflected, refracted, diffracted, and can produce interference patterns.

These properties may be demonstrated using a ripple tank similar to that shown in Figure 7.6. As the motor turns, the wooden bar vibrates, creating ripples on the surface of the water. The ripples are lit from above. This creates shadows on the viewing screen. The shadows show the shape and movement of the waves. Each shadow corresponds to a particular point on the wave, and is referred to as a **wavefront**.

Figure 7.6 *A ripple tank.*

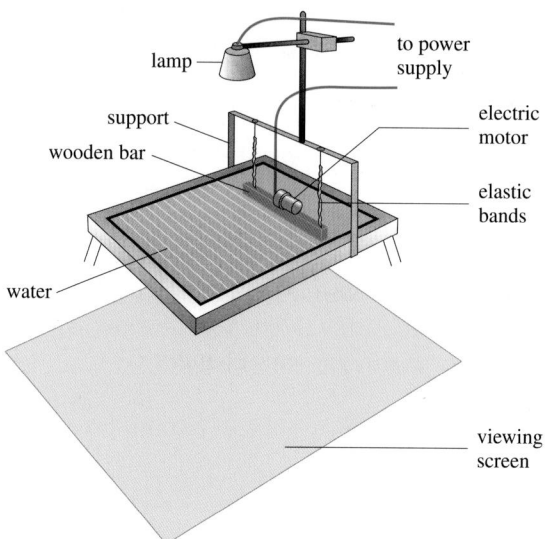

Figure 7.7 illustrates the pattern of wavefronts produced by a low-frequency vibrator and one of higher frequency. Note that for the higher frequency the wavelength is less, since wave speed is constant and $c = f\lambda$.

Figure 7.7 *Ripple tank patterns for low- and high-frequency vibrations.*

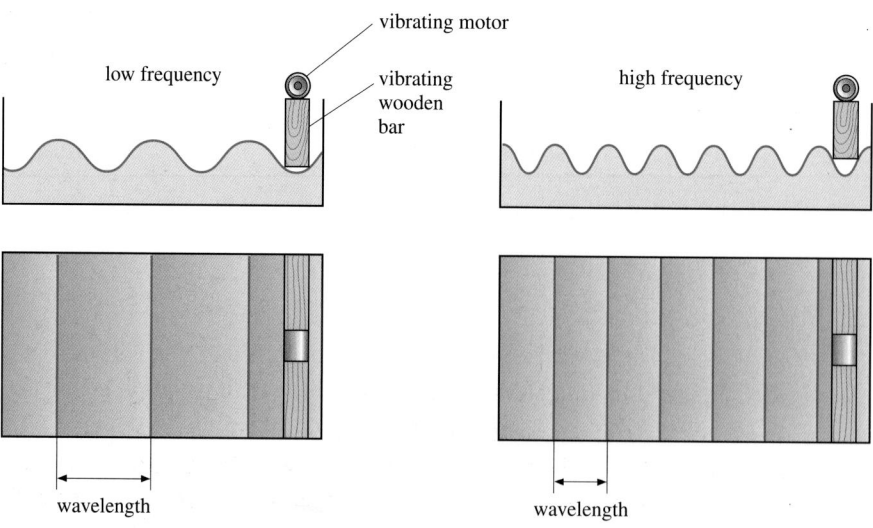

Circular waves may be produced by replacing the vibrating bar with a small dipper, or by allowing drops of water to fall into the ripple tank. A circular wave is illustrated in Figure 7.8. This pattern is characteristic of waves spreading out from a point source.

7.1 Transverse and longitudinal waves

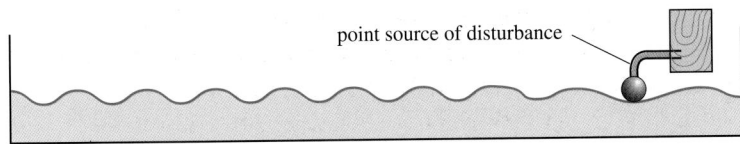

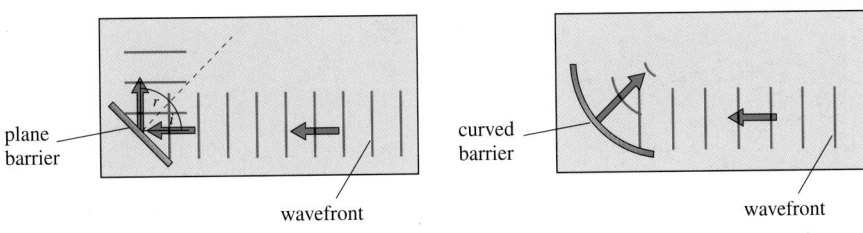

point source of disturbance

Figure 7.8 *Ripple tank pattern for a point source.*

circular waves

We shall now see how the ripple tank may be used to demonstrate the wave properties of reflection, refraction, diffraction and interference.

Reflection

As the waves strike a plane barrier placed in the water, they are reflected. The angle of reflection equals the angle of incidence, and there is no change in wavelength (see Figure 7.9). If a curved barrier is used, the waves can be made to converge or diverge.

plane barrier

wavefront

curved barrier

wavefront

Figure 7.9 *Ripple tank pattern showing reflection at a plane surface and a curved one.*

Refraction

If a glass block is submerged in the water, this produces a sudden change in the depth of the water. The speed of surface ripples on water depends on the depth of the water: the shallower the water, the slower the speed. Thus, the waves move more slowly as they pass over the glass block. The frequency of the waves remains constant, and so the wavelength decreases. If the waves are incident at an angle to the submerged block, they will change direction, as shown in Figure 7.10.

★ The change in direction of a wave due to a change in speed is called **refraction**.

Figure 7.10 *Ripple tank pattern showing refraction.*

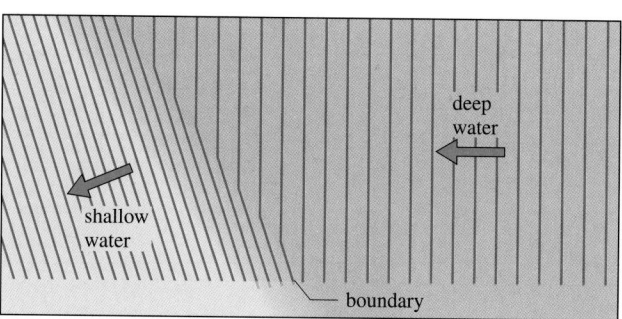

As the waves re-enter the deeper water, their speed increases to the former value and they change direction once again.

Diffraction

When waves pass through a narrow gap, they spread out. This spreading out is called **diffraction**. The extent of diffraction depends on the size of the gap compared with the wavelength. It is most noticeable if the size of the gap is approximately equal to the wavelength. Diffraction is illustrated in Figure 7.11. Note that diffraction may also occur at an edge.

 Diffraction is defined as the spreading of a wave into regions where it would not be seen if it moved only in straight lines.

Figure 7.11 *Ripple tank pattern showing diffraction at (a) a wide gap, (b) a narrow gap.*

a)
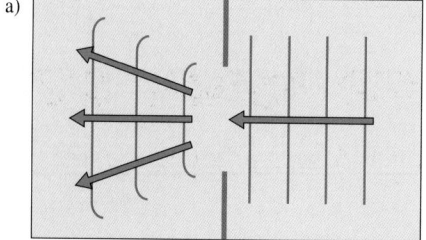
b)
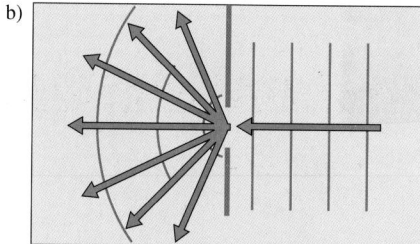

Interference

If two or more waves overlap, the resultant displacement is the sum of the individual displacements. Remember that displacement is a vector quantity. The overlapping waves are said to **interfere**. This may lead to a resultant wave of either a larger or a smaller displacement than either of the two component waves, as shown in Figure 7.12. These are examples of **constructive interference**, in which the resultant wave has a greater amplitude, and **destructive interference**, in which the resultant wave has a smaller amplitude.

Interference can be demonstrated in the ripple tank by using two point sources. Figure 7.13 shows such an interference pattern.

7.1 Transverse and longitudinal waves

a)

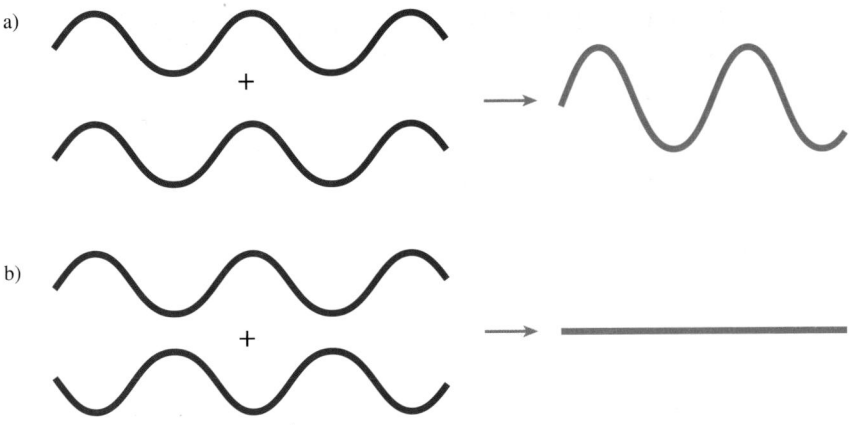

Figure 7.12 a) *Constructive interference and* b) *destructive interference.*

b)

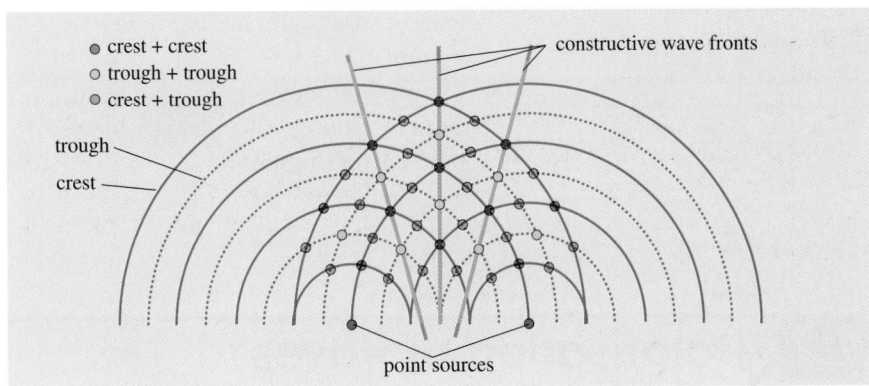

Figure 7.13 *Ripple tank pattern showing two-source interference.*

* A progressive wave travels outwards from the source, carrying energy but without transferring matter.

* In a transverse wave, the oscillations are at right angles to the direction in which the wave carries energy.

* In a longitudinal wave, the oscillations are along the direction in which the wave carries energy.

* The speed c, frequency f and wavelength λ of a wave are related by $c = f\lambda$.

* The intensity of a wave is the energy passing through unit area per unit time. Intensity is proportional to the square of the amplitude and to the square of the frequency.

* For a wave spreading uniformly in all directions, without absorption, from a point source, the intensity at a distance r from the source is proportional to $1/r^2$.

Section 7.1 questions

1   A certain sound wave in air has a speed
    340 m s^{-1} and wavelength 1.7 m. For this
    wave, calculate:
    (a) the frequency,
    (b) the period.

2   The speed of electromagnetic waves in
    vacuum (or air) is $3.0 \times 10^8 \text{ m s}^{-1}$.

    (a) The visible spectrum extends from a
         wavelength of 400 nm (blue light) to
         700 nm (red light). Calculate the range
         of frequencies of visible light.
    (b) A typical frequency for v.h.f. television
         transmission is 250 MHz. Calculate the
         corresponding wavelength.

3   Two waves travel with the same speed and
    have the same amplitude, but the first has
    twice the wavelength of the second.
    Calculate the ratio of the intensities
    transmitted by the waves.

4   A student stands at a distance of 5.0 m from
    a point source of sound, which is radiating
    uniformly in all directions. The intensity of
    the sound wave at her ear is
    $6.3 \times 10^{-6} \text{ W m}^{-2}$.
    (a) The receiving area of the student's ear
         canal is 1.5 cm^2. Calculate how much
         energy passes into her ear in one
         minute.
    (b) The student moves to a point 1.8 m
         from the source. Calculate the new
         intensity of the sound.

7.2   Polarisation of waves

Think about generating waves in a rope by moving your hand holding the
stretched rope up and down, or from side to side. The transverse vibrations
of the rope will be in just one plane – the vertical plane if your hand is
moving up and down, or the horizontal plane if it moves from side to side.
The vibrations are said to be **plane-polarised** in either a vertical plane or a
horizontal plane. However, if the rope passes through a vertical slit, then only
if you move your hand up and down will the wave pass through the slit. If the
rope passes through a horizontal slit, only waves generated by a side-to-side
motion of the hand will be transmitted. This is illustrated in Figure 7.14. If
both orientations of slit are used, the transmission of both types of waves
would be blocked.

> ★ The condition for a wave to be plane-polarised is for the vibrations to
>   be in just one plane, which contains the direction in which the wave
>   is travelling.

Clearly, polarisation can apply only to transverse waves. Longitudinal waves
vibrate along the direction of wave travel, and whatever the orientation of
the slit, it would make no difference to the transmission of the waves.

The fact that light can be polarised was understood only in the early 1800s.
This was a most important discovery. It showed that light is a transverse

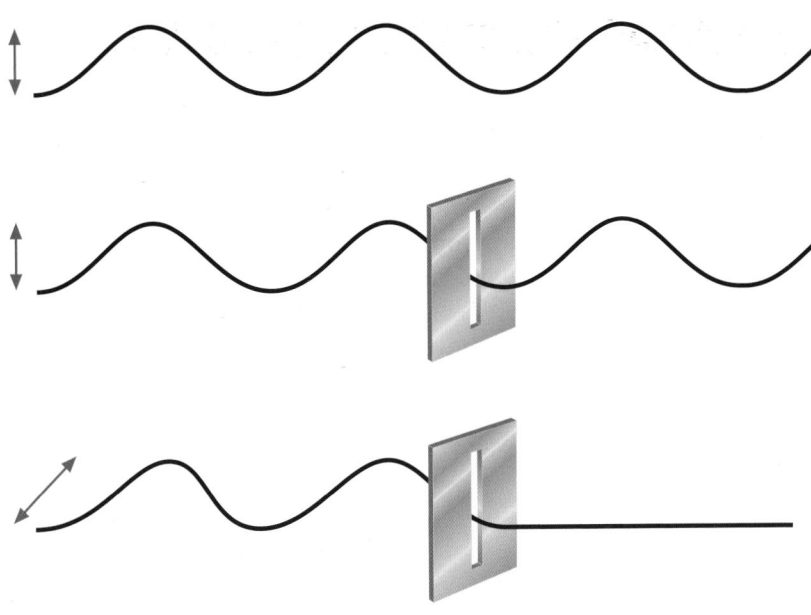

Figure 7.14 Transverse waves on a rope.

wave motion, and opened the way, fifty years later, to Maxwell's theory of light as electromagnetic radiation. Maxwell described light in terms of oscillating electric and magnetic fields, at right angles to each other and at right angles to the direction of travel of the wave energy. When we talk about the direction of polarisation of a light wave, we refer to the direction of the electric field component of the electromagnetic wave.

The Sun and domestic light bulbs emit **unpolarised** light, that is the vibrations take place in many directions at once, instead of in the single plane associated with plane-polarised radiation (Figure 7.15).

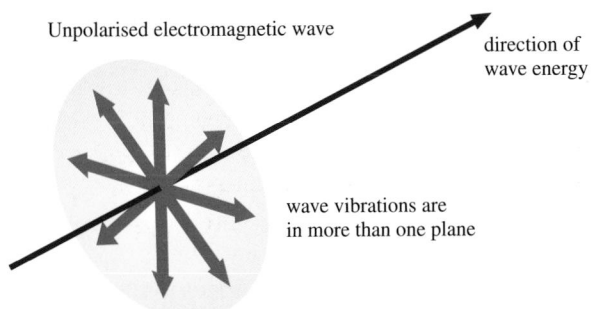

Unpolarised electromagnetic wave

direction of wave energy

wave vibrations are in more than one plane

Polarised electromagnetic wave

direction of wave energy

wave vibrations are in one plane

Figure 7.15 Unpolarised and polarised electromagnetic waves.

Polarising light waves

Some transparent materials, such as a Polaroid sheet, allow vibrations to pass through in one plane only. A Polaroid sheet contains long chains of organic molecules aligned parallel to each other. When unpolarised light arrives at the sheet, the component of the electric field of the incident radiation which is parallel to the molecules is strongly absorbed, whereas radiation with its electric field perpendicular to the molecules is transmitted through the sheet. The Polaroid sheet acts as a **polariser**, producing plane-polarised light from light which was originally unpolarised. Figure 7.16 illustrates unpolarised light entering the polariser, and polarised light leaving it.

Figure 7.16 Action of a polariser.

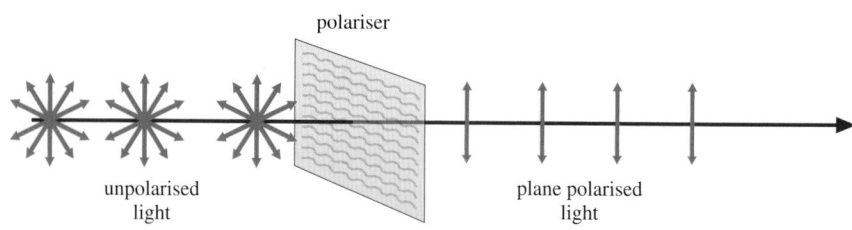

If you try to view plane-polarised light through a second sheet of Polaroid which is placed so that its polarising direction is at right angles to the polarising direction of the first sheet, it will be found that no light is transmitted. In this arrangement, the Polaroids are said to be crossed. The second Polaroid sheet is acting as an **analyser**. If the two Polaroids have their polarising directions parallel, then plane-polarised light from the first Polaroid can pass through the second. These two situations are illustrated in Figure 7.17. Although the action of the Polaroid sheet is not that of a simple slit, the arrangement of the crossed Polaroids has the same effect as the crossed slits in the rope-and-slits experiment.

Applications of polarisation

See the light but not the glare!

A very familiar application of the polarisation of light arises in the use of Polaroid sunglasses. Light reflected from a non-metallic surface, such as the smooth surface of a swimming-pool, is partially polarised parallel to the surface. Wearing sunglasses with Polaroid lenses, arranged with the polarising direction vertical, will prevent the transmission of much of the light reflected from the water. Reduction of the glare of the reflected light allows you to see objects more clearly. Fishermen find that Polaroid sunglasses eliminate glare from the surface of a stream or pond, so that they can see beneath the water more easily.

Wearing Polaroid sunglasses helps fishermen catch their prey!

a)

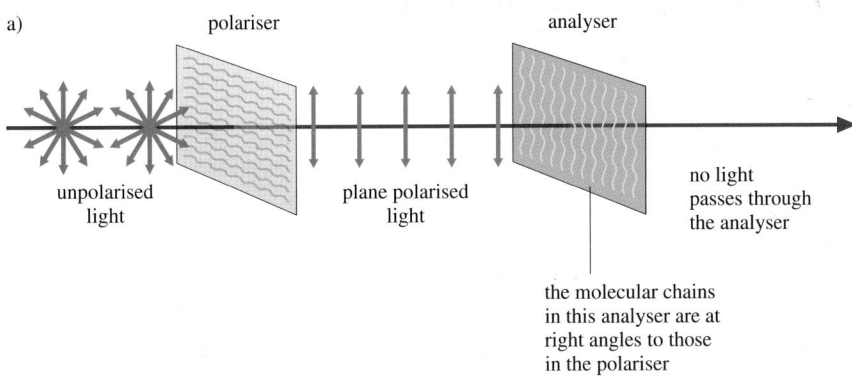

b)

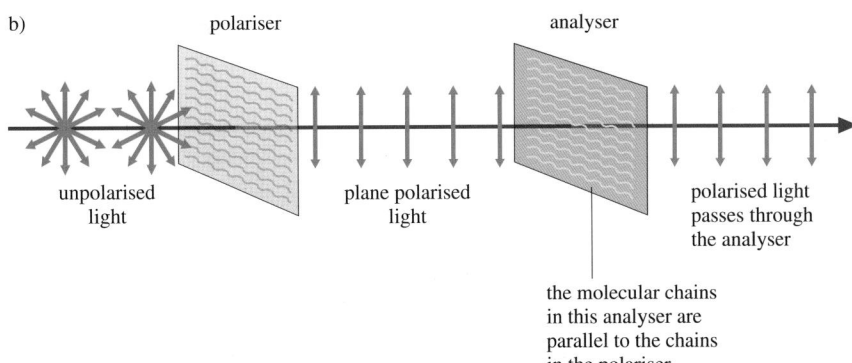

Figure 7.17 *Polariser and analyser in a) crossed and b) parallel situations.*

Stress analysis

When polarised light passes through some transparent materials, the plane of polarisation is rotated. If the material is put under stress, the amount of stress affects the degree of rotation. If the incident polarised light is white, each of its component colours is rotated by a different amount, creating a pattern of coloured fringes when viewed through the analyser Polaroid. Engineers make use of this phenomenon by making models of structural components in a suitable transparent material, such as Perspex, stressing the model, and examining the pattern of fringes to identify regions of high stress. Where possible, the structure is re-designed to remove these danger points. The photograph illustrates a Perspex model being examined between crossed Polaroids.

Poor TV reception?

In certain parts of the United Kingdom television reception is poor. To correct this problem the signal is often boosted by a local relay station before being re-transmitted. Sometimes this introduces further difficulties because of interference between the incident and the re-transmitted signal. This can be avoided if the main transmitter emits waves that are vertically polarised, and the relay station emits waves that are horizontally polarised. This is illustrated in Figure 7.18.

Stress analysis by polarised light.

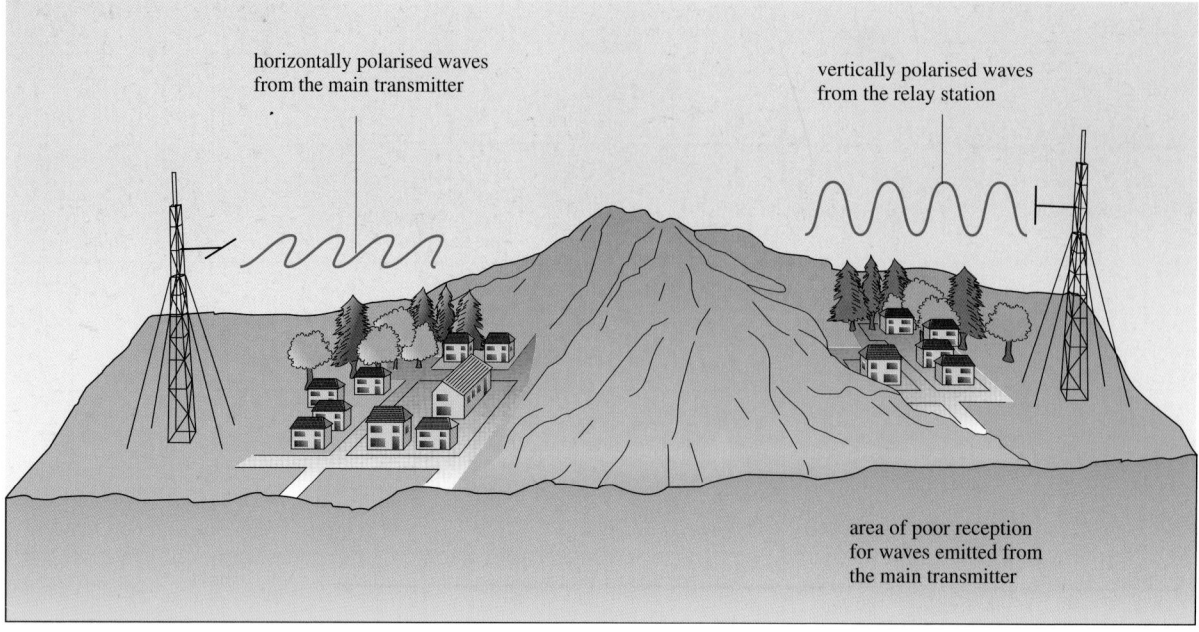

horizontally polarised waves
from the main transmitter

vertically polarised waves
from the relay station

area of poor reception
for waves emitted from
the main transmitter

Figure 7.18 *Avoiding interference between television signals.*

Section 7.2 summary

★ In a plane-polarised wave, the vibrations of the wave are in one direction only, which contains the direction of travel of the wave.

★ Transverse waves can be polarised – longitudinal waves cannot.

★ Plane-polarised light can be produced from unpolarised light by using a polariser, such as a sheet of Polaroid.

★ Rotating an analysing Polaroid in a beam of plane-polarised light prevents transmission of the polarised light. This occurs when the polarising directions of polariser and analyser are at right angles.

7.3 Interference

Any moment now the unsuspecting fisherman in Figure 7.19 is going to experience the effects of **interference**. The amplitude of oscillation of his boat will be significantly affected by the two approaching waves and their interaction when they reach his position.

We have already met the idea of interference through demonstrations of wave properties with the ripple tank. In this section we shall look at the phenomenon more closely and obtain some important results.

Figure 7.19 *Fisherman in a boat.*

Figure 7.20 shows two waves arriving together. If they arrive *in phase*, that is if their crests arrive at exactly the same time, they will interfere *constructively*. A resultant wave will be produced which has crests much higher than either of the two individual waves, and troughs which are much deeper. If the two incoming waves have the same frequency and equal amplitude A, the resultant wave produced by constructive interference has an amplitude of $2A$. The frequency of the resultant is the same as that as the incoming waves.

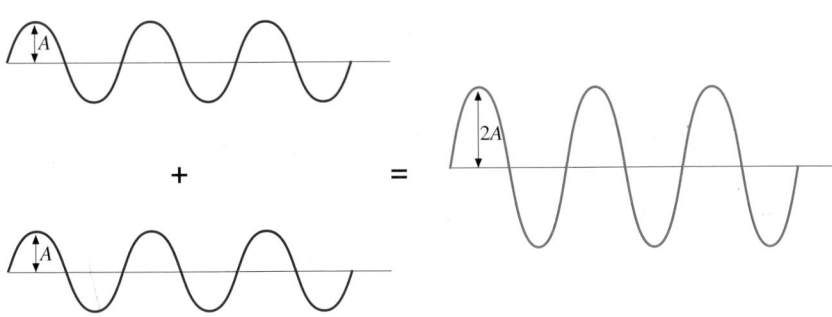

Figure 7.20 *Constructive interference.*

If the two waves arrive *out of phase* (with a phase difference of π radians or 180°), that is if the peaks of one wave arrive at the same time as the troughs from the other, they will interfere *destructively*. The resultant wave will have a smaller amplitude. In the case shown in Figure 7.21, where the incoming waves have equal amplitude, the resultant wave has zero amplitude.

Figure 7.21 *Destructive interference.*

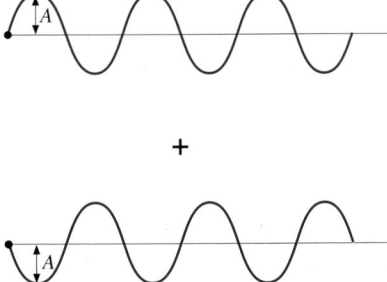

This situation is an example of the **principle of superposition of waves**. The principle describes how waves, which meet at the same point in space, interact.

> ★ The principle of superposition states that the resultant displacement at a point is equal to the sum of the displacements of the individual waves at that point.

Because displacement is a vector, we must remember to add the individual displacements taking account of their directions. The principle applies to all types of wave.

Interference patterns

If we consider the effect of superposition at a number of points in space, we can build up an interference pattern – a pattern showing some areas where there is constructive interference, and hence a large wave disturbance, and other areas where the interference is destructive, and there is little or no wave disturbance.

Figure 7.22 illustrates the production of an interference pattern by two point sources A and B. The point C is equidistant from A and B. A wave travelling to C from A covers the same distance as a wave travelling to C from B. If the waves started out in phase at A and B, they will arrive in phase at C. They combine constructively, producing a maximum disturbance at C.

Figure 7.22 Producing an interference pattern.

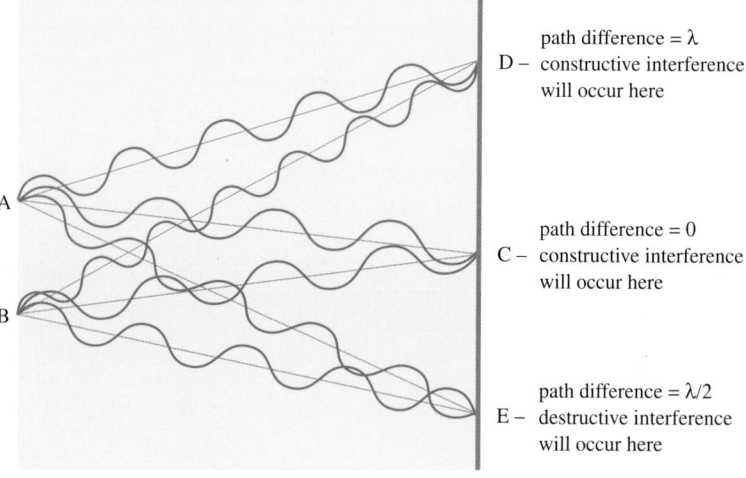

path difference = λ
D – constructive interference will occur here

path difference = 0
C – constructive interference will occur here

path difference = $\lambda/2$
E – destructive interference will occur here

At other places, such as D, the waves will have travelled different distances from the two sources. There is a **path difference** between the waves arriving at D. If this path difference is a whole number of wavelengths (λ, 2λ, 3λ and so on) the waves arrive in phase and interfere constructively, producing maximum disturbance again. However, at places such as E the path difference is an odd number of half-wavelengths ($\lambda/2$, $3\lambda/2$, $5\lambda/2$ and so on). The waves arrive at E out of phase, and interference is destructive, producing a minimum resultant disturbance. This collection of maxima and minima produced by the superposition of overlapping waves is called an **interference pattern**.

7.3 Interference

Producing an interference pattern with sound waves

Figure 7.23 shows an experimental arrangement to demonstrate interference with sound waves from two loudspeakers connected to the same signal generator and amplifier, so that each speaker produces a note of the same frequency. The demonstration is best carried out in the open air (on playing-fields, for example) to avoid reflections from walls. Moving about in the space around the speakers, you pass through places where the waves interfere constructively and you can hear a loud sound. In places where the waves interfere destructively, the note is much quieter than elsewhere in the pattern.

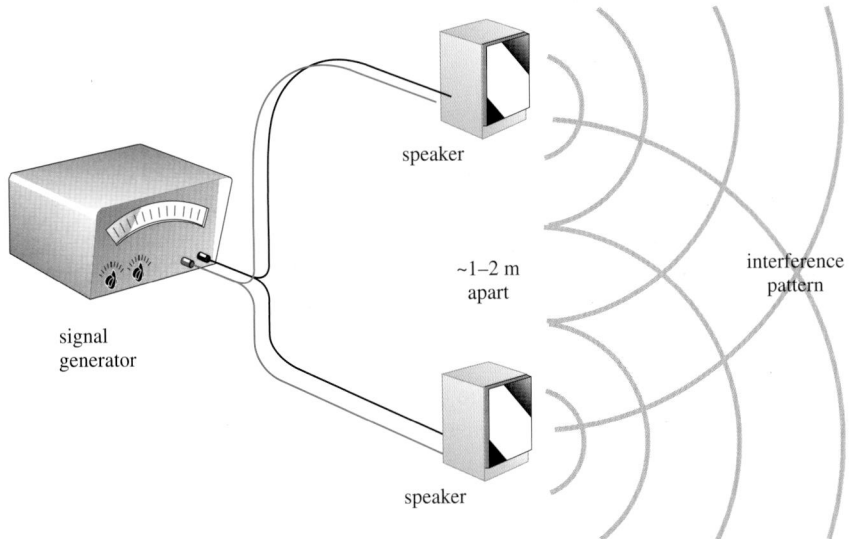

Figure 7.23 *Demonstration of interference with sound waves.*

Producing an interference pattern with light waves

If you try to set up a demonstration with two separate light sources, such as car headlights, you will find that it is not possible to produce an observable interference pattern (Figure 7.24).

Figure 7.24 *Failure of an interference demonstration with light.*

A similar demonstration works with sound waves from two separate loudspeakers. What has gone wrong?

To produce an observable interference the two wave sources must have a **constant phase relationship**. In the sound experiment, the waves from the two loudspeakers have a constant phase relationship because the loudspeakers are connected to the same oscillator and amplifier. If the waves emitted from the speakers are in phase when the experiment begins, they stay in phase for the whole experiment.

 Wave sources which maintain this constant phase relationship are described as **coherent** sources.

This is illustrated in Figure 7.25.

Figure 7.25 *Coherent wave trains.*

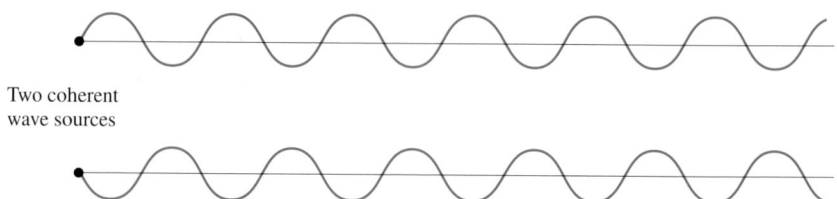

Two coherent
wave sources

Light is emitted from sources as a series of pulses or packets of energy. These pulses last for a very short time, about a nanosecond (10^{-9} s). Between each pulse there is an abrupt change in the phase of the waves. Waves from two separate sources may be in phase at one instant, but out of phase in the next nanosecond. The human eye cannot cope with such rapid changes, so the pattern is not observable. Separate light sources produce incoherent waves (Figure 7.26). To obtain observable patterns the frequencies of the two sources must be the same but it is not essential for the amplitudes of the waves from the two sources to be the same. If the amplitudes are not equal, a completely dark fringe will never be obtained, and the contrast of the pattern is reduced.

Figure 7.26 *Incoherent light trains.*

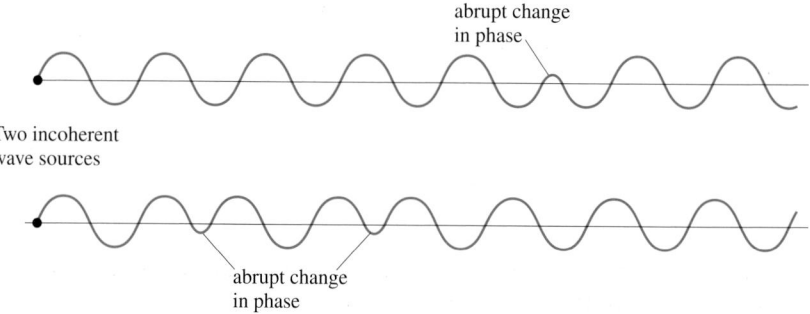

abrupt change
in phase

Two incoherent
wave sources

abrupt change
in phase

Young's double-slit experiment

In 1801 Thomas Young demonstrated how light waves could produce an interference pattern. The experimental arrangement is shown in Figure 7.27. A monochromatic light source (a source of one colour, and hence one

Figure 7.27 *Young's double-slit experiment.*

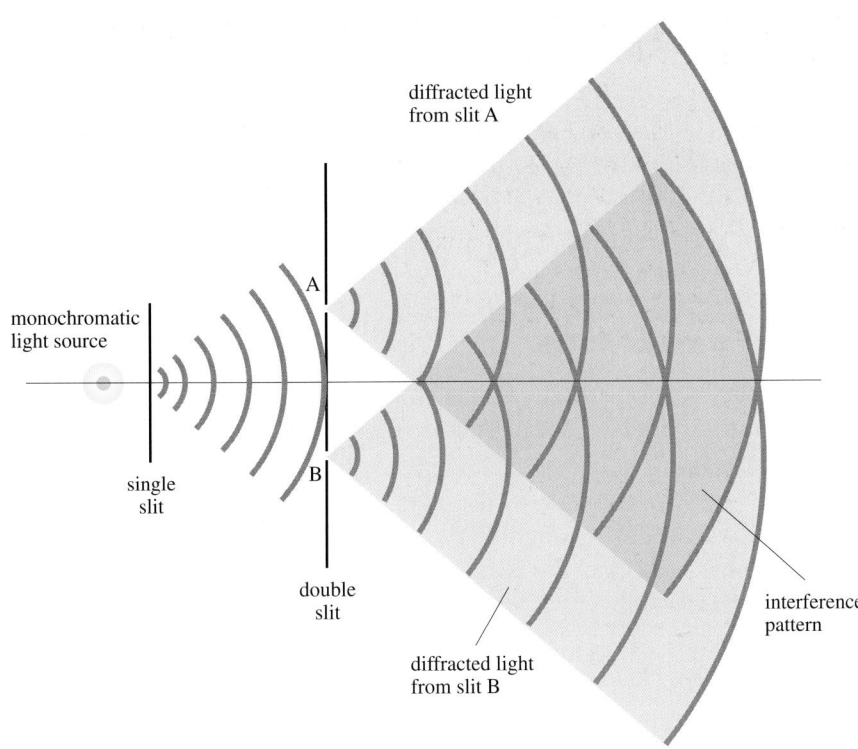

Figure 7.27 *Young's double-slit experiment.*

wavelength λ) is placed behind a single slit to create a small, well-defined source of light. Light from this source is diffracted at the slits, producing two light sources at the slits A and B. Because these two light sources originate from the same primary source, they are coherent and create a sustained and observable interference pattern as seen in the photograph of the dark and bright interference fringes. Bright fringes are seen where constructive interference occurs, that is where the path difference between the two diffracted waves from the sources A and B is $n\lambda$, where n is a whole number. Dark fringes are seen where destructive interference occurs. The condition for a dark fringe is that the path difference should be $(n + \frac{1}{2})\lambda$.

The distance w on the screen between successive bright fringes is called the fringe width. The fringe width is related to the wavelength λ of the light source by the equation

$$\bigstar \quad w = \frac{\lambda D}{d},$$

where D is the distance from the sources to the screen and d is the distance between the centres of the slits.

Although Young's original double-slit experiment was carried out with light, the conditions for constructive and destructive interference apply for any two-source situation. The same formula applies for all types of wave, provided that the fringes are detected at a distance of many wavelengths from the two sources.

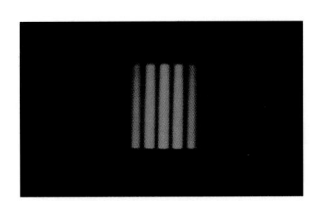

Fringe pattern in Young's experiment.

Calculate the observed fringe width for a Young's double-slit experiment using light of wavelength 600 nm and slits 0.50 mm apart. The distance from the slits to the screen is 0.80 m.

Using $w = \lambda D/d$, $w = 600 \times 10^{-9} \times 0.80/0.50 \times 10^{-3} = 9.6 \times 10^{-4}$ m
$$= \textbf{0.96 mm}$$

Now it's your turn

1 Calculate the wavelength of light which produces fringes of width 0.50 mm on a screen 60 cm from two slits 0.75 mm apart.
   Ans: 625 nm

2 Radar waves of wavelength 50 mm create a fringe pattern 1.0 m from the aerials. Calculate the distance between the aerials if the fringe spacing is 80 cm.
   Ans: 0.63 m

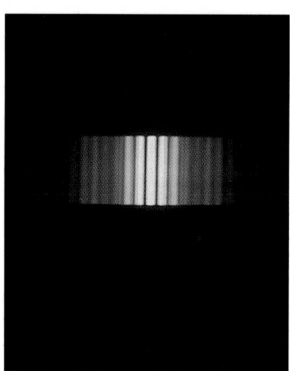

White-light interference fringes.

White-light fringes

If the two slits in Young's experiment are illuminated with white light, each of the different wavelengths making up the white light produces its own fringe pattern. At the centre of the pattern, where the path difference for all waves is zero, there will be a white maximum with a black fringe on each side. Thereafter, the maxima and minima of the different colours overlap in such a way as to produce a pattern of coloured fringes. Only a few will be visible; a short distance from the centre so many wavelengths overlap that they combine to produce what is effectively white light again. The effect is illustrated in the photograph opposite.

Section 7.3 summary

★ The principle of superposition of waves states that when waves meet at the same point in space, the resultant displacement is given by the sum of the displacements of the individual waves.

★ Constructive interference is obtained when the waves which meet are completely in phase, so that the resultant wave is of greater amplitude than any of its constituents.

★ Destructive interference is obtained when the waves which meet are completely out of phase.

★ To produce a sustained and observable interference pattern the sources must be coherent (have a constant phase relationship).

★ For light sources to be coherent, they must originate from the same single source.

★ Young's double-slit experiment:

Condition for constructive interference: path difference $= n\lambda$

Condition for destructive interference: path difference $= (n + \frac{1}{2})\lambda$

fringe width $w = \lambda D/d$, where d is the separation of the source slits and D is the distance of the screen from the slits.

Sections 7.2–7.3 questions

1 You have been sold a pair of sunglasses, and are not sure whether or not they have Polaroid lenses. How could you find out?

2 Compare a two-source experiment to demonstrate the interference of sound waves with a Young's double-slit experiment using light. What are the similarities and differences between the two experiments?

3 **(a)** Explain the term *coherence* as applied to waves from two sources.
   (b) Describe how you would produce two coherent sources of light.
   (c) A double-slit interference pattern is produced using slits separated by 0.45 mm, illuminated with light of wavelength 633 nm from a laser. The pattern is projected on a wall 2.50 m from the slits. Calculate the fringe separation.

4 Figure 7.28 shows the arrangement for obtaining interference fringes in Young's double-slit experiment. Describe and explain what will be seen on the screen if the arrangement is altered in each of the following ways:
   (a) the slit separation is halved,
   (b) the distance from slits to screen is doubled,
   (c) the monochromatic light source is replaced with a white-light source,
   (d) a piece of Polaroid is placed in front of each slit, the polarising directions of the Polaroids being vertical for one slit and horizontal for the other.

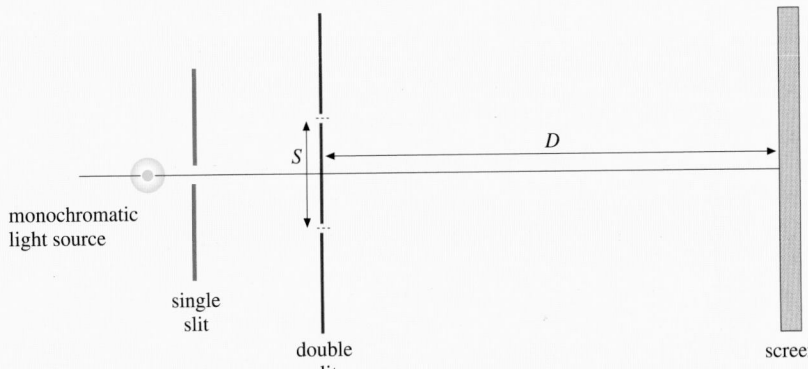

S – slit separation

D – distance from slits to screen

monochromatic light source

single slit

double slit

screen

Figure 7.28 *Young's double-slit experiment.*

Photograph of cello being bowed.

7.4 Standing waves

The notes we hear from this cello are created by the vibrations of its strings. The wave patterns on the vibrating strings are called **standing waves** (or **stationary waves**). The waves which carry the sound to our ears transfer energy, and so are called **progressive waves**.

Figure 7.29 shows a single transverse pulse travelling along a 'slinky' spring. The pulse is reflected when it reaches the fixed end. If a second pulse is sent along the slinky, the reflected pulse will pass through the outward-going pulse, creating a new pulse shape (Figure 7.30). Interference has taken place between the outward and reflected pulses.

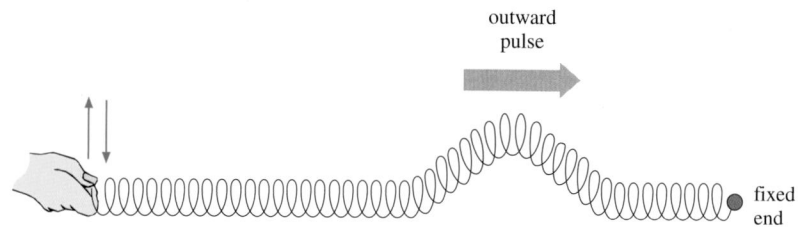

Figure 7.29 *Single transverse pulse travelling along a slinky.*

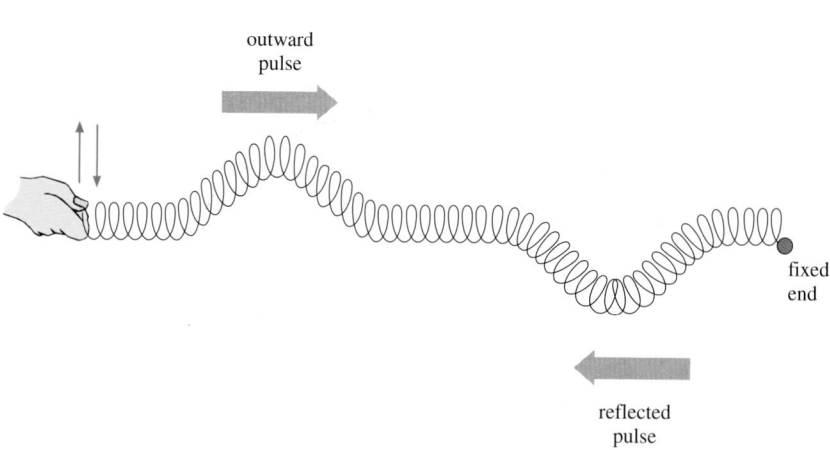

Figure 7.30 *Reflected pulse about to meet an outward-going pulse.*

If the interval between pulses is reduced, a progressive wave is generated. When the wave reaches the fixed end, it is reflected. We now have two progressive waves of equal frequency and amplitude travelling in opposite directions on the same spring. The waves interfere, producing a wave pattern (Figure 7.31) in which the crests and troughs do not move. This pattern is called a **standing wave** (or a **stationary wave**, because it does not move). A standing wave is the result of interference between two waves of equal frequency and amplitude, travelling along the same line with the same speed but in opposite directions.

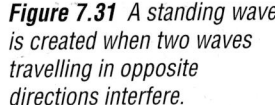

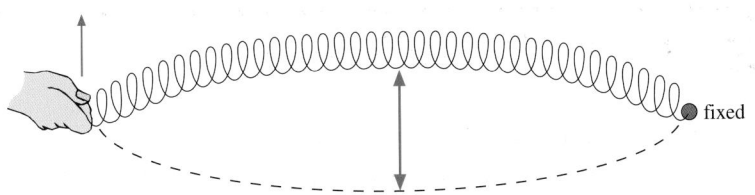

Standing waves in strings

If a string is plucked and allowed to vibrate freely, there are certain frequencies at which it will vibrate. The amplitude of vibration at these frequencies is large. This is known as a **resonance** effect.

It is possible to investigate standing waves in a more controlled manner using a length of string under tension and a vibrator driven by a signal generator. As the frequency of the vibrator is changed, different standing wave patterns are formed. Some of these are shown opposite.

Figure 7.32 shows the simplest way in which a stretched string can vibrate. The wave pattern has a single loop. This is called the **fundamental mode** of vibration, or the **first harmonic**. At the ends of the string there is no vibration. These points are called **nodes**. At the centre of the string, the amplitude is a maximum. A point of maximum amplitude is called an **antinode**. Nodes and antinodes do not move along the string.

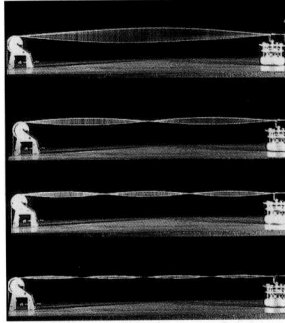

First four modes of vibration of a string.

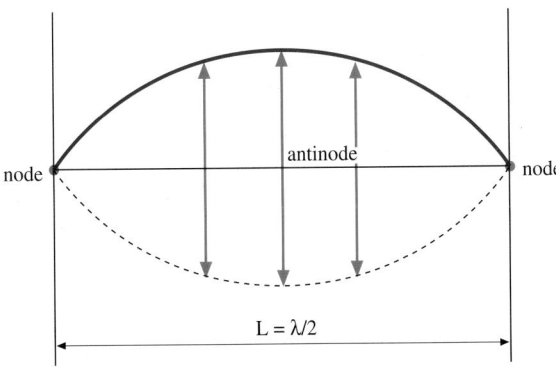

node   antinode   node

$L = \lambda/2$

Figure 7.32 *Fundamental mode of vibration of a stretched string.*

The wavelength λ of the standing wave in the fundamental mode is $2L$. From the wave equation $c = f\lambda$, the frequency f_1 of the note produced by the string vibrating in its fundamental mode is given by $f_1 = c/2L$, where c is the speed of the progressive waves which have interfered to produce the standing wave.

Figure 7.33 shows the second mode of vibration of the string. The standing wave pattern has two loops. This mode is sometimes called the **first overtone**, or the **second harmonic** (don't be confused!). The wavelength of this second mode is L. Applying the wave equation, the frequency f_2 is found to be c/L.

Figure 7.33 *Second mode of vibration of a stretched string.*

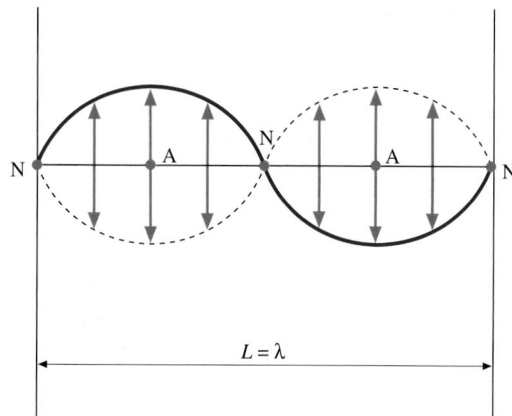

$$L = \lambda$$

Figure 7.34 shows the third mode (the second overtone, or third harmonic). This is a pattern with three loops. The wavelength is $2L/3$, and the frequency f_3 is $3c/2L$.

Figure 7.34 *Third mode of vibration of a stretched string.*

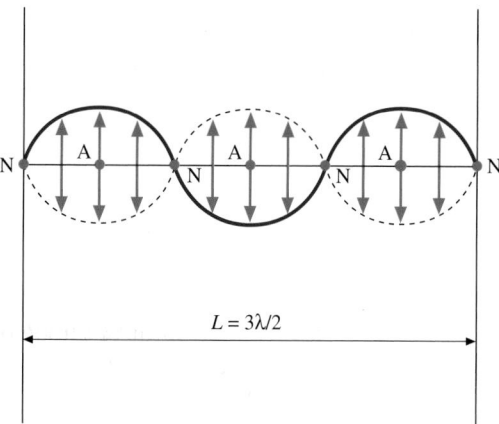

$$L = 3\lambda/2$$

The general expression for the frequency f_n of the nth mode (or the nth harmonic, or $(n - 1)$th overtone) is

$$\bigstar \quad f_n = \frac{cn}{2L} \qquad n = 1, 2, 3, \dots.$$

The key features of a standing wave pattern on a string, which distinguish it from a progressive wave, are as follows. The nodes and antinodes do not move along the string, whereas in a progressive wave, the crests and troughs do move along it. The amplitude of vibration varies with position along the string: it is zero at a node, and maximum at an antinode. In a progressive wave, all points have the same amplitude. Between adjacent nodes, all points of the standing wave vibrate in phase. That is, all particles of the string are at their maximum displacement at the same instant. In a progressive wave, phase varies continuously along the wave.

This type of situation, where a system responds to particular frequencies of oscillation, is an example of **resonance**. We shall return to this in Chapter 9.

Standing waves in air

Figure 7.35 shows an experiment to demonstrate the formation of standing waves in air. A fine, dry powder (such as cork dust or lycopodium powder) is sprinkled evenly along the tube. A loudspeaker powered by a signal generator is placed at the open end. The frequency of the sound from the loudspeaker is gradually increased. At certain frequencies, the powder forms itself into evenly-spaced heaps along the tube. A standing wave has been set up in the air, caused by the interference of the sound wave from the loudspeaker and the wave reflected from the closed end of the tube. At nodes (positions of zero amplitude) there is no disturbance, and the powder can settle into a heap. At antinodes the disturbance is at a maximum, and the powder is dispersed.

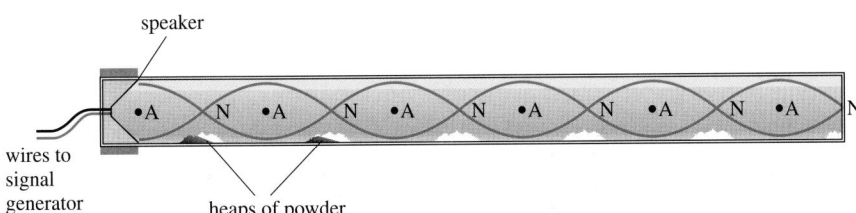

Figure 7.35 Demonstration of standing waves in air.

For standing waves in a closed pipe, the air cannot move at the closed end, and so this must always be a node N. However, the open end is a position of maximum disturbance, and this is an antinode A. (In fact, the antinode is slightly outside the open end. The distance of the antinode from the end of the tube is called the end-correction. The value of the end-correction depends on the diameter of the tube.)

Figure 7.36 shows the simplest way in which the air in a pipe, closed at one end, can vibrate. Figure 7.36a illustrates the motion of some of the air particles in the tube. Their amplitude of vibration is zero at the closed end, and increases with distance up the tube to a maximum at the open end. This representation is tedious to draw, and Figure 7.36b is the conventional way of showing the amplitude of vibration. The amplitudes along the axis in Figure 7.36a are plotted as a continuous curve in Figure 7.36b. One danger of using diagrams like Figure 7.36b is that they give the impression that the sound wave is transverse rather than longitudinal. So be warned! The mode illustrated in Figure 7.36 is the fundamental mode (the first harmonic). The wavelength of this standing wave is $4L$, where L is the length of the pipe. Using the wave equation, the frequency f_1 of the fundamental mode is given by $f_1 = c/4L$, where c is the speed of the sound in air.

Other modes of vibration are possible. Figures 7.37 and 7.38 show the second mode (the first overtone, or second harmonic) and the third mode (the second overtone, or third harmonic). The corresponding wavelengths are $4L/3$ and $4L/5$, and the frequencies are $f_2 = 3c/4L$ and $f_3 = 5c/4L$.

Figure 7.36 *Fundamental mode of vibration of air in a closed pipe.*

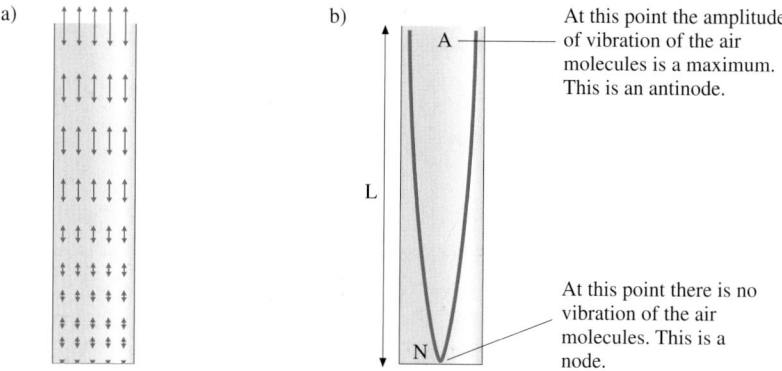

a)

b)

At this point the amplitude of vibration of the air molecules is a maximum. This is an antinode.

At this point there is no vibration of the air molecules. This is a node.

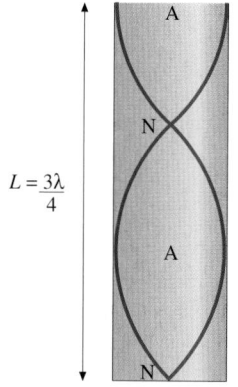

$L = \frac{3\lambda}{4}$

Figure 7.37 *Second mode of vibration of air in a closed pipe.*

The general expression for the frequency f_n of the nth mode of vibration of the air in the closed tube (the nth harmonic, or the $(n-1)$th overtone) is

$$\bigstar \; f_n = \frac{(2n-1)c}{4L}.$$

This is another example of resonance. The particular frequencies at which standing waves are obtained in the pipe are the resonant frequencies of the pipe.

Measuring the speed of sound using standing waves

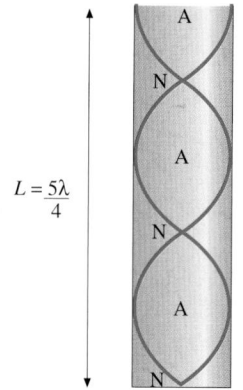

$L = \frac{5\lambda}{4}$

Figure 7.38 *Third mode of vibration of air in a closed pipe.*

The principle of resonance in a tube closed at one end can be used to measure the speed of sound in air. A glass tube is placed in a cylinder of water. By raising the tube, the length of the column of air can be increased. A vibrating tuning fork of known frequency f is held above the open end of the glass tube, causing the air in it to vibrate. The tube is gradually raised, increasing the length of the air column. At a certain position the note becomes much louder. This is known as the first position of resonance, and occurs when a standing wave corresponding to the fundamental mode is established inside the tube. The length L_1 of the air column is noted. The tube is raised further until a second resonance position is found. This corresponds to the second mode of vibration. The length L_2 at this position is also noted. The two resonance positions are illustrated in Figure 7.39.

At the first position of resonance, $\lambda/4 = L_1 + e$, where e is the end-correction of the tube (to allow for the fact that the antinode is slightly above the open end of the tube). At the second position of resonance, $3\lambda/4 = L_2 + e$. By subtracting these equations, we can eliminate e to give

$$\lambda/2 = L_2 - L_1.$$

From the wave equation, the speed of sound c is given by $c = f\lambda$. Thus,

$$c = 2f(L_2 - L_1).$$

Figure 7.39 *Speed of sound by the resonance tube method.*

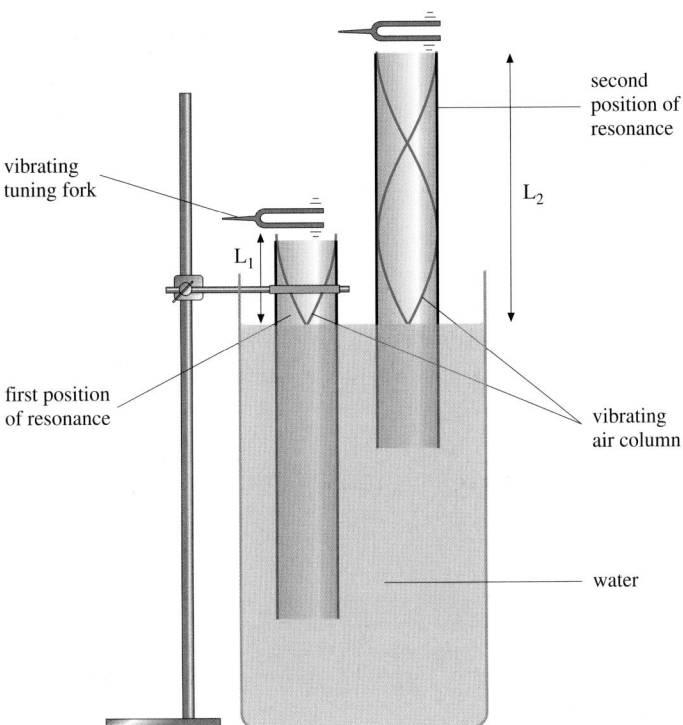

second
position of
resonance

L_2

vibrating
tuning fork

L_1

first position
of resonance

vibrating
air column

water

Figure 7.40 illustrates a method of measuring the speed of sound using standing waves in free air, rather than in a resonance tube. The signal generator and loudspeaker produce a note of known frequency f. The reflector is moved slowly back and forth until the trace on the oscilloscope has a minimum amplitude. When this happens, a standing wave has been set up with one of its nodes in the same position as the microphone. The microphone is now moved along the line between the loudspeaker and the reflector. The amplitude of the trace on the oscilloscope will increase to a maximum, and then decrease to a minimum. The microphone has been moved from one node, through an antinode, to the next node. The distance d between these positions is measured. We know that the distance between nodes is $\lambda/2$. The speed of sound can then be calculated using $c = f\lambda$, giving $c = 2fd$.

Figure 7.40 *Speed of sound from standing waves in free air.*

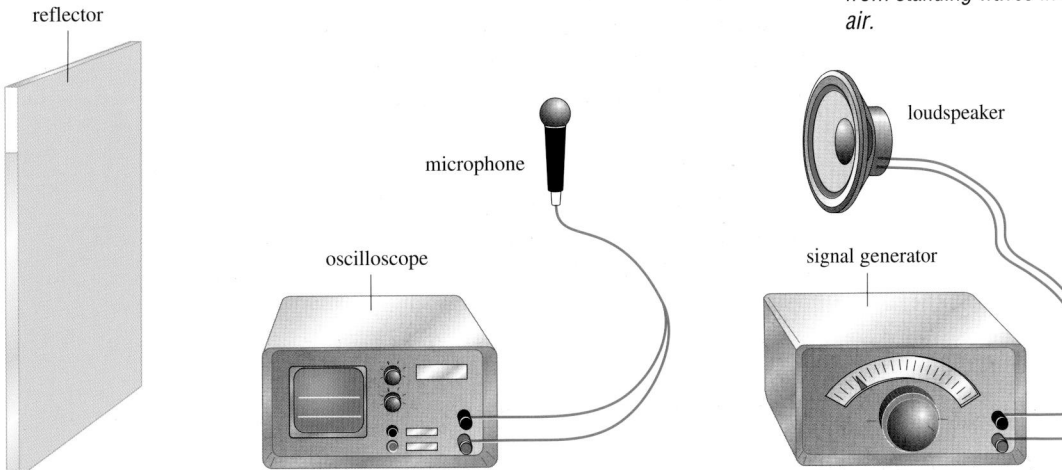

reflector

microphone

loudspeaker

oscilloscope

signal generator

7.4 Standing waves

Examples

1 A string 75 cm long is fixed at one end. The other end is vibrated up and down with a frequency of 15 Hz. This frequency gives a standing-wave pattern with three complete loops on the string. Calculate the speed of the progressive waves which have interfered to produce the standing wave.

   The three-loop pattern corresponds to the situation where the length L of the string is $3\lambda/2$ (see Figure 7.34). The wavelength λ is thus $2 \times 0.75/3 = 0.50$ m. The frequency of the wave is 15 Hz, so by the wave equation $c = f\lambda = 15 \times 0.50 = $ **7.5 m s^{-1}**.

2 Find the fundamental frequency and first two overtones for an organ pipe which is 0.17 m long and closed at one end. The speed of sound in air is 340 m s^{-1}.

   The frequencies of the fundamental and first two overtones of a tube of length L, closed at one end, are $c/4L$, $3c/4L$ and $5c/4L$ (see Figures 7.36–38). The frequencies are thus $340/4 \times 0.17 = $ **500 Hz**, $3 \times 340/4 \times 0.17 = $ **1500 Hz** and $5 \times 340/4 \times 0.17 = $ **2500 Hz**.

Now it's your turn

1 A violin string vibrates with a fundamental frequency of 440 Hz. What are the frequencies of its first two overtones?
   Ans: 880 Hz; 1320 Hz

2 The speed of waves on a certain stretched string is 48 m s^{-1}. When the string is vibrated at a frequency of 64 Hz, standing waves are set up. Find the separation of successive nodes in the standing-wave pattern.
   Ans: 0.38 m

3 You can make an empty lemonade bottle resonate by blowing across the top. What fundamental frequency of vibration would you expect for a bottle 25 cm deep? The speed of sound in air is 340 m s^{-1}.
   Ans: 340 Hz

4 A certain closed organ pipe can resonate at consecutive frequencies of 640 Hz, 896 Hz and 1152 Hz. Deduce its fundamental frequency.
   Ans: 128 Hz

Section 7.4 summary

★ A standing wave is the result of interference between two progressive waves of equal frequency and amplitude travelling along the same line with the same speed, but in opposite directions.

★ Points of zero amplitude on a standing wave are called nodes; points of maximum amplitude are called antinodes.

★ For standing waves on a stretched string, frequency f_n of nth mode is given by $f_n = cn/2L$, where c is the speed of progressive waves on the string and L is the length of the string.

★ For standing waves in air in a tube closed at one end, frequency f_n of nth mode is given by $f_n = (2n - 1)c/4L$, where c is the speed of sound in air and L is the length of the tube.

Section 7.4 questions

1 State and explain four ways in which standing waves differ from progressive waves.

2 A source of sound of frequency 2000 Hz is placed in front of a flat wall. When a microphone is moved away from the source towards the wall, a series of maxima and minima are detected.

   (a) Explain what has happened to create these maxima and minima.
   (b) The speed of sound in air is 340 m s^{-1}. Calculate the distance between successive minima.

3 A string is stretched between two fixed supports separated by 1.20 m. Stationary waves are generated on the string. It is observed that two stationary wave frequencies are 180 Hz and 135 Hz; there is no resonant frequency between these two. Calculate:
   (a) the speed of progressive waves on the stretched string,
   (b) the lowest resonant frequency of the string.

7.5 Diffraction

Although we often hear the statement 'light travels in straight lines', there are occasions when this appears not to be the case. Newton tried to explain the fact that when light travels through an aperture, or passes the edge of an obstacle, it deviates from the straight-on direction and appears to spread out. We have seen from the ripple tank demonstration (Figure 7.11) that water waves spread out when they pass through an aperture. This effect is called **diffraction**. The fact that light undergoes diffraction is powerful evidence that light has wave properties. (Newton's attempt to explain diffraction was not, in fact, based on a wave theory of light. The Dutch scientist Christian Huygens, a contemporary of Newton, favoured the wave theory, and used it to account for reflection, refraction and diffraction. Not until 1815 did the French scientist Augustin Fresnel develop the wave theory of light so as to explain diffraction in detail.)

Figure 7.11 showed that the degree to which waves are diffracted depends upon the size of the obstacle or aperture and the wavelength of the wave. The greatest effects occur when the wavelength is about the same size as the aperture. The wavelength of light is very small (green light has wavelength 5×10^{-7} m), and therefore diffraction effects can be difficult to detect.

Huygens' explanation of diffraction

If we let a single drop of water fall into a ripple tank it will create a circular wavefront which will spread outwards from the disturbance (Figure 7.8). Huygens put forward a wave theory of light which was based on the way in which circular wavefronts advance. He suggested that at any instant all points on a circular wavefront could be regarded as secondary disturbances, giving rise to their own outward-spreading circular wavelets. The envelope, or tangent curve, of the wavefronts produced by the secondary sources gives the new position of the original wavefront. This construction is illustrated in Figure 7.41.

Figure 7.41 Huygens' construction for a circular wavefront.

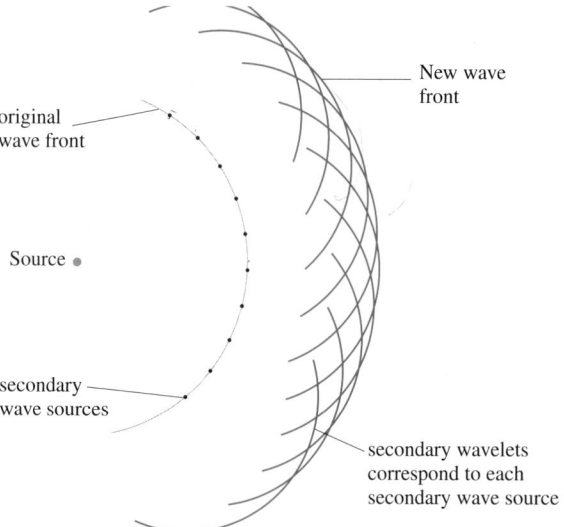

original wave front

New wave front

Source

secondary wave sources

secondary wavelets correspond to each secondary wave source

If a wavefront is restricted in any way, for example by passing through an aperture in the form of a slit, some of the wavelets are removed, causing the edges of the wavefront to be curved. If the wavelength is small compared with the size of the aperture, the wavefronts which pass through the aperture show curvature only at their ends, and the diffraction effect is relatively small. If the aperture is comparable with the wavelength, the diffracted wavefronts become circular, centred on the slit. Note that there is no change of wavelength on diffraction. This effect was illustrated in Figure 7.11.

Diffraction of light at a single slit.

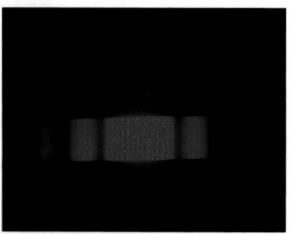

This photograph shows the diffraction pattern created by a single slit illuminated by monochromatic light. The central region of the pattern is a broad, bright area with narrow, dark fringes on either side. Beyond these is a further succession of bright and dark areas. The bright areas become less and less intense as we move away from the centre.

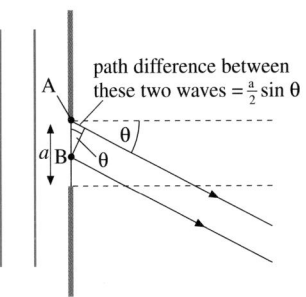

Figure 7.42 *Light leaving a single slit.*

This single-slit diffraction pattern has many features that we associate with an interference pattern. But how can a single slit produce an interference-type pattern? The explanation that follows is based on Huygens' wavelet idea.

Figure 7.42 shows plane wavefronts arriving at a single slit of width a. Each point on the wavefront passing through the slit can be considered to be a source of secondary wavelets. One such source is at A, at the top edge of the slit, and a second is at B, at the centre of the slit, a distance $a/2$ along the wavefront from A. These two sources behave like the sources in a two-source interference experiment. The wavelets spreading out from these points overlap and create an interference pattern. In the straight-on direction, there is no path difference between the waves from A and B. Constructive interference occurs in this direction, giving a bright fringe in the centre of the pattern. To either side of the central fringe there are directions where the path difference between the waves from A and B is an odd number of half-wavelengths. This is the condition for destructive interference, resulting in dark fringes. The condition for constructive interference is that the path difference should be a whole number of wavelengths. Thus, the dark fringes alternate with bright fringes.

This argument can be applied to the whole of the slit. Every wavelet spreading out from a point in the top half of the slit can be paired with one coming from a point $a/2$ below it in the lower half of the slit. When wavelets from points right across the aperture are added up, we find that there are certain directions in which constructive interference occurs, and other directions in which the interference is destructive. Figure 7.43 is a graph of the intensity of the diffraction pattern as a function of the angle θ at which the light is viewed. It shows that most of the intensity is in the central area, and that this is flanked by dark and bright fringes.

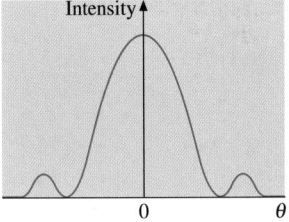

Figure 7.43 *Light intensity graph for single slit diffraction.*

We can use Figure 7.42 to derive an expression for the angle at which the first dark fringe is obtained. Remember that the condition for destructive interference is that the path difference between the two rays should be half a wavelength. The path difference between the two rays shown in Figure 7.42 is $\frac{1}{2}a \sin \theta$. If this is to be $\frac{1}{2}\lambda$, we have

$$\bigstar \quad \sin \theta = \frac{\lambda}{a}.$$

This is the condition to observe the first dark fringe at angle θ. More generally,

$$\bigstar \quad \sin \theta = \frac{n\lambda}{a},$$

where n is a whole number called the *order* of the dark fringe being considered, counting outwards from the centre.

Although we have been concentrating on a diffraction pattern obtained with light, the derivation above applies to any type of wave passing through an aperture. The wavelength of light is generally small compared with the size of slits of other apertures, so the diffraction angle θ is also small. Provided that θ is only a few degrees (less than about 10°), the approximation $\sin \theta = \theta$ may be used (remember that θ must be in radians!). Very often, the single-slit diffraction equation for light is expressed in the form

$$\star \quad \theta = \frac{n\lambda}{a},$$

making use of the $\sin \theta$ approximation. But take care! This approximate form may not apply for the diffraction of other types of wave, such as sound or water waves, where the wavelength may be closer in magnitude to the size of the aperture, and diffraction angles are larger.

Example

1 Calculate the angle between the centre of the diffraction pattern and the first minimum when light of wavelength 600 nm passes through a slit 0.10 mm wide.

Using $a \sin \theta = n\lambda$, we have $\sin \theta = n\lambda/a$. Substituting, $\sin \theta = 1 \times 6.0 \times 10^{-7}/1.0 \times 10^{-4}$ (don't forget to convert the nm and mm to m!) $= 0.0060$, and $\theta = \mathbf{0.34°}$. Using the $\sin \theta = \theta$ approximation, we would have obtained $\theta = n\lambda/a = 1 \times 6.0 \times 10^{-7}/1.0 \times 10^{-4} = \mathbf{0.0060\ rad}$ (which is equal to 0.34°).

Now it's your turn

1 Calculate the angle between the centre of the diffraction pattern and the first minimum when a sound wave of wavelength 1.0 m passes through a door 1.2 m wide.
Ans: 56°

2 Calculate the wavelength of water waves which, on passing through a gap 50 cm wide, create a diffraction pattern such that the angle between the centre of the pattern and the second-order minimum is 60°.
Ans: 22 cm

The diffraction grating

A **diffraction grating** is a plate on which there is a very large number of parallel, identical, very closely-spaced slits. If monochromatic light is incident on this plate, a pattern of narrow bright fringes is produced, as shown in Figure 7.44.

parallel beam of
monochromatic light
or laser light

diffraction
grating

screen

Figure 7.44 *Arrangement for obtaining a fringe pattern with a diffraction grating.*

Although the device is called a *diffraction* grating, we shall use straightforward superposition and interference ideas in obtaining an expression for the angles at which the maxima of intensity are obtained.

Figure 7.45 shows a parallel beam of light incident normally on a diffraction grating in which the spacing between adjacent slits is d. Consider first rays 1 and 2 which are incident on adjacent slits. The path difference between these rays when they emerge at an angle θ is $d \sin \theta$. To obtain constructive interference in this direction from these two rays, the condition is that the path difference should be an integral number of wavelengths. The

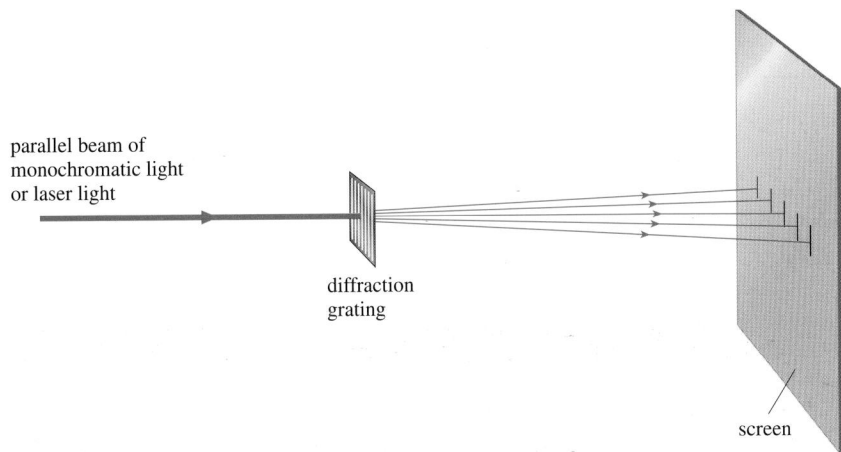

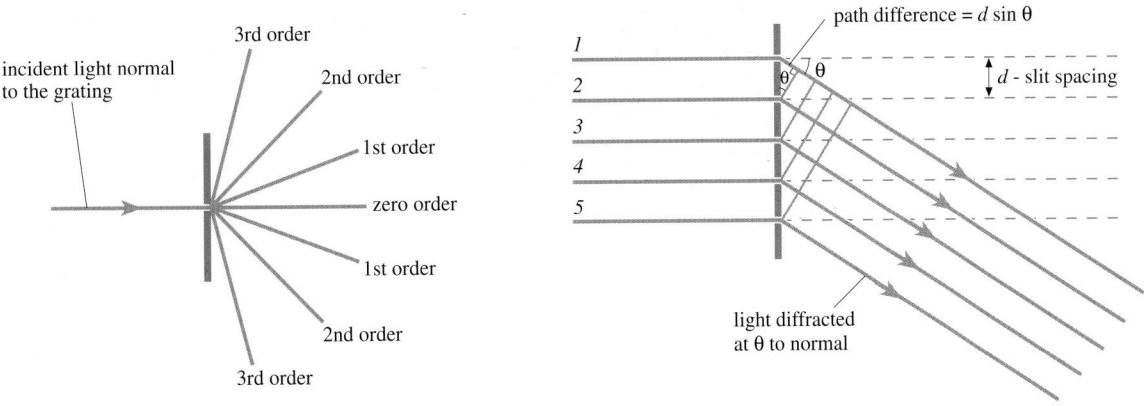

Figure 7.45 *The diffraction grating.*

path difference between rays 2 and 3, 3 and 4, and so on, will also be $d \sin \theta$. The condition for constructive interference is the same. Thus, the condition for a maximum of intensity at angle θ is

$$\star \quad d \sin \theta = n\lambda,$$

where λ is the wavelength of the monochromatic light used, and n is a whole number.

When $n = 0$, $\sin \theta = 0$ and θ is also zero; this gives the straight-on direction, or what is called the zero-order pattern. When $n = 1$, we have the first order diffraction pattern, and so on.

Example

Monochromatic light is incident normally on a grating with 7.00×10^5 lines per metre. A second-order maximum is observed at an angle of diffraction of $40.0°$. Calculate the wavelength of the incident light.

The slits on a diffraction grating are created by drawing parallel lines on the surface of the plate. The relationship between the slit spacing d and the number N of lines per metre is $d = 1/N$. For this grating, $d = 1/7.00 \times 10^5 = 1.43 \times 10^{-6}$ m. Using $n\lambda = d \sin \theta$, $\lambda = (d/n) \sin \theta = (1.43 \times 10^{-6}/2) \sin 40.0° = \textbf{460 nm}$.

Now it's your turn

1   Monochromatic light is incident normally on a grating with 5.00×10^5 lines per metre. A third-order maximum is observed at an angle of diffraction of $78.0°$. Calculate the wavelength of the incident light.
    Ans: 652 nm

2   Light of wavelength 5.90×10^{-7} m is incident normally on a diffraction grating with 8.00×10^5 lines per metre. Calculate the diffraction angles of the first- and second-order diffraction images.
    Ans: 28.2°, 70.7°

3   Light of wavelength 590 nm is incident normally on a grating with spacing 1.67×10^{-6} m. How many orders of diffraction maxima can be obtained?
    Ans: 2

The diffraction grating with white light

If white light is incident on a diffraction grating, each wavelength λ making up the white light is diffracted by a different amount, as described by the equation $n\lambda = d \sin \theta$. Red light, because it has the longest wavelength in the visible spectrum, is diffracted through the largest angle. Blue light has the shortest wavelength, and is diffracted the least. Thus, the white light is split into its component colours, producing a continuous spectrum (Figure 7.46). The spectrum is repeated in the different orders of diffraction. Depending on the grating spacing, there may be some overlapping of different orders. For example, the red component of the first-order image may overlap with the blue end of the second-order spectrum.

An important use of the diffraction grating is in a **spectrometer**, a piece of apparatus used to measure spectra. By measuring the angle at which a particular diffracted image appears, the wavelength of the light producing that image may be determined.

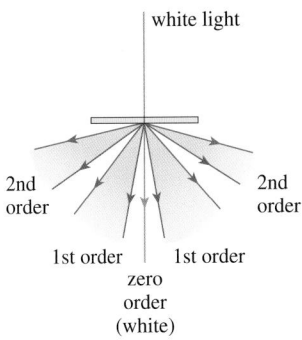

Figure 7.46 Production of the spectrum of white light with a diffraction grating.

Section 7.5 summary

★ Diffraction is the spreading out of waves after passing through an aperture or meeting an obstacle. It is most obvious when the size of the aperture and the wavelength of the wave are approximately the same.

★ The condition for a diffraction maximum in a diffraction grating is $d \sin \theta = n\lambda$, where d is the grating spacing, θ is the angle at which the diffraction maximum is observed, n is an integer (the order of the image), and λ is the wavelength of the light.

★ Single slit diffraction: the condition for minima is $a \sin \theta = n\lambda$, where a is the width of the slit.

Section 7.5 questions

1 Blue and red light, with wavelengths 450 nm and 650 nm respectively, is incident normally on a diffraction grating which has 4.0×10^5 lines per metre.

   (a) Calculate the grating spacing.
   (b) Calculate the angle between the second-order maxima for these wavelengths.
   (c) For each wavelength, find the maximum order that can be observed.

2 Discuss any difference between the interference patterns formed by (a) two parallel slits 1 μm apart, (b) a diffraction grating with grating spacing 1 μm, when illuminated with monochromatic light.

3 Light of wavelength 633 nm passes through a slit 50 μm wide. Calculate the angular separation between the central maximum and the first minimum of the diffraction pattern.

Exam Questions

1 Assume that waves spread out uniformly in all directions from the epicentre of an earthquake. The intensity of a particular earthquake wave was measured as 5.0×10^6 W m^{-2} at a distance of 40 km from the epicentre. What was the intensity at a distance of only 2 km from the epicentre?

2 A string is stretched between two fixed supports 3.5 m apart. Stationary waves are generated by disturbing the string. One possible mode of vibration of the stationary waves is shown in Figure 7.47. The nodes and antinodes are labelled N and A respectively.

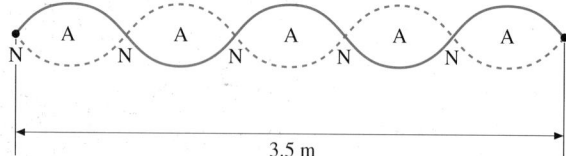

Figure 7.47

(a) Distinguish between a node and an antinode in a stationary wave.
(b) State the phase difference between the vibrations of particles of the string at any two consecutive antinodes.
(c) Calculate the ratio of the frequency of the mode of vibration shown in Figure 7.47 to the frequency of the fundamental mode of vibration of the string.
(d) The frequency of the mode of vibration shown in Figure 7.47 is 160 Hz. Calculate the speed of the progressive waves which produced this stationary wave.

3 Explain why light will not pass through two sheets of Polaroid when they are arranged in contact with their polarising directions at right angles. What would happen if a third sheet of Polaroid, with its polarising direction at 45° to the polarising directions of the other two, were placed between them? (If you can get hold of Polaroid filters, it is worth trying this experiment, starting with thinking about how you would determine the polarising direction of each!)

4 A vibrating tuning fork of frequency 320 Hz is held over the open end of a resonance tube. The other end of the tube is immersed in water. The length of the air column is gradually increased until resonance first occurs. Taking the speed of sound in air as 340 m s^{-1}, calculate the length of the air column. (Neglect any end-correction.)

5 We can hear sounds round corners. We cannot see round corners. Both sound and light are waves. Explain why sound and light seem to behave differently.

6 Figure 7.48 shows a narrow beam of monochromatic laser light incident normally on a diffraction grating. The central bright spot is formed at O.

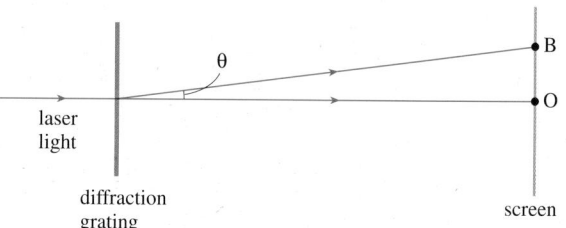

Figure 7.48

(a) Write down the relationship between the wavelength λ of the light and the angle θ for the first diffraction image formed at B. Identify any other symbol used.
(b) The screen is 1.1 m from the diffraction grating and the grating has 300 lines per mm. The laser light has wavelength 6.3×10^{-7} m. Find the distance OB from the central spot to the first bright image at B.
(c) The diffraction grating is now replaced by one which has 600 lines per mm. For this second grating, calculate the distance from the central spot to the first bright image.

CHAPTER EIGHT
Photons, electrons and atoms

In the last chapter we spent a long time in describing properties of waves, and in emphasising experiments that showed that light is a wave motion. We now look at phenomena which, during the last hundred years, have completely revolutionised physics. A number of experiments around the 1900s showed that light behaves more like a stream of particles than as a wave. Later, it was shown experimentally that electrons, which had always been thought of as particles, could also behave as waves. This idea of **wave-particle duality** is one of the most important concepts of modern physics. We start by describing the photoelectric effect, the phenomenon that proved that light had a particle nature. This is the work which Einstein first explained in 1905, and for which he received the 1921 Nobel Prize. We shall then describe the diffraction of electrons. Remember that diffraction is a characteristic of waves, yet electrons have always been regarded as particles!

This picture was taken with an image-intensifying camera. This is a special camera capable of taking clear pictures in extremely low light levels. It makes use of the fact that some metals emit electrons when light shines on them. This effect is known as photoelectric emission. Each electron liberated by a photon of light is made to produce a pulse of current which in turn is used to produce the image in the camera.

CHAPTER 8

8.1 The photoelectric effect

Some of the electrons in a metal are free to move around in it. (It is these free electrons that form the electric current when a potential difference is applied across the ends of a metal wire.) However, to remove free electrons from a metal requires energy, because they are held in the metal by the electrostatic attraction of the positively-charged nuclei. If an electron is to escape from the surface of a metal, work must be done on it. The electron must be given energy. When this energy is in the form of light energy, the phenomenon is called **photoelectric emission**.

 Photoelectric emission is the release of electrons from the surface of a solid when electromagnetic radiation is incident on its surface.

Demonstration of photoelectric emission

A clean zinc plate is placed on the cap of a gold-leaf electroscope. The electroscope is then charged negatively, and the gold leaf deflects, proving that the zinc plate is charged. This is illustrated in Figure 8.1. If visible light of any colour is shone on to the plate, the leaf does not move. Even when the intensity (the brightness) of the light is increased, the leaf remains in its deflected position. However, when ultra-violet radiation is shone on the plate the leaf falls, showing that it is losing negative charge. This means that electrons are being emitted from the zinc plate. These electrons are called **photoelectrons**. If the intensity of the ultra-violet radiation is increased, the leaf falls more quickly, showing that the rate of emission of electrons has increased.

The difference between ultra-violet radiation and visible light is that ultra-violet radiation has a shorter wavelength and a higher frequency than visible light.

Figure 8.1 *Demonstration of photoelectric emission.*

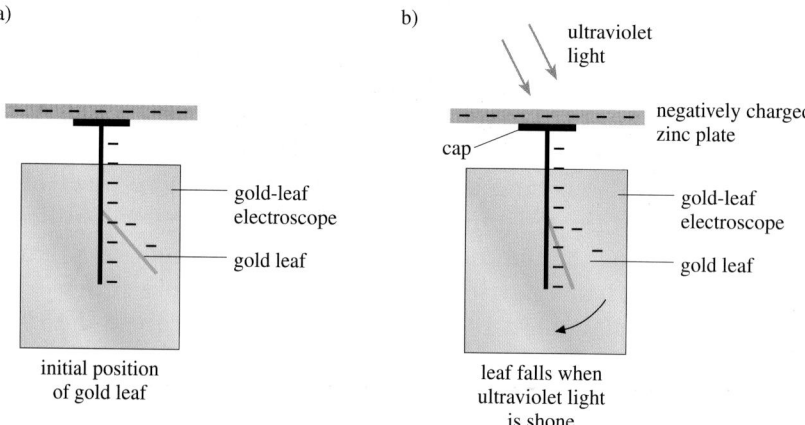

Further investigations with apparatus like this lead to the following conclusions:

1 If photoemission takes place, it does so instantaneously. There is no delay between illumination and emission.

2 Photoemission takes place only if the frequency of the incident radiation is above a certain minimum value called the threshold frequency f_0.

3 Different metals have different threshold frequencies.

4 Whether or not emission takes place depends only on whether the frequency of radiation used is above the threshold for that surface. It does not depend on the intensity of the radiation.

5 For a given frequency, the rate of emission of photoelectrons is proportional to the intensity of the radiation.

Another experiment, using the apparatus shown in Figure 8.2, can be carried out to investigate the energies of the photoelectrons. If ultra-violet radiation of a fixed frequency (above the threshold) is shone on to the metal surface it emits photoelectrons. Some of these electrons travel from A to B. Current is detected using the microammeter. If a potential difference is applied between A and B, with B negative with respect to A, any electron going from A to B will gain potential energy as it moves against the electric field. The gain in potential energy is at the expense of the kinetic energy of the electron. That is,

$$loss\ in\ kinetic\ energy = gain\ in\ potential\ energy$$
$$= charge\ of\ electron \times potential\ difference$$

Please see Chapter 6 to clarify this result.

Figure 8.2 Experiment to measure the maximum kinetic energy of photoelectrons.

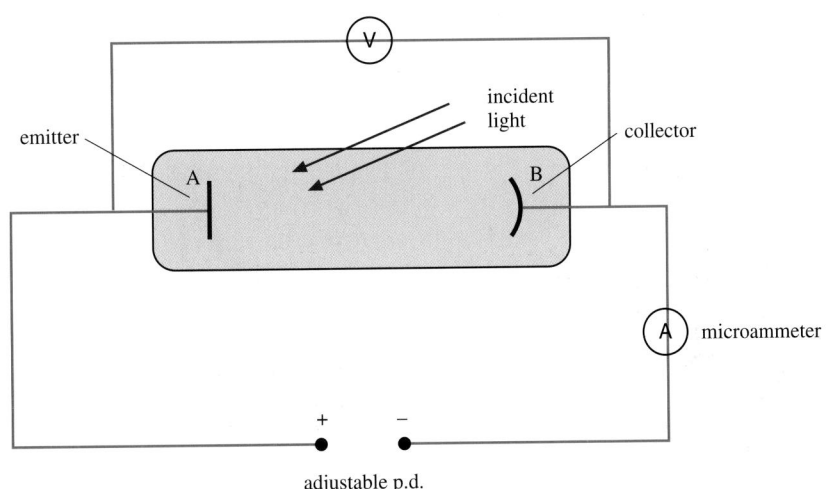

incident light

emitter

collector

A

B

V

A microammeter

+ −

adjustable p.d.

If the voltage between A and B is gradually increased, the current registered on the microammeter decreases and eventually falls to zero. The minimum value of the potential difference necessary to stop the electron flow is known as the stopping potential. It measures the maximum kinetic energy with which the photoelectrons are emitted. The fact that there is a current in the microammeter at voltages less than the stopping potential indicates that there is a range of kinetic energies for these electrons.

If the experiment is repeated with radiation of greater intensity but the same frequency, the maximum current in the microammeter increases, but the value of the stopping potential is unchanged.

The experiment can be repeated using ultra-violet radiation of different frequencies, measuring the stopping potential for each frequency. When the maximum kinetic energy of the photoelectrons is plotted against the frequency of the radiation, the graph of Figure 8.3 is obtained.

Figure 8.3 *Graph of maximum kinetic energy of photoelectrons against frequency of radiation.*

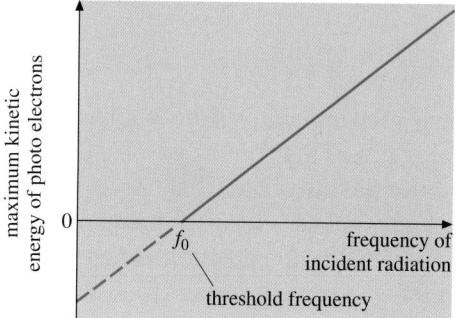

The following conclusions are drawn from this experiment:

1   The photoelectrons have a range of kinetic energies, from zero up to some maximum value. If the frequency of the incident radiation is increased the maximum kinetic energy of the photoelectrons also increases.

2   For constant frequency of the incident radiation, the maximum kinetic energy is unaffected by the intensity of the radiation.

3   If the graph of Figure 8.3 is extrapolated to the point where the maximum kinetic energy of the photoelectrons is zero, the minimum frequency required to cause emission from the surface (the threshold frequency) may be found.

At the time when the photoelectric effect was first being studied, it was fully accepted that light was a wave motion. The conclusions of experiments on photoemission produced doubt as to whether light is a continuous wave. One of the main problems concerns the existence of a threshold frequency.

Classical wave theory predicts that when an electromagnetic wave (that is, light) interacts with an electron, the electron will absorb energy from it. So, if an electron absorbs enough energy it should be able to escape from the metal. Remember from Chapter 7 that the energy carried by a wave depends on its amplitude and its frequency. Thus, even if we have a low-frequency wave, its energy can be boosted by increasing the amplitude (that is, by increasing the brightness of the light). So, according to wave theory, we ought to be able to cause photoemission by using any frequency of light, provided we make it bright enough. Alternatively, we could use less bright light and shine it on the metal for a longer time, until enough energy to cause emission has been delivered. But this does not happen. The experiments we have described earlier showed conclusively that radiation of frequency below the threshold, no matter how intense or for how long it is used, does not produce photoelectrons. The classical wave theory of electromagnetic radiation leads to the following predictions:

1 Whether an electron is emitted or not should depend on the energy of the incident wave, that is on its intensity. A very intense wave, of any frequency, should cause photoemission.

2 The maximum kinetic energy of the photoelectrons should be greater if the radiation intensity is greater.

3 There is no reason why photoemission should be instantaneous.

These predictions, based on wave theory, do not match the observations. A new approach, based on an entirely new concept, the **quantum theory**, was used to explain these findings.

Einstein's theory of photoelectric emission

In 1901 the German physicist Max Planck had suggested that the energy carried by electromagnetic radiation might exist as discrete packets called **quanta**. The energy E carried in each quantum is given by

$$\bigstar \quad E = hf,$$

where f is the frequency of the radiation and h is a constant called the Planck constant. The value of the Planck constant is $6.63 \times 10^{-34}\,\text{J s}$.

In 1905 Albert Einstein developed the theory of quantised energy to explain all the observations associated with photoelectric emission. He proposed that light radiation consists of a stream of energy packets called **photons**.

Albert Einstein.

★ A photon is the special name given to a quantum of energy when the energy is in the form of electromagnetic radiation.

When a photon interacts with an electron, it transfers all its energy to the electron. It is only possible for a single photon to interact with a single electron; the photon cannot share its energy between several electrons. This transfer of energy is instantaneous.

The photon theory of photoelectric emission is as follows. If the frequency of the incident radiation is less than the threshold frequency for the metal, the energy carried by each photon is insufficient for an electron to escape the surface of the metal. If the photon energy is insufficient for electron escape it is converted to thermal energy in the metal.

★ The minimum amount of energy necessary for an electron to escape is called the **work function energy** ϕ.

Some values for the work function energy and threshold frequencies f_0 of different metals are given in Figure 8.4.

Figure 8.4

metal	Work function energies and threshold frequencies		
	θ/J	θ/eV	f_0/Hz
sodium	3.8×10^{-19}	2.4	5.8×10^{14}
calcium	4.6×10^{-19}	2.9	7.0×10^{14}
zinc	5.8×10^{-19}	3.6	8.8×10^{14}
silver	6.8×10^{-19}	4.3	1.0×10^{15}
platinum	9.0×10^{-19}	5.6	1.4×10^{15}

Remember! 1 eV = 1.6×10^{-19} J.

If the frequency of the incident radiation is equal to the threshold frequency, the energy carried by each photon is just sufficient for electrons at the surface to escape. If the frequency of the incident radiation is greater than the threshold frequency, surface electrons will escape and have surplus energy in the form of kinetic energy. These electrons will have the maximum kinetic energy. If a photon interacts with an electron below the surface, some energy may be lost as the electron moves towards the surface, so that it is emitted with less than the maximum kinetic energy. This gives rise to a range of values of kinetic energy.

Einstein used the principle of conservation of energy to derive the photoelectric equation

8.1 The photoelectric effect

★ *photon energy = work function energy + maximum kinetic energy of photoelectron*

or

$$\bigstar \quad hf = \phi + \tfrac{1}{2}m_e v_{max}^{2}.$$

For radiation incident at the threshold frequency, $\tfrac{1}{2}m_e v_{max}^{2} = 0$, so that $hf_0 = \phi$. The photoelectric equation can then be written

$$\bigstar \quad hf = hf_0 + \tfrac{1}{2}m_e v_{max}^{2}.$$

Example

The work function energy of platinum is 9.0×10^{-19} J. Calculate **(a)** the threshold frequency for the emission of photoelectrons from platinum, **(b)** the maximum kinetic energy of a photoelectron when radiation of frequency 2.0×10^{15} Hz is incident on a platinum surface.

(a) Using $hf_0 = \phi, f_0 = \phi/h$, so
   $f_0 = 9.0 \times 10^{-19}/6.6 \times 10^{-34} = \mathbf{1.4 \times 10^{15}}$ **Hz**.
(b) Using $hf = hf_0 + \tfrac{1}{2}m_e v_{max}^{2}, hf - hf_0 = \tfrac{1}{2}m_e v_{max}^{2}$ and
   $\tfrac{1}{2}m_e v_{max}^{2} = 6.6 \times 10^{-34}(2.0 \times 10^{15} - 1.4 \times 10^{15}) = \mathbf{4.0 \times 10^{-19}}$ **J**.

Now it's your turn

1 The work function energy of silver is 4.3 eV. Show that the threshold frequency is about 1.0×10^{15} Hz.

2 Electromagnetic radiation of frequency 3.0×10^{15} Hz is incident on the surface of sodium metal. The emitted photoelectrons have a maximum kinetic energy of 1.6×10^{-18} J. Calculate the threshold frequency for photoemission from sodium.
   Ans: 5.8×10^{14} Hz

Section 8.1 summary

★ Electrons may be emitted from metal surfaces if the metal is illuminated by electromagnetic radiation. This phenomenon is called photoelectric emission.

★ Photoelectric emission cannot be explained by the wave theory of light. It is necessary to use the quantum theory, in which electromagnetic radiation is thought of as consisting of packets of energy called photons.

★ The energy E of a photon is given by $E = hf$, where h is the Planck constant and f is the frequency of the radiation.

★ The work function energy ϕ of a metal is the minimum energy needed to free an electron from the surface of the metal.

★ The Einstein photoelectric equation is $hf = \phi + \frac{1}{2}m_e v_{max}^2$.

★ The threshold frequency f_0 is given by $hf_0 = \phi$.

Section 6.1 questions

1 Calculate the energy range, in eV, of photons in the visible region of the electromagnetic spectrum (wavelengths 400 nm–700 nm). (Planck constant $h = 6.6 \times 10^{-34}$ J s; speed of light $c = 3.0 \times 10^8$ m s^{-1}; electron charge $e = -1.6 \times 10^{-19}$ C.)

2 The threshold frequency for photoemission from a certain metal is 8.8×10^{14} Hz.

  (a) To what wavelength does this frequency correspond?

  (b) Calculate the maximum kinetic energy of the emitted photoelectrons when the metal is illuminated with ultra-violet radiation of wavelength 240 nm.

(Planck constant $h = 6.6 \times 10^{-34}$ J s; speed of light $c = 3.0 \times 10^8$ m s^{-1}.)

3 In a stopping-potential experiment using ultra-violet radiation of wavelength 265 nm, the photocurrent from a certain metal is reduced to zero at an applied potential difference of 1.18 V. Calculate the work function energy of the metal. (Planck constant $h = 6.63 \times 10^{-34}$ J s; speed of light $c = 3.00 \times 10^8$ m s^{-1}; electron charge $e = -1.60 \times 10^{-19}$ C.)

8.2   Wave-particle duality

If light waves can behave like particles (photons), perhaps moving particles can behave like waves?

If a beam of X-rays of a single wavelength is directed at a thin metal foil, a diffraction pattern is produced. This is a similar effect to the diffraction pattern produced when light passes through a diffraction grating (see Chapter 7). An X-ray diffraction pattern is shown in the first photograph. The foil contains many tiny crystals. The gaps between neighbouring planes of atoms in the crystals act as slits, creating a diffraction pattern.

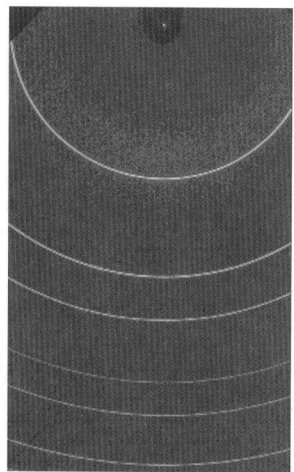

X-ray diffraction pattern of a metal foil.

8.2 Wave-particle duality

If a beam of electrons is directed at a thin metal foil, a similar diffraction pattern is produced, as shown in the second photograph. The electrons, which we normally consider to be particles, are exhibiting a property we would normally associate with waves. Remember that, to observe diffraction, the wavelength of the radiation should be comparable with the size of the aperture. The separation of planes of atoms in crystals is of the order of 10^{-10} m. The fact that diffraction is observed with electrons suggest that they have a wavelength of about the same magnitude.

In 1924 the French physicist Louis de Broglie suggested that all moving particles have a wave-like nature. Using ideas based upon the quantum theory and Einstein's theory of relativity, he suggested that the momentum p of a particle and its wavelength λ are related by the equation

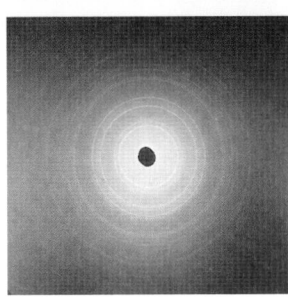

Electron diffraction pattern of a metal foil.

$$\star \ \ \lambda = \frac{h}{p},$$

where h is the Planck constant. λ is known as the de Broglie wavelength.

Example

Calculate the de Broglie wavelength of an electron travelling with a speed of 1.0×10^7 m s^{-1}.
(Planck constant $h = 6.6 \times 10^{-34}$ J s; electron mass $m_e = 9.1 \times 10^{-31}$ kg)

Using $\lambda = h/p$ and $p = mv$,
$\lambda = 6.6 \times 10^{-34}/9.1 \times 10^{-31} \times 1.0 \times 10^7 = \mathbf{7.3 \times 10^{-11}}$ **m**.

Now it's your turn

1 Calculate the de Broglie wavelength of an electron travelling with a speed of 5.5×10^7 m s^{-1}.
   (Planck constant $h = 6.6 \times 10^{-34}$ J s; electron mass $m_e = 9.1 \times 10^{-31}$ kg)
   Ans: 1.3×10^{-12} m

2 Calculate the de Broglie wavelength of an electron which has been accelerated from rest through a potential difference of 100 V.
   (Planck constant $h = 6.6 \times 10^{-34}$ J s; electron mass $m_e = 9.1 \times 10^{-31}$ kg; electron charge $e = -1.6 \times 10^{-19}$ C)
   Ans: 1.2×10^{-10} m

Section 8.2 questions

1 Calculate the de Broglie wavelength of an electron with energy 1.0 keV.
(Planck constant $h = 6.6 \times 10^{-34}$ J s; electron mass $m_e = 9.1 \times 10^{-31}$ kg; 1 eV = 1.6×10^{-19} J)

2 Calculate the speed of a neutron with de Broglie wavelength 1.5×10^{-10} m.
(Planck constant $h = 6.6 \times 10^{-34}$ J s; electron mass $m_n = 1.7 \times 10^{-27}$ kg.)

8.3 Emission spectra

Continuous and line spectra

If white light from a tungsten filament lamp is passed through a prism, the light is dispersed into its component colours, as illustrated in Figure 8.5. The band of different colours is called a **continuous spectrum**. A continuous spectrum has all colours (and wavelengths) between two limits. In the case of white light, the colour and wavelength limits are violet (about 400 nm) and red (about 700 nm). Since this spectrum has been produced by the emission of light from the tungsten filament lamp, it is referred to as an **emission spectrum**. Finer detail of emission spectra than is obtained using a prism may be achieved using a diffraction grating.

Figure 8.5 *Continuous spectrum of white light from a tungsten filament lamp.*

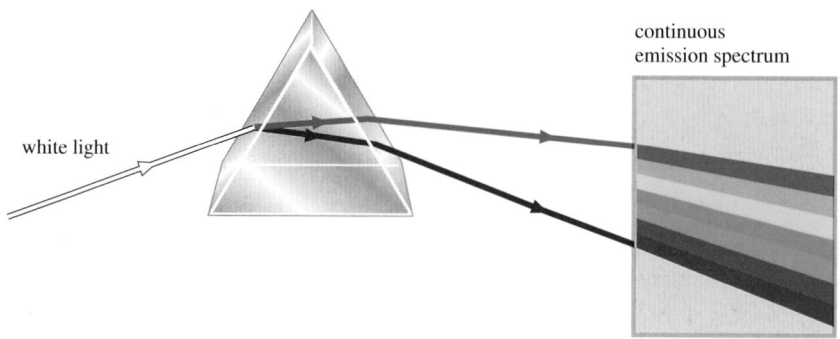

white light

continuous emission spectrum

A discharge tube is a transparent tube containing a gas at low pressure. When a high potential difference is placed across two electrodes in the tube, light is emitted. Examination of the light with a diffraction grating shows that the emitted spectrum is no longer continuous, but consists of a number of bright lines.

Line spectrum of hydrogen
from a discharge tube.

Such a spectrum is illustrated in the photograph. It is known as a **line spectrum**. A line spectrum consists of a number of separate colours, each colour being seen as the image of the slit in front of the source. The wavelength corresponding to the lines of the spectrum are characteristic of the gas which is in the discharge tube.

Electron energy levels in atoms

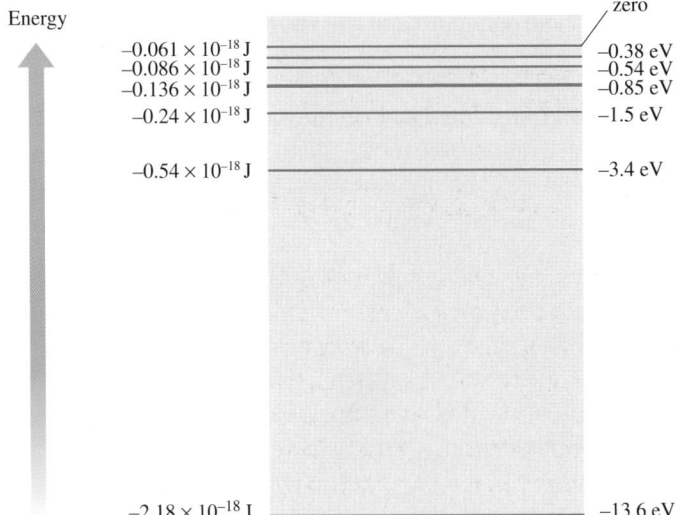

Figure 8.6 *Electron energy levels for the hydrogen atom.*

To explain how line spectra are produced we need to understand how electrons in atoms behave. Electrons in an atom can have only certain specific energies. These energies are called the **electron energy levels** of the atom. The energy levels may be represented as a series of lines against a vertical scale of energy, as illustrated in Figure 8.6. The electron in the hydrogen atom can have any of these energy values, but cannot have energies between them.

Normally electrons occupy the lowest energy levels available. Under these conditions the atom and its electrons are said to be in the **ground state**. Figure 8.7 represents a hydrogen atom with its single electron in the lowest energy state.

If, however, the electron absorbs energy, perhaps by being heated, or by collision with another electron, it may be promoted to a higher energy level. The energy absorbed is exactly equal to the difference in energy of the two levels. Under these conditions the atom is described as being in an **excited state**. This is illustrated in Figure 8.8.

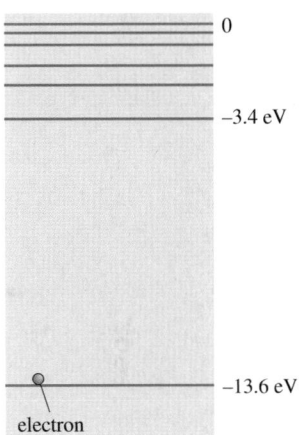

Figure 8.7 *Hydrogen atom in its ground state.*

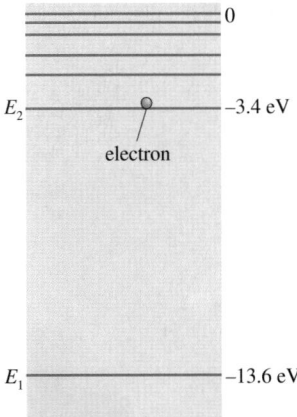

Figure 8.8 *Hydrogen atom in an excited state.*

8.3 Emission spectra

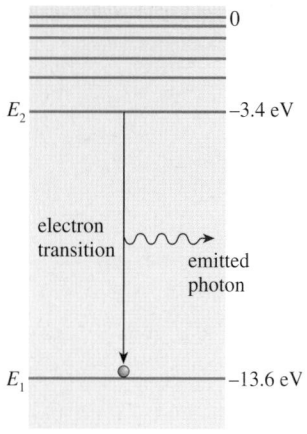

Figure 8.9 *Return of atom to ground state with photon emission.*

An excited atom is unstable. After a short time, the excited electron will return to a lower level. To achieve this, the electron must lose energy. It does so by emitting a photon of electromagnetic radiation, as illustrated in Figure 8.9. The energy hf of the photon is given by

$$\bigstar \quad hf = E_2 - E_1,$$

where E_2 is the energy of the higher level and E_1 that of the lower, and h is the Planck constant. Using the relation between the speed of light c, frequency f and wavelength λ, the wavelength of the emitted radiation is given by

$$\bigstar \quad \lambda = \frac{hc}{\Delta E},$$

where $\Delta E = E_2 - E_1$. This movement of an electron between energy levels is called an **electron transition**. Note that the larger the energy of the transition, the higher the frequency (and the shorter the wavelength) of the emitted radiation.

Note that this downward transition results in the **emission** of a photon. The atom can be raised to an excited state by the **absorption** of a photon, but the photon must have just the right energy, corresponding to the difference in energy of the excited state and the initial state. So, a downward transition corresponds to photon emission, and an upward transition to photon absorption.

Spectroscopy

Figure 8.10 shows some of the possible transitions that might take place when electrons in an excited atom return to lower energy levels. Each of the transitions results in the emission of a photon with a particular wavelength. For example, the transition from E_4 to E_1 results in light with the highest frequency and shortest wavelength. On the other hand, the transition from E_4 to E_3 gives the lowest frequency and longest wavelength.

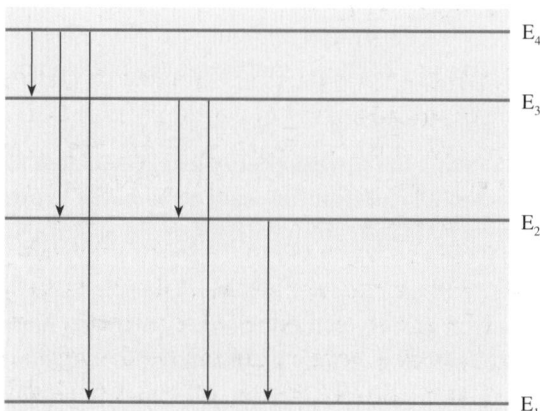

Figure 8.10 *Some energy levels of electrons in an atom.*

8.3 Emission spectra

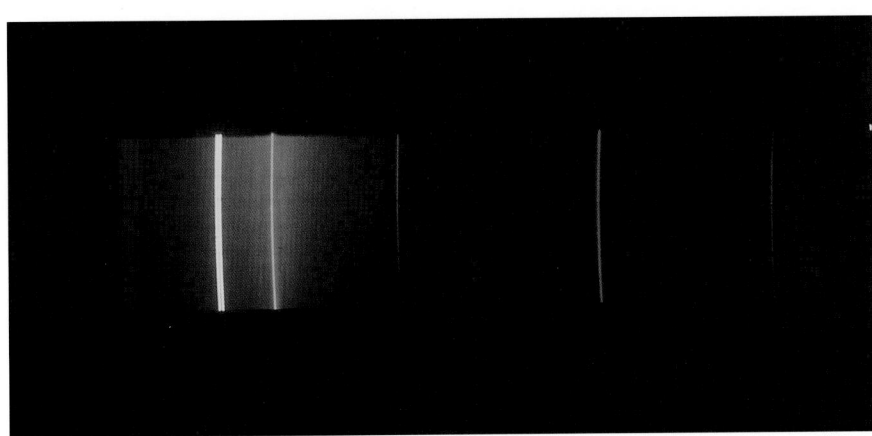

Mercury line spectrum.

Because all elements have different energy levels, the energy differences are unique to each element. Consequently, each element produces a different and characteristic line spectrum. Spectra can be used to identify the presence of a particular element. The line spectrum of mercury is shown in the photograph. The study of spectra is called **spectroscopy**, and instruments used to measure the wavelengths of spectra are **spectrometers**. Spectrometers for accurate measurement of wavelength make use of diffraction gratings (see Chapter 7) to disperse the light.

Continuous spectra

Whilst the light emitted by isolated atoms such as those in low-pressure gases produces line spectra, the light emitted by atoms in a solid, a liquid, or a gas at high pressure produces a continuous spectrum. This happens because of the proximity of the atoms to each other. Interaction between the atoms results in a broadening of the electron energy levels. Consequently, transitions of a wide range of magnitudes of energy are possible, and light of a broad spread of wavelengths may be emitted. This is seen as a **continuous spectrum**.

Absorption spectra

If white light passes through a low-pressure gas and the spectrum of the white light is then analysed, it is found that light of certain wavelengths is missing. In their place are dark lines. This type of spectrum is called an **absorption spectrum** and is illustrated in the photograph on page 210. As the white light passes through the gas, some electrons absorb energy and make transitions to higher energy levels. The wavelengths of the light they absorb correspond exactly to the energies needed to make the particular upward transitions. When these excited electrons return to lower levels, the photons are emitted in all directions, rather than in the original direction of the white light. Thus, some wavelengths appear to be missing. It follows that the wavelengths missing from an absorption spectrum are those present in the emission spectrum of the same element. This is illustrated overleaf.

Absorption spectrum.

Relation between an absorption spectrum and the emission spectrum of the same element. (a) Spectrum of white light (b) absorption spectrum of element (c) emission spectrum of same element.

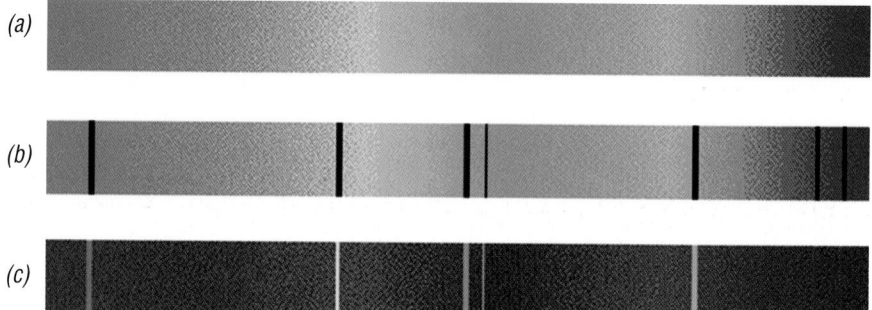

(a)

(b)

(c)

The electromagnetic spectrum

Visible light is just a small region of the **electromagnetic spectrum**. All electromagnetic waves are transverse waves, consisting of electric and magnetic fields which oscillate at right angles to each other and to the direction in which the wave is travelling. This is illustrated in Figure 8.11. Electromagnetic waves show all the properties common to wave motions: they can be reflected, refracted, and diffracted; they obey the principle of superposition and produce interference patterns (see Chapter 7). Because they are transverse waves, they can, in addition, be polarised (see Chapter 7). In a vacuum all electromagnetic waves travel at the same speed, 3.00×10^8 m s^{-1}.

Figure 8.11 *Oscillating electric and magnetic fields in an electromagnetic wave.*

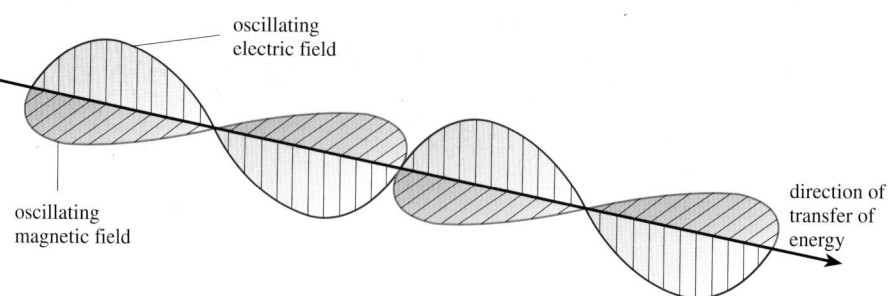

oscillating electric field

oscillating magnetic field

direction of transfer of energy

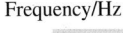

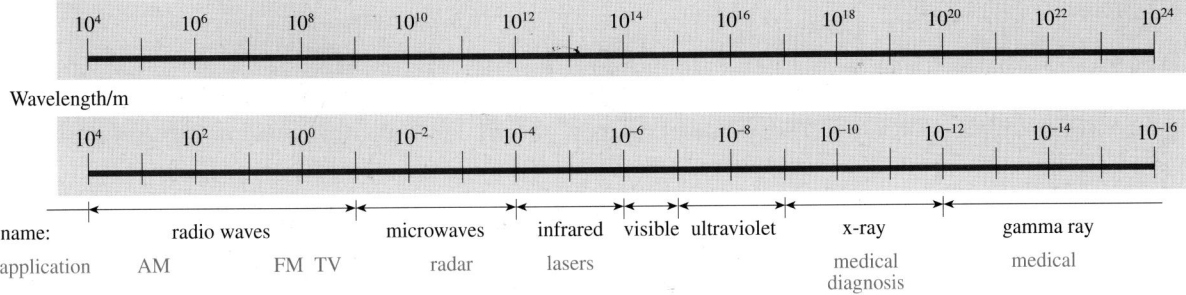

Frequency/Hz

Wavelength/m

| name: | radio waves | microwaves | infrared | visible | ultraviolet | x-ray | gamma ray |

| application | AM | FM TV | radar | lasers | | medical diagnosis | medical |

The complete electromagnetic spectrum is divided into a series of regions based on the properties of electromagnetic waves in these regions, as illustrated in Figure 8.12. It should be noted that there is no clear boundary between regions.

Figure 8.12 *The electromagnetic spectrum.*

Example

Calculate the wavelength of the radiation emitted when the electron in a hydrogen atom makes a transition from the energy level at -0.54×10^{-18} J to the level at -2.18×10^{-18} J.
(Planck constant $h = 6.62 \times 10^{-34}$ J s; speed of light $c = 3.00 \times 10^8$ m s^{-1}.)

Here $\Delta E = E_2 - E_1 = -0.54 \times 10^{-18} - (-2.18 \times 10^{-18}) = 1.64 \times 10^{-18}$ J.
Using $\lambda = hc/\Delta E$,
$\lambda = 6.62 \times 10^{-34} \times 3.00 \times 10^8/1.64 \times 10^{-18} = 1.21 \times 10^{-7}$ m = **121 nm**.

Now it's your turn

1 Calculate the wavelength of the radiation emitted when the electron in a hydrogen atom makes a transition from the energy level at -3.4×10^{-18} J to the level at -8.5×10^{-18} J.
   (Planck constant $h = 6.6 \times 10^{-34}$ J s; speed of light $c = 3.0 \times 10^8$ m s^{-1}.)
   Ans: 3.9×10^{-8} m

2 The electron in a hydrogen atom makes a transition from the energy level at -13.58 eV to the level at -0.38 eV when a photon is absorbed. Calculate the frequency of the radiation absorbed.
   (Planck constant $h = 6.62 \times 10^{-34}$ J s; 1 eV = 1.60×10^{-19} J.)
   Ans: 3.19×10^{15} Hz

Section 8.3 summary

★ Electrons in atoms can have only certain energies. These energies may be represented in an energy level diagram.

★ Electrons in a given energy level may absorb energy and make a transition to a higher energy level.

★ Excited electrons may return to a lower level with the emission of a photon. The frequency f of the emitted radiation is given by $E_2 - E_1 = hf$, where E_2 and E_1 are the energies of the upper and lower levels and h is the Planck constant; the wavelength λ is given by $\lambda = c/f$, where c is the speed of light.

★ When an electron absorbs energy from white light and moves to a higher energy level, a line absorption spectrum is produced.

★ All wavelengths of electromagnetic radiation have the same speed $c = 3.00 \times 10^8$ m s^{-1} in a vacuum.

Section 8.3 questions

1 Calculate the wavelength of the radiation emitted when the electron in the hydrogen atom makes a transition from the energy level at −0.54 eV to the level at −3.39 eV. (Planck constant $h = 6.62 \times 10^{-34}$ J s; speed of light $c = 3.00 \times 10^8$ m s^{-1}; 1 eV = 1.60×10^{-19} J.)

2 The energy required to completely remove an electron in the ground state from an atom is called the **ionisation energy**. This energy may be supplied by the absorption of a photon, in which case the process is called **photoionisation**. Use information from Figure 8.6 to deduce the wavelength of radiation required to achieve photoionisation of hydrogen. (Planck constant $h = 6.62 \times 10^{-34}$ J s; speed of light $c = 3.00 \times 10^8$ m s^{-1}.)

Exam Questions

1 A zinc plate is placed on the cap of a gold leaf electroscope and charged negatively. The gold leaf is seen to deflect. Explain fully the following observations.

   (a) When the zinc plate is illuminated with red light, the gold leaf remains deflected.
   (b) When the zinc plate is irradiated with ultra-violet radiation, the leaf collapses.
   (c) When the intensity of the ultra-violet radiation is increased, the leaf collapses more quickly.

   (d) If the zinc plate is initially charged positively, the gold leaf remains deflected regardless of the nature of the incident radiation.

2 A beam of monochromatic light of wavelength 630 nm transports energy at the rate of 0.25 mW. Calculate the number of photons passing a given cross-section of the beam each second. (Planck constant $h = 6.6 \times 10^{-34}$ J s; speed of light $c = 3.0 \times 10^8$ m s^{-1}.)

3 The work function energy of a certain metal is 4.0×10^{-19} J.

  (a) Calculate the longest wavelength for which photoemission is obtained.

  (b) This metal is irradiated with ultra-violet radiation of wavelength 250 nm. Calculate, for the emitted electrons:

    i) the maximum kinetic energy,

    ii) the maximum velocity.

(Planck constant $h = 6.6 \times 10^{-34}$ J s; speed of light $c = 3.0 \times 10^8$ m s^{-1}; electron mass $m_e = 9.1 \times 10^{-31}$ kg.)

4 Calculate the de Broglie wavelengths of:

  (a) a ball of mass 0.30 kg moving at 50 m s^{-1},

  (b) a bullet of mass 50 g moving at 500 m s^{-1},

  (c) an electron of mass 9.1×10^{-31} kg moving at 3.0×10^7 m s^{-1},

  (d) a proton of mass 1.7×10^{-27} kg moving at 3.0×10^6 m s^{-1}.

(Planck constant $h = 6.6 \times 10^{-34}$ J s.)

5 Atoms in the gaseous state (for example, a gas in a discharge tube) produce an emission spectrum consisting of a series of separate lines. The wavelengths of these lines are characteristic of the particular atoms involved. Hot atoms in the solid state (for example, a hot metal filament in an electric light bulb) produce a continuous emission spectrum which is characteristic of the temperature of the filament rather than of the atoms involved. Suggest reasons for this difference.

6 Figure 8.13 shows some transitions which may take place as excited electrons return to lower energy levels. Calculate the wavelengths of the light emitted in each transition.
(Planck constant $h = 6.6 \times 10^{-34}$ J s; speed of light $c = 3.0 \times 10^8$ m s^{-1}.)

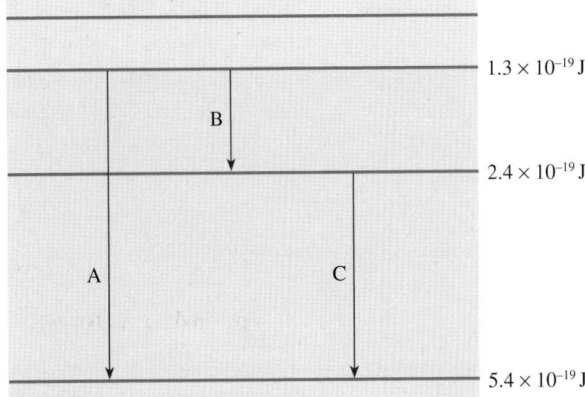

Figure 8.13

7 When the visible spectrum emitted by the Sun is observed closely it is observed that light of certain frequencies is missing and in their place are dark lines.

  (a) Explain how the cool outer gaseous atmosphere of the Sun could be responsible for the absence of these frequencies.

  (b) Suggest how an analysis of this spectrum could be used to determine which gases are present in the Sun's atmosphere.

CHAPTER NINE

Circular motion and oscillations

Understanding the physics of circular motion has enabled scientists to launch satellites into orbit around the Earth, to identify particles from their curved tracks in bubble chambers, and to design fairground rides!

In Chapter 2 we dealt with uniformly-accelerated motion. In this chapter we shall study two important examples of motion in which the acceleration is not uniform – circular motion and simple harmonic oscillations. Both have very wide applications in physics and in everyday life.

9.1   Going round in circles

Centripetal force

Newton's first law of motion (Chapter 4) tells us that an object with a resultant force of zero acting on it will either not be moving at all, or will be moving in a straight line at constant speed (that is, its velocity does not change). The object is said to be in equilibrium. (The full conditions of equilibrium require there to be no resultant force or resultant moment acting on the body.)

An object travelling in a circle may have a constant speed, but it is not travelling in a straight line. Therefore its velocity is changing, since velocity is speed in a certain direction. A change in velocity means the object is accelerating.

In Chapter 2 we dealt with objects moving along a straight line, where acceleration related to changes of speed. In circular motion, objects accelerate because their *direction* changes.

The acceleration a of an object moving in a circle of radius r with speed v is given by the expression

$$\star \quad a = \frac{v^2}{r}.$$

This acceleration is towards the centre of the circle. In order to make an object accelerate, there must be a resultant force acting on it. This force is called the **centripetal force**. From Newton's second law of motion (see Chapter 4), the force F to give a mass m an acceleration a is

$$F = ma.$$

By substituting for a, the centripetal force is

$$\star \quad F = \frac{mv^2}{r}.$$

$\star$ The centripetal force acts towards the centre of the circle

This means that the centripetal force acts at right angles to the instantaneous velocity of the object.

Consider a ball on a string which is being swung in a horizontal circle. The tension in the string provides the centripetal force. At any instant the direction of ball's velocity is along the tangent to the circle, as shown in Figure 9.1. If the string breaks or is released, there is no longer any centripetal force. The ball will travel in the direction of the tangent to the circle at the moment of release.

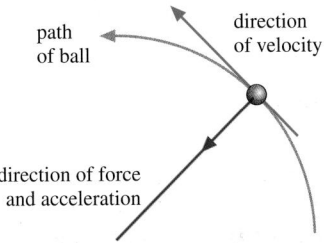

path of ball

direction of velocity

direction of force and acceleration

Figure 9.1 *A ball swung in a circle on the end of a string.*

Example

An aircraft in a display team makes a turn in a horizontal circle of radius 500 m. It is travelling at a speed of 100 m s^{-1} (230 mph).

(a) Calculate the centripetal acceleration of the aircraft.
(b) The mass of the pilot in the aircraft is 80 kg. Calculate the centripetal force acting on the pilot.

(a) From $a = v^2/r$, the centripetal acceleration $a = 100^2/500 = \textbf{20 m s}^{-2}$.
(b) From $F = ma$, the centripetal force on the pilot
   $F = 80 \times 20 = \textbf{1600 N}$.

Now it's your turn

1 In a model of the hydrogen atom, the electron moves round the nucleus in a circular orbit of radius 5.3×10^{-11} m. The speed of the electron is 2.2×10^6 m s^{-1}. Calculate the centripetal acceleration of the electron.
   Ans: 9.1×10^{22} m s^{-2}

2 A satellite orbits the Earth 200 km above its surface. The acceleration towards the centre of the Earth is 9.2 m s^{-2}. The radius of the Earth is 6400 km. Calculate (a) the speed of the satellite, (b) the time to complete one orbit.
   Ans: (a) 7.8 km s^{-1}; (b) 5.3×10^3 s

Radians and degrees

In circular motion, it is convenient to measure angles in **radians** rather than degrees. One degree is by tradition equal to the angle of a circle divided by 360.

> ★ One radian (rad) is defined as the angle subtended at the centre of a circle by an arc equal in length to the radius.

Thus, to obtain an angle in radians, we divide the length of the arc by the radius of the circle (see Figure 9.2).

$$\theta = \frac{length\ of\ arc}{radius\ of\ circle}.$$

The angle in radians in a complete circle would be

$$\theta = \frac{circumference\ of\ the\ circle}{radius} = \frac{2\pi r}{r} = 2\pi.$$

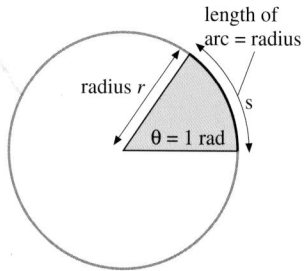

Figure 9.2 θ in radians = arc/radius.

Since the angle of a complete circle is 360°, then

$$\bigstar \quad 2\pi \text{ rad} = 360°$$

or

$$\bigstar \quad 1 \text{ rad} = 57.3°.$$

Angular velocity

For an object moving in a circle,

> $\bigstar$ the **angular speed** is defined as the angle swept out by the radius per second.

The **angular velocity** is the angular speed in a given direction (for example, clockwise). The unit of angular speed and angular velocity is the radian per second (rad s^{-1}).

$$\bigstar \quad angular\ speed\ \omega = \frac{\Delta\theta}{\Delta t}.$$

Figure 9.3 shows an object travelling at constant speed v in a circle of radius r. In a time Δt the object moves along an arc of length Δs and sweeps out an angle $\Delta\theta$.

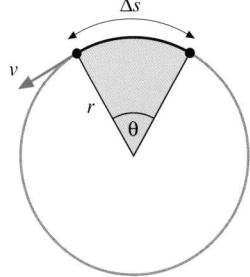

Figure 9.3 *Angular velocity* $\omega = v/r$.

From the definition of the radian,

$$\Delta\theta = \Delta s/r,$$

or

$$\Delta s = r\Delta\theta.$$

Dividing both sides of this equation by Δt,

$$\Delta s/\Delta t = r\Delta\theta/\Delta t.$$

By definition, $\Delta s/\Delta t$ is the linear speed v of the object, and $\Delta\theta/\Delta t$ is the angular speed ω. Hence

$$\bigstar \quad v = r\omega.$$

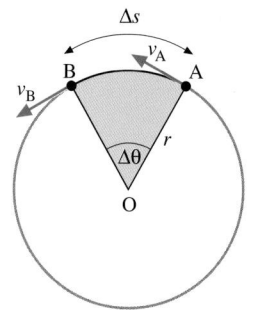

Figure 9.4 *Diagram for proof of* $a = v^2/r$.

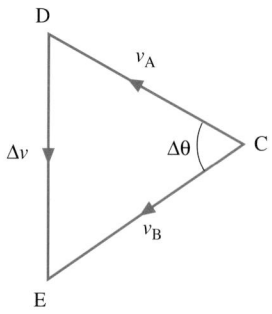

Figure 9.5 *Vector diagram for proof of* $a = v^2/r$.

Derivation of expression for centripetal acceleration

Figure 9.4 shows an object which has travelled at constant speed v in a circular path from A to B in time Δt. At A, its velocity is v_A, and at B the velocity is v_B. Both v_A and v_B are vectors.

The change in velocity Δv may be seen in the vector diagram of Figure 9.5. A vector Δv must be added to v_A in order to give the new velocity v_B.

The angle between the two radii OA and OB is $\Delta\theta$. This angle is also equal to the angle between the vectors v_A and v_B, because triangles OAB and CDE are similar. Consider angle $\Delta\theta$ to be so small that the arc AB may be approximated to a straight line. Then, using similar triangles, DE/CD = AB/OA, and $\Delta v/v_A = \Delta s/r$ or

$$\Delta v = \Delta s(v_A/r).$$

The time to travel the distance Δs or the angle $\Delta\theta$ is Δt. Dividing both sides of the equation by Δt,

$$\Delta v/\Delta t = (\Delta s/\Delta t)(v/r),$$

and from the definitions of acceleration ($a = \Delta v/\Delta t$) and speed ($v = \Delta s/\Delta t = v_A = v_B$) we have

$$a = v(v/r),$$

or

$$a = v^2/r.$$

This expression can be written in terms of angular speed ω. Since $v = r\omega$,

$$\bigstar \quad \text{centripetal acceleration} = \frac{v^2}{r} = r\omega^2$$

and since $F = ma$,

$$\bigstar \quad \text{centripetal force} = \frac{mv^2}{r} = mr\omega^2.$$

Example

The drum of a spin dryer has a radius of 20 cm and rotates at 600 revolutions per minute.

(a) Show that the angular speed of the drum is about 63 rad s^{-1}.
(b) Calculate, for a point on the edge of the drum:
   i) its linear speed,
   ii) its acceleration towards the centre of the drum.

(a) 600 revolutions per minute is 10 revolutions per second. The time for one revolution is thus 0.10 s. Each revolution is 2π rad, so the angular speed $\omega = \theta/t = 2\pi/0.10 = \textbf{63 rad s}^{-1}$.
(b) **i)** Using $v = r\omega$, $v = 0.20 \times 63 = \textbf{13 m s}^{-1}$.
   ii) Using $a = v^2/r$, $a = (13)^2/0.20 = \textbf{850 m s}^{-2}$.

Now it's your turn

A toy train moves round a circular track of diameter 0.70 m, completing one revolution in 10 seconds. Calculate, for this train:

(a) the linear speed,
(b) the angular speed,
(c) the centripetal acceleration.
   Ans: (a) 0.22 m s^{-1}; (b) 0.63 rad s^{-1}; (c) 0.14 m s^{-2}

Examples of circular motion

When a ball is whirled round on the end of a string, you can see clearly that the string is making the ball accelerate towards the centre of the circle. However, in other examples it is not always so easy to see what force is providing the centripetal acceleration.

A satellite in Earth orbit experiences gravitational attraction towards the centre of the Earth. This attractive force provides the centripetal force and causes the satellite to accelerate towards the centre of the Earth, and so moves in a circle. We shall return to this in detail in Chapter 10.

A charged particle moving at right angles to a magnetic field experiences a force at right angles to its direction of motion, and therefore moves in the arc of a circle. This will be considered in more detail in Chapter 11.

For a car travelling in a curved path, the frictional force between the tyres and the road surface provides the centripetal force. If this frictional force is not large enough, for example if the road is icy, then the car carries on moving in a straight line – it skids.

A passenger in a car that is cornering appears to be flung away from the centre of the circle. The centripetal force required to maintain the passenger in circular motion is provided through the seat of the car. This force is below the centre of mass of the passenger, causing rotation about the centre of mass (Figure 9.6). The effect is that the upper part of the passenger moves outwards unless another force acts on the upper part of the body, preventing rotation.

Figure 9.6 Passenger in a car rounding a corner.

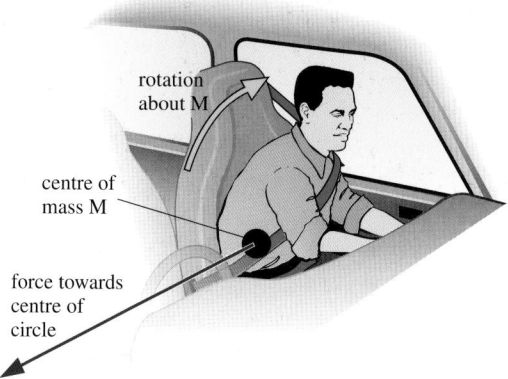

For cornering which does not rely only on friction, the road can be banked as in some car and cycle racetracks, illustrated in Figure 9.7. The road provides a resultant force normal to its surface through contact between the tyres and the road. This resultant force F is at an angle to the vertical, and can be resolved into a vertical component F_v and a horizontal component F_h. F_v is equal to the weight of the vehicle, thus maintaining equilibrium in the vertical direction. The horizontal component F_h provides the centripetal force towards the centre of the circle. Many roads are banked for greater road safety, so as to reduce the chance of loss of control of vehicles due to skidding outwards on the corner, and for greater passenger comfort.

Figure 9.7 Cornering on a banked track.

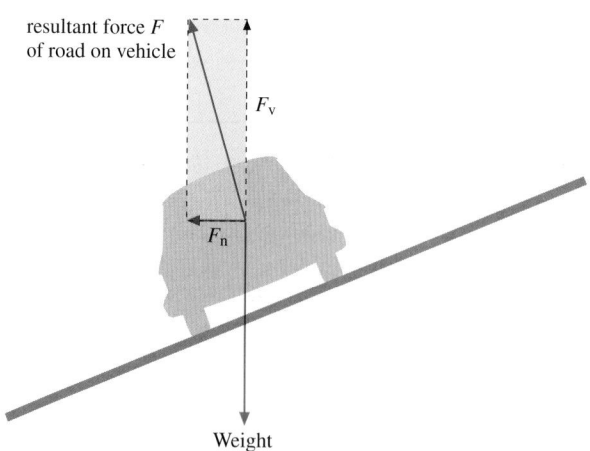

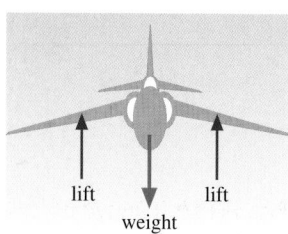

Figure 9.8 An aircraft in straight, level flight.

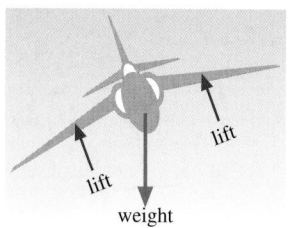

Figure 9.9 An aircraft banking.

An aircraft has a lift force caused by the different rates of flow of air above and below the wings. The lift force balances the weight of the aircraft when it flies on a straight, level path (Figure 9.8). In order to change direction, the aircraft is banked so that the wings are at an angle to the horizontal (Figure 9.9). The lift force now has a horizontal component which provides a centripetal force to change the aircraft's direction.

9.1 Going round in circles

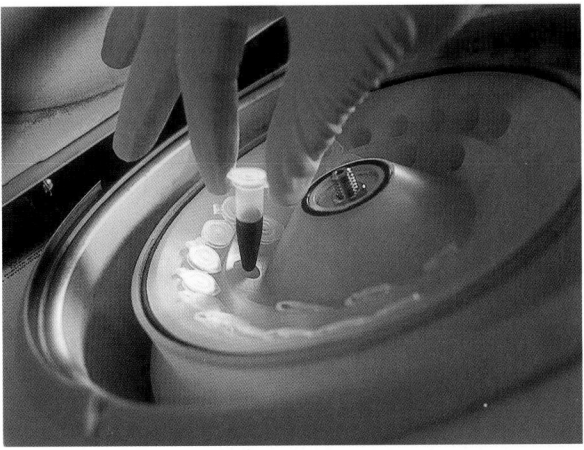

A centrifuge is a device used to spin objects at high speed about an axis. They are used to separate particles in mixtures. More massive particles require larger centripetal forces in order to maintain circular motion than do less massive ones. As a result, the more massive particles tend to separate from less massive particles, collecting further away from the axis of rotation. Space research centres, such as NASA, use centrifuges which are large enough to rotate a person. Their purpose is to investigate the effects of large accelerations on the human body.

Separation of a solid from a liquid in a laboratory centrifuge.

A centrifuge testing the effect of acceleration on the human body.

Motion in a vertical circle

Some pleasure park rides, such as the one in the photograph at the beginning of the chapter, involve rotation in a vertical circle. A person on such a ride must have a resultant force acting towards the centre of the circle.

The forces acting on the person are the person's weight, which always acts vertically downwards, and the reaction from the seat, which acts at right angles to the seat.

Consider a person moving round a vertical circle at speed v. At the bottom of the ride, the reaction R_b from the seat must provide the centripetal force as well as overcoming the weight W of the person. Figure 9.10 illustrates the

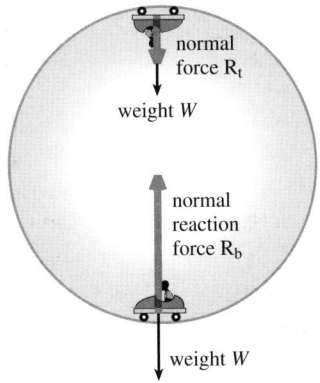

normal
force R_t

weight W

normal
reaction
force R_b

weight W

Figure 9.10 Forces on a person on a circular ride.

situation. Hence the centripetal force is given by

$$mv^2/r = R_b - W.$$

At the top of the ride, the weight W and the reaction force F_t both act downwards towards the centre of the circle. The centripetal force is now given by

$$mv^2/r = R_t + W.$$

This means that the force R_t from the seat at the top of the ride is less than the force R_b at the bottom. If the speed v is not large, then at the top of the circle the weight may be greater than the centripetal force. The person would lose contact with the seat and fall inwards.

Example

A rope is tied to a bucket of water, and the bucket is swung in a vertical circle of radius 1.2 m. What must be the minimum speed of the bucket at the highest point of the circle if the water is to stay in the bucket throughout the motion?

This example is similar to the problem of the pleasure park ride. Water will fall out of the bucket if its weight is greater than the centripetal force. The critical speed v is given by $mv^2/r = mg$, or $v^2 = gr$. Here $v = (9.8 \times 1.2)^{\frac{1}{2}} = \textbf{3.4 m s}^{-1}$.

Now it's your turn

At an air show, an aircraft diving at a speed of 170 m s^{-1} pulls out of the dive by moving in the arc of a circle at the bottom of the dive. Calculate the minimum radius of this circle if the centripetal acceleration of the aircraft is not to exceed five times the acceleration of free fall. The pilot has mass 85 kg. What is his apparent weight (i.e. the sum of the forces acting vertically downwards on him) at the instant when the aircraft is at its lowest point?

Ans: 590 m; 5000 N

Section 9.1 summary

★ Angles may be measured in radians (rad). One radian is the angle subtended at the centre of a circle by an arc of the circle equal in length to its radius.

★ Angular speed ω is the angle swept out per unit time by a line rotating about a point.

★ A particle moving along a circle of radius r with linear speed v has angular velocity ω given by $v = r\omega$.

★ An object moving along a circle of radius r with linear speed v and angular speed ω has an acceleration a towards the centre (the centripetal acceleration) given by $a = v^2/r = r\omega^2$.

★ A resultant force acting towards the centre of the circle, called the centripetal force, is required to make an object move in the circle.

★ For an object of mass m moving along a circle of radius r with linear speed v and angular speed ω, the centripetal force F is given by $F = mv^2/r = mr\omega^2$.

Section 9.1 questions

1 State how the centripetal force is provided in the following examples:
(a) a planet orbiting the Sun,
(b) a child on a playground roundabout,
(c) a train on a curved track,
(d) a passenger in a car going round a corner.

2 NASA's 20-G centrifuge is used for testing space equipment and the effect of acceleration on humans. The centrifuge consists of an arm of length 17.8 m, rotating at constant speed and producing an acceleration equal to 20 times the acceleration of free fall. Calculate:
(a) the angular speed required to produce a centripetal acceleration of 20 g,
(b) the rate of rotation of the arm.
($g = 9.8$ m s^{-2}).

9.2 Oscillations

Some movements involve repetitive to-and-fro motion, such as a pendulum, the beating of a heart, the motion of a child on a swing, and the vibrations of a guitar string.

Another example would be a mass bouncing up and down on a spring, as illustrated in Figure 9.11. One complete movement from the starting or rest position, up, then down and finally back up to the rest position is known as an **oscillation**.

★ The time taken for one complete oscillation is referred to as the **period T** of the oscillation

The oscillations repeat themselves.

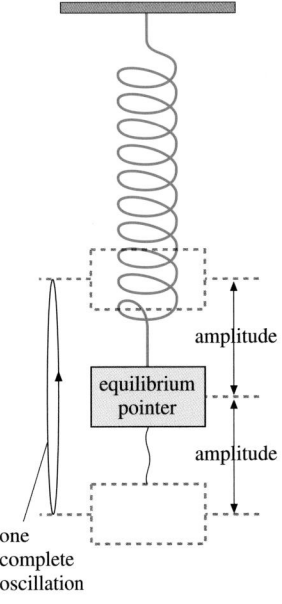

Figure 9.11 Oscillation of a mass on a spring.

⭐ The number of oscillations per unit time is the **frequency** f

Frequency may be measured in hertz (Hz), where one hertz is one oscillation per second (1 Hz = 1 s^{-1}). However, frequency may also be measured in min^{-1}, hour^{-1}, and so on. For example, it would be appropriate to measure the frequency of the tides in h^{-1}. Since period T is the time for one oscillation then

⭐ frequency $f = 1/T$.

As the mass oscillates, it moves from its rest position.

⭐ The distance from the equilibrium position is known as the **displacement**.

This is a vector quantity, since the displacement may be on either side of the equilibrium position.

⭐ The **amplitude** (a scalar quantity) is the maximum displacement.

Some oscillations maintain a constant period even when the amplitude of the oscillation changes. Galileo discovered this fact for a pendulum. He timed the swings of an oil lamp in Pisa Cathedral, using his pulse as a measure of time. Oscillators that have a constant time period are called isochronous, and may be made use of in timing devices. For example, in quartz watches the oscillations of a small quartz crystal provide constant time intervals. Galileo's experiment was not precise, and we now know that a pendulum swinging with a large amplitude is not isochronous.

The quantities period, frequency, displacement and amplitude should be familiar from our study of waves in Chapter 7. It should not be a surprise to meet them again, as the idea of oscillations (for example, of a small piece of a rope) is vital to the understanding of waves.

Displacement-time graphs

It is possible to plot displacement-time graphs (see Chapter 2) for oscillators. One experimental method is illustrated in Figure 9.12. A mass on a spring oscillates above a position sensor that is connected to a computer through a datalogging interface causing a trace to appear on the monitor.

The graph describing the variation of displacement with time may have different shapes, depending on the oscillating system. For many oscillators the graph is approximately a sine (or cosine) curve. A sinusoidal displacement-time graph is a characteristic of an important type of

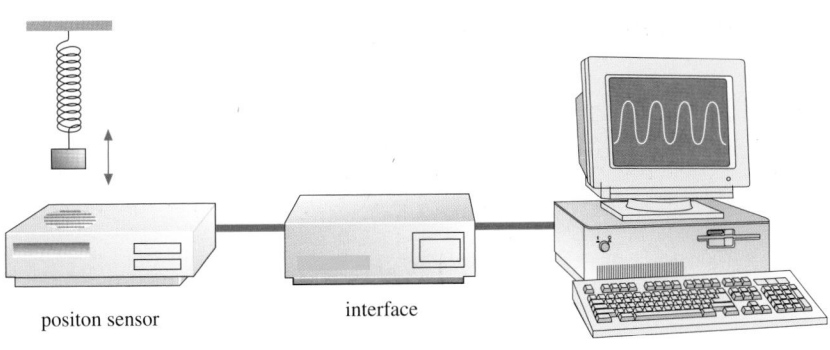

positon sensor   interface

oscillation called **simple harmonic motion** (s.h.m.). Oscillators which move in s.h.m. are called **harmonic** oscillators. We shall analyse simple harmonic motion in some detail, because it successfully describes many oscillating systems, both in real life and in theory. Fortunately, the mathematics of s.h.m. can be approached through a simple defining equation. The properties of the motion can be deduced from the relations between graphs of displacement against time, and velocity against time, which we have already met in Chapter 2.

9.3 Simple harmonic motion

★ Simple harmonic motion is defined as the motion of a particle about a fixed point such that its acceleration a is proportional to its displacement x from the fixed point, and is directed towards the point.

Mathematically, we write this definition as

$$\star \quad a = -\omega^2 x,$$

where ω^2 is a constant. We take the constant as a squared quantity, because this will ensure that the constant is always positive (the square of a positive number, or of a negative number, will always be positive). Why worry about keeping the constant positive? This is because the minus sign in the equation must be preserved. It has a special significance, because it tells us that the acceleration a is always in the opposite direction to the displacement x. (Remember that both acceleration and displacement are vector quantities, so the minus sign is shorthand for the idea that the acceleration is always directed towards the fixed point from which the displacement is measured. This is illustrated in Figure 9.13.)

The defining equation is represented in a graph of a against x as a straight line, of negative gradient, through the origin, as shown in Figure 9.14. The

Figure 9.13 *Directions of displacement and acceleration are always opposite.*

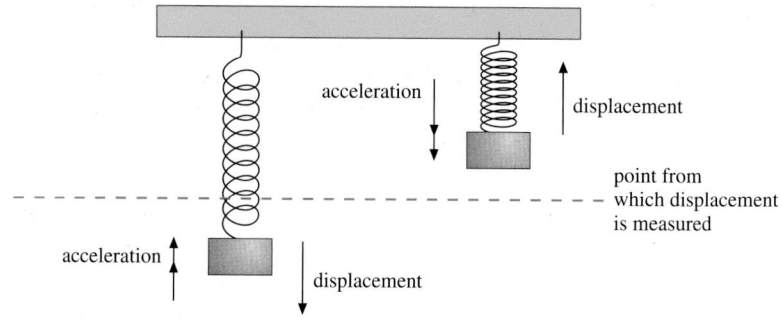

Figure 9.14 *Graph of the defining equation for simple harmonic motion.*

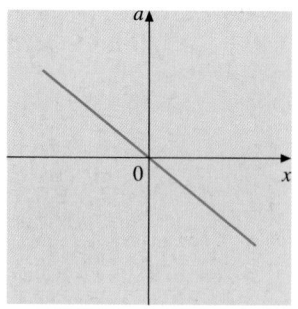

gradient is negative because of the minus sign in the equation. Note that both positive and negative values for the displacement should be considered.

The square root of the constant ω^2 (that is, ω) is known as the **angular frequency** of the oscillation. This angular frequency ω is related to the frequency f of the oscillation by the expression

$$\star \quad \omega = 2\pi f.$$

By Newton's second law, the force acting on a body is proportional to the acceleration of the body. The defining equation for simple harmonic motion can thus be related to the force acting on the particle. If the acceleration of the particle is proportional to its displacement from a fixed point, the resultant force acting on the particle is also proportional to the displacement. We can bring in the idea of the direction of the acceleration by specifying that the force is always acting towards the fixed point, or by calling it a **restoring** force.

Solution of equation for simple harmonic motion

In order to find the displacement-time relation for a particle moving in a simple harmonic motion, we need to solve the equation $a = -\omega^2 x$. To derive the solution requires mathematics which is beyond the requirements of AS/A2 Physics. However, you need to know the form of the solution. This is

$$\star \quad x = A \sin \omega t$$

or

$$\star \quad x = A \cos \omega t,$$

where A is the amplitude of the oscillation. The solution $x = A \sin \omega t$ is used

9.3 Simple harmonic motion

when, at time $t = 0$, the particle is at its equilibrium position where $x = 0$. Conversely, if at time $t = 0$ the particle is at its maximum displacement, $x = A$, the solution is $x = A \cos \omega t$. The variation with time t of the displacement x for the two solutions is shown in Figure 9.15. The period T of the oscillations is also shown.

In Chapter 2 it was shown that the gradient of a displacement-time graph may be used to determine velocity. Referring to Figure 9.15, it can be seen that at each time at which $x = A$, the gradient of the graph is zero (this applies to both solutions). Thus, the velocity is zero whenever the particle has its maximum displacement. If we think about a mass vibrating up and down on a spring, this means that when the spring is fully stretched and the mass has its maximum displacement, the mass stops moving downwards and has zero velocity. Also from Figure 9.15, we can see that the gradient of the graph is at a maximum whenever $x = 0$. This means that when the spring is neither under or over-stretched the speed of the mass is at a maximum. After passing this point the spring forces the mass to slow down until it changes direction.

If a full analysis is carried out, it is found that the variation of velocity with time is cosinusoidal if the sinusoidal displacement solution is taken, and siunusoidal if the cosinusoidal displacement solution is taken. This is illustrated in Figure 9.16.

The velocity v of the particle is given by the expressions

$$\bigstar \quad v = A\omega \cos \omega t \qquad \text{when} \qquad x = A \sin \omega t$$

and

$$\bigstar \quad v = -A\omega \sin \omega t \qquad \text{when} \qquad x = A \cos \omega t.$$

In each case, there is a phase difference between velocity and displacement. The velocity curve is $\pi/2$ rad ahead of the displacement curve. (If the phase angle is not considered, the variation with time of the velocity is the same in each case.) The maximum speed v_0 is given by

$$\bigstar \quad v_0 = A\omega.$$

For completeness, Figure 9.16 also shows the variation with time of the acceleration a of the particle. This could be derived from the velocity-time graph by taking the gradient. The equations for the acceleration are

$$\bigstar \quad a = -A\omega^2 \sin \omega t \qquad \text{when} \qquad x = A \sin \omega t$$

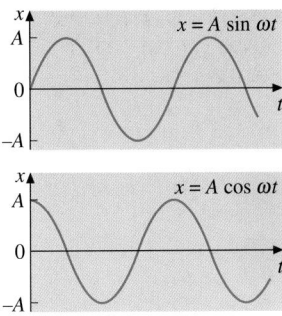

Figure 9.15 *Displacement-time curves for the two solutions to the s.h.m. equation.*

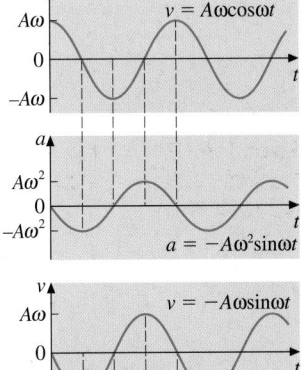

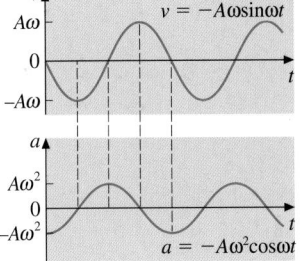

Figure 9.16 *Velocity-time and acceleration-time graphs for the two solutions to the s.h.m. equation.*

and

$$\bigstar \quad a = -A\omega^2 \cos \omega t \qquad \text{when} \qquad x = A \cos \omega t.$$

Note that, for both solutions, these equations are consistent with the defining equation for simple harmonic motion, $a = -\omega^2 x$. You can easily prove this by eliminating $\sin \omega t$ from the first set of equations, and $\cos \omega t$ from the second.

Example

The displacement x at the time t of a particle moving in simple harmonic motion is given by $x = 0.25 \cos 7.5t$, where x is in metres and t is in seconds.

(a) Use the equation to find the amplitude, frequency and period for the motion.

(b) Find the displacement when $t = 0.50$ s.

(a) Compare the equation with $x = A \cos \omega t$. The amplitude $A = \mathbf{0.25 \ m}$. The angular frequency $\omega = 7.5$ rad s^{-1}. Remember that $\omega = 2\pi f$, so the frequency $f = \omega/2\pi = 7.5/2\pi = \mathbf{1.2 \ Hz}$. The period $T = 1/f = 1/1.2 = \mathbf{0.84 \ s}$.

(b) Substitute $t = 0.50$ s in the equation, remembering that the angle ωt is in radians and not degrees. $\omega t = 7.5 \times 0.50 = 3.75$ rad $= 215°$. So $x = 0.25 \cos 215° = \mathbf{-0.20 \ m}$.

Now it's your turn

A mass oscillating on a spring has an amplitude of 0.10 m and a period of 2.0 s.

(a) Deduce the equation for the displacement x if timing starts at the instant when the mass has its maximum displacement.

(b) Calculate the time interval from $t = 0$ before the displacement is 0.08 m.

Ans: (a) $x = 0.10 \cos \pi t$; (b) 0.20 s

Sections 9.2–9.3 summary

★ The period of an oscillation is the time taken to complete one oscillation.

★ Frequency is the number of oscillations per unit time.

★ Frequency f is related to period T by the expression $f = 1/T$.

★ The displacement of a particle is the distance from the equilibrium position.

9.3 Simple harmonic motion

★ Amplitude is the maximum displacement.

★ Simple harmonic motion (s.h.m.) is defined as the motion of a particle about a fixed point such that its acceleration a is proportional to its displacement x from the fixed point, and is directed towards the fixed point: $a \propto -x$ or $a = -\omega^2 x$.

★ The constant ω in the defining equation for simple harmonic motion is known as the angular frequency.

★ For a particle oscillating in s.h.m. with frequency f, then $\omega = 2\pi f$.

★ Simple harmonic motion is described in terms of displacement x, amplitude A, frequency f, angular frequency ω by the relations

displacement: $x = A \sin \omega t$ or $x = A \cos \omega t$

velocity: $v = A\omega \cos \omega t$ or $v = -A\omega \sin \omega t$

acceleration: $a = -A\omega^2 \sin \omega t$ or $a = -A\omega^2 \cos \omega t$

Remember that $\omega = 2\pi f$, and equations may appear in either form.

Section 9.3 questions

1 A particle is oscillating in simple harmonic motion with period 4.5 ms and amplitude 3.0 cm. At time $t = 0$, the particle is at the equilibrium position. Calculate, for this particle:
   (a) the frequency,
   (b) the angular frequency,
   (c) the maximum speed,
   (d) the magnitude of the maximum acceleration,
   (e) the speed at time $t = 1.0$ ms.

2 A particle is oscillating in simple harmonic motion with frequency 50 Hz and amplitude 15 mm. Calculate the speed when the displacement from the equilibrium position is 12 mm.

9.4 Examples of simple harmonic motion

By definition, an object whose acceleration is proportional to the displacement from the equilibrium position, and which is always directed towards the equilibrium position, is undergoing simple harmonic motion. We now look at two simple examples of oscillatory motion which approximate to this definition. They may easily be set up as demonstrations of s.h.m.

CHAPTER 9

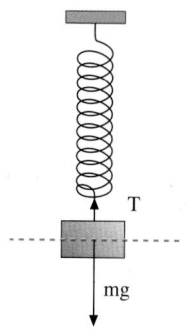

Figure 9.17 *Mass on a helical spring.*

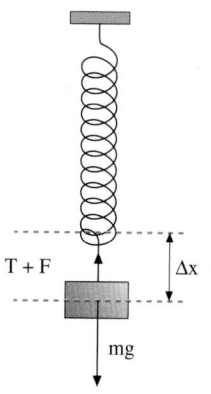

Figure 9.18 *Mass on a helical spring: additional extension* Δx.

Mass on a helical spring

Figure 9.17 illustrates a mass m suspended from a spring. (This sort of spring is called a *helical* spring because it has the shape of a helix.)

The weight mg of the mass is balanced by the tension T in the spring. When the spring is extended by an amount Δx as in Figure 9.18, there is an additional upward force in the spring given by

$$F = -k\Delta x,$$

where k is a constant for a particular spring, known as the spring constant. The spring constant is a measure of the stiffness of the spring. A stiff spring has a large value of k; a more flexible spring has a smaller value of k and, for the same force, would have a larger extension than one with a large spring constant. The spring constant k is given by the expression

$$\bigstar \quad k = \frac{F}{\Delta x},$$

and is measured, in the SI system, in newtons per metre (N m^{-1}).

When the mass is released, the restoring force F pulls the mass towards the equilibrium position. (The minus sign in the expression for F shows the direction of this force.) The restoring force is proportional to the displacement. This means that the acceleration of the mass is proportional to the displacement from the equilibrium position and is directed towards the equilibrium position. This is the condition for simple harmonic motion.

The full theory shows that, for a mass m suspended from a light spring having spring constant k, the period T of the oscillations is given by

$$\bigstar \quad T = 2\pi\sqrt{\frac{m}{k}}.$$

For oscillations to be simple harmonic, the spring must obey Hooke's law throughout, that is the extensions must not exceed the limit of proportionality. Furthermore, for large amplitude oscillations, the spring may become slack. Ideally, the spring would have no mass, but if the suspended mass is more than about twenty times the mass of the spring, the error involved in assuming that the spring has no mass is less than 1%.

This example of simple harmonic motion is particularly useful in modelling the vibrations of molecules. A molecule containing two atoms oscillates as if the atoms were connected by a tiny spring. The spring constant of this spring depends upon the type of bonding between the atoms. The frequency of oscillation of the molecule can be measured experimentally using spectroscopy, and this gives direct information about the bonding. This

9.4 Examples of simple harmonic motion

model can be extended to solids, where atoms are often thought of as being connected to their neighbours by springs. Again, this leads to an experimental way of obtaining information about interatomic forces in the solid.

The simple pendulum

Figure 9.19 illustrates a simple pendulum. Ideally, a simple pendulum is a point mass m on a light, inelastic string. In real experiments we use a pendulum bob of finite size. When the bob is pulled aside through an angle θ and then released, there will be a restoring force acting in the direction of the equilibrium position.

Because the simple pendulum moves in the arc of a circle, the displacement will be an angular displacement θ rather than the linear displacement x we have been using so far.

The two forces on the bob are its weight mg and the tension T in the string. The component of the weight along the direction of the string, $mg \cos \theta$, is equal to the tension in the string. The component of the weight at right angles to the direction of the string, $mg \sin \theta$, is the restoring force F. This makes the bob accelerate towards the equilibrium position.

The restoring force depends on $\sin \theta$. As θ increases, the restoring force is not proportional to the displacement (in this case θ), and so the motion is oscillatory but not simple harmonic. However, the situation is different if the angle θ is kept small (less than about 5°). For these small angles, θ is proportional to $\sin \theta$. In fact, if θ is measured in radians, then

$$\theta \text{ in radians} \approx \sin \theta$$

You can check this using your calculator. Some values of $\theta/°$, θ/rad and $\sin \theta$ are given in Figure 9.20.

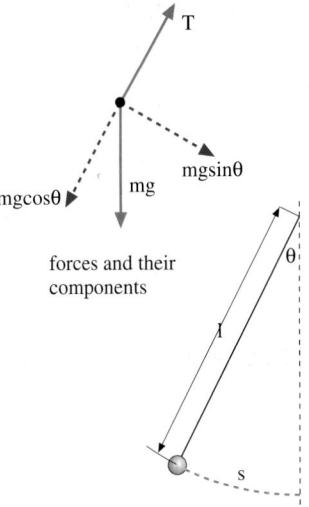

forces and their components

Figure 9.19 *The simple pendulum.*

Figure 9.20

$\theta/°$	θ/rad	$\sin \theta$
1.00	0.0175	0.0175
2.00	0.0349	0.0349
3.00	0.0524	0.0523
5.00	0.0873	0.0871
10.00	0.1745	0.1736

This means that, for small-amplitude oscillations (the angle of the string to the vertical should be less than about 5°), the pendulum bob oscillates with simple harmonic motion.

A full treatment of the theory shows that the period T is related to the length l of the pendulum (l is the distance between the centre of mass of the pendulum bob and the point of suspension – see Figure 9.19) by the expression

$$\star \quad T = 2\pi\sqrt{\frac{l}{g}},$$

where g is the acceleration of free fall. An experiment in which the period of a simple pendulum is measured can be used as to determine the acceleration of free fall. The experiment is repeated for different lengths of pendulum, and the gradient of a graph of T^2 against l is $4\pi^2/g$. This provides an alternative to dynamics experiments in which the time for a body to fall through measured distances is determined, and g is calculated from the equation of motion (see Chapter 2).

Example

A helical spring is clamped at one end and hangs vertically. It extends by 10 cm when a mass of 50 g is hung from its free end. Calculate:
(a) the spring constant of the spring,
(b) the period of small amplitude oscillations of the mass.

(a) Using $k = F/\Delta x$, the spring constant
$k = 50 \times 10^{-3} \times 9.8/10 \times 10^{-2} = $ **4.9 N m^{-1}**.
(b) Using $T = 2\pi(m/k)^{\frac{1}{2}}$, the period $T = 2\pi(50 \times 10^{-3}/4.9)^{\frac{1}{2}} = $ **0.63 s**.

Now it's your turn

(a) The acceleration of free fall at the Earth's surface is 9.8 m s^{-2}. Calculate the length of a simple pendulum which would have a period of 1.0 s.
(b) The acceleration of free fall on the Moon's surface is 1.6 m s^{-2}. Calculate, for the pendulum in (a), its period of oscillation on the Moon.

Ans: 0.25 m; 2.5 s

Section 9.4 summary

$\star$ The period T of small oscillations of a mass m suspended on a light helical spring of spring constant k is $T = 2\pi(m/k)^{\frac{1}{2}}$.

$\star$ The period T of small oscillations of a simple pendulum of length l is $T = 2\pi(l/g)^{\frac{1}{2}}$, where g is the acceleration of free fall.

Section 9.4 questions

1 Geologists use the fact that the period of oscillation of a simple pendulum depends on the acceleration of free fall to map variations of g. A geologist determines the frequency of oscillation of a test pendulum of effective length 515.6 mm to be 0.6948 Hz. Calculate the acceleration of free fall at this locality.

2 A spring stretches by 85 mm when a mass of 50 g is hung from it. The spring is then stretched a further distance of 15 mm from the equilibrium position, and the mass is released at time $t = 0$. Calculate:
   (a) the spring constant,
   (b) the amplitude of the oscillations,
   (c) the period,
   (d) the displacement at time $t = 0.20$ s.
      ($g = 9.8$ m s^{-2}.)

<table><tr><td>9.5</td><td></td></tr></table>

9.5 Energy changes in simple harmonic motion

Kinetic energy

In Section 9.3 we saw that the velocity of a particle vibrating with simple harmonic motion varies with time and, consequently, with the displacement of the particle. For the case where displacement x is zero at time $t = 0$, displacement and velocity are given by

$$x = A \sin \omega t$$

and

$$v = A\omega \cos \omega t.$$

There is a trigonometrical relation between the sine and the cosine of an angle θ, which is $\sin^2 \theta + \cos^2 \theta = 1$. Applying this relation, we have

$$x^2/A^2 + v^2/A^2\omega^2 = 1,$$

which leads to

$$v^2 = A^2\omega^2 - x^2\omega^2.$$

(If we had taken the displacement and velocity equations for the case when the displacement is a maximum at $t = 0$, exactly the same relation would have been obtained. Try it!)

The kinetic energy of the particle (of mass m) oscillating with simple harmonic motion is $\frac{1}{2}mv^2$. Thus, the kinetic energy E_k at displacement x is

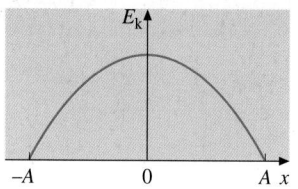

Figure 9.21 *Variation of kinetic energy in s.h.m.*

given by

$$\star \quad E_k = \tfrac{1}{2}m\omega^2(A^2 - x^2).$$

The variation with displacement of the kinetic energy is shown in Figure 9.21.

Potential energy

In Section 9.3 we stated that the defining equation for simple harmonic motion could be expressed in terms of the restoring force acting on the particle. Since $F = ma$ and $a = -\omega^2 x$ then at displacement x, this force is

$$F_{res} = -m\omega^2 x,$$

where m is the mass of the particle. To find the change in potential energy of the particle when the displacement increases by Δx, we need to find the work done against the restoring force.

The work done in moving the point of application of a force F by a distance Δx is $F\Delta x$. In the case of the particle undergoing simple harmonic motion, we know that the restoring force is directly proportional to displacement. To calculate the work done against the restoring force in giving the particle a displacement x, we take account of the fact that F_{res} depends on x by taking the average value of F_{res} during this displacement. The average value is just $\tfrac{1}{2}m\omega^2 x$, since the value of F_{res} is zero at $x = 0$ and increases linearly to $m^2 x$ at displacement x. Thus, the potential energy E_p at displacement x is given by (*average restoring force*) $\times$ (*displacement*), or

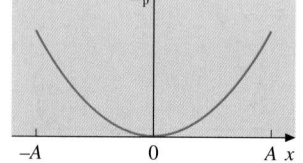

Figure 9.22 *Variation of potential energy in s.h.m.*

$$\star \quad E_p = \tfrac{1}{2}m\omega^2 x^2.$$

The variation with displacement of the potential energy is shown in Figure 9.22.

Total energy

The total energy E_{tot} of the oscillating particle is given by

$$E_{tot} = E_k + E_p$$

$$= \tfrac{1}{2}m\omega^2(A^2 - x^2) + \tfrac{1}{2}m\omega^2 x^2$$

$$\star \quad = \tfrac{1}{2}m\omega^2 A^2.$$

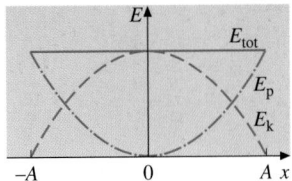
This total energy is constant since m, ω and A are all constant. We might have expected this result, as it merely expresses the law of conservation of energy.

The variations with displacement x of the total energy E_{tot}, the kinetic energy E_k and the potential energy E_p are shown in Figure 9.23.

Figure 9.23 *Energy variations in s.h.m.*

Example

A particle of mass 60 g oscillates in simple harmonic motion with angular frequency 6.3 rad s^{-1} and amplitude 15 mm. Calculate:
(a) the total energy,
(b) the kinetic and potential energies at half-amplitude (at displacement $x = 7.5$ mm).

(a) Using $E_{tot} = \frac{1}{2}m\omega^2 A^2$, $E_{tot} = \frac{1}{2} \times 60 \times 10^{-3} \times 6.3^2 \times (15 \times 10^{-3})^2$
(don't forget to convert g to kg and mm to m!) = **2.7 × 10^{-4} J**.
(b) Using $E_k = \frac{1}{2}m\omega^2(A^2 - x^2)$,
$E_k = \frac{1}{2} \times 60 \times 10^{-3} \times 6.3^2 \times [(15 \times 10^{-3})^2 - (7.5 \times 10^{-3})^2]$
$= $ **2.0 × 10^{-4} J**.

Using $E_p = \frac{1}{2}m\omega^2 x^2$, $E_p = \frac{1}{2} \times 60 \times 10^{-3} \times (7.5 \times 10^{-3})^2$
$= $ **0.7 × 10^{-4} J**.

Note that $E_{tot} = E_k + E_p$, as expected.

Now it's your turn

A particle of mass 0.40 kg oscillates in simple harmonic motion with frequency 5.0 Hz and amplitude 12 cm. Calculate, for the particle at displacement 10 cm:
(a) the kinetic energy,
(b) the potential energy,
(c) the total energy.

Ans: (a) 0.85 J; (b) 1.92 J; (c) 2.77 J

Section 9.5 summary

★ The kinetic energy E_k of a particle of mass m oscillating in simple harmonic motion with angular frequency ω and amplitude A is $E_k = \frac{1}{2}m\omega^2(A^2 - x^2)$, where x is the displacement.

★ The potential energy E_p of a particle of mass m oscillating in simple harmonic motion with angular frequency ω is $E_p = \frac{1}{2}m\omega^2 x^2$, where x is the displacement.

★ The total energy E_{tot} of a particle of mass m oscillating in simple harmonic motion with angular frequency ω and amplitude A is $E_{tot} = \frac{1}{2}m\omega^2 A^2$.

★ For a particle oscillating in simple harmonic motion, $E_{tot} = E_k + E_p$; this expresses the law of conservation of energy.

Section 9.5 questions

1   One particle oscillating in simple harmonic motion has ten times the total energy of a second, but the frequencies and masses are the same. Calculate the ratio of the amplitudes of the two motions.

2   **(a)** Calculate the displacement, expressed as a fraction of the amplitude A, of a particle moving in simple harmonic motion with a speed equal to half the maximum value.

   (b) Calculate the displacement at which the energy of the particle has equal amounts of kinetic and of potential energy.

9.6 Free and damped oscillations

A particle is said to be undergoing **free oscillations** when the only force acting on it is the restoring force.

There are no forces to dissipate energy and so the oscillations have constant amplitude. Total energy remains constant. This is the situation we have been considering so far. Simple harmonic oscillations are free oscillations.

In real situations, however, frictional and other resistive forces cause the oscillator's energy to be dissipated, and this energy is converted eventually into heat energy. The oscillations are said to be **damped**.

The total energy of the oscillator decreases with time. The damping is said to be **light** when the amplitude of the oscillations decreases gradually with time. This is illustrated in Figure 9.24. The decrease in amplitude is, in fact, exponential with time. The period of the oscillations is slightly greater than that of the corresponding free oscillations.

Figure 9.24 *Lightly-damped oscillations.*

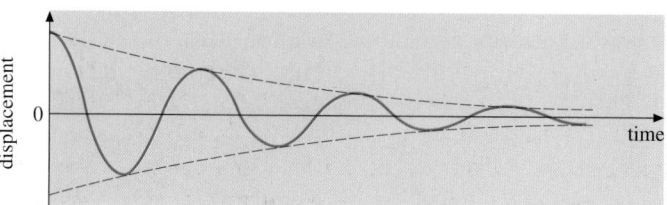

9.6 Free and damped oscillations

Heavy damping causes the oscillations to die away more quickly. If the damping is increased further, then the system reaches **critical damping** point. Here the displacement decreases to zero in the shortest time, without any oscillation (Figure 9.25).

Any further increase in damping produces **overdamping**. The displacement decreases to zero in a longer time than for critical damping (Figure 9.25).

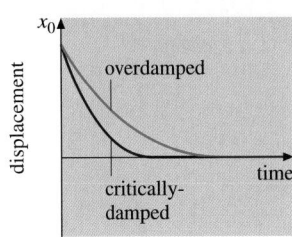

Figure 9.25 *Critical damping and overdamping.*

Damping is often useful in an oscillating system. For example, vehicles have springs between the wheels and the frame to give a smoother and more comfortable ride. If there was no damping, a vehicle would move up and down for some time after hitting a bump in the road. Dampers (shock absorbers) are connected in parallel with the springs so that the suspension has critical damping and comes to rest in the shortest time possible. Dampers often work through hydraulic action. When the spring is compressed, a piston connected to the vehicle frame forces oil through a small hole in the piston, so that the energy of the oscillation is dissipated as heat energy in the oil.

Many swing doors have a damping mechanism fitted to them. The purpose of the damper is so that the open door, when released, does not overshoot the closed position with the possibility of injuring someone approaching the door. Most door dampers operate in the overdamped mode.

Vehicle suspension system showing springs and dampers.

Forced oscillations and resonance

When a vibrating body undergoes free (undamped) oscillations, it vibrates at its **natural frequency**. We have already met the idea of a natural frequency in Chapter 7, when talking about standing waves on strings. The natural frequency of such a system is the frequency of the first mode of vibration, that is, the fundamental frequency. A practical example is a guitar string, plucked at its centre, which oscillates at a particular frequency which depends on the speed of progressive waves on the string and the length of the string. The speed of progressive waves on the string depends on the mass per unit length of the string and the tension in the string.

Vibrating objects may have periodic forces acting on them. These periodic forces will make the object vibrate at the frequency of the applied force, rather than at the natural frequency of the system. The object is then said to be undergoing **forced vibrations**. Figure 9.26 illustrates apparatus which may be used to demonstrate the forced vibrations of a mass on a helical spring. The oscillator provides the forcing (driving) frequency.

As the frequency of the oscillator is gradually increased from zero, the mass begins to oscillate. At first the amplitude of the oscillations is small, but it increases with increasing frequency. When the driving frequency equals the natural frequency of oscillation of the mass-spring system, the amplitude of the oscillations reaches a maximum. The frequency at which this occurs is

Figure 9.26 *Demonstration of forced oscillations.*

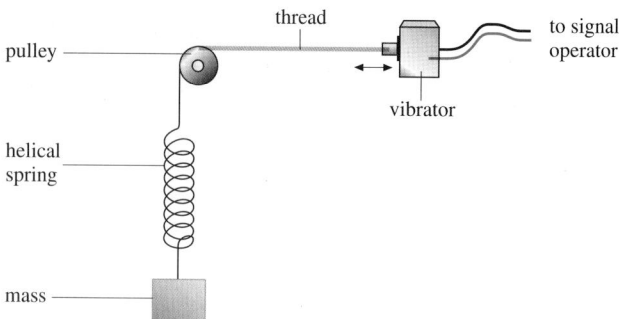

called the **resonant frequency**, and **resonance** is said to occur. If the driving frequency is increased further, the amplitude of oscillation of the mass decreases. The variation with driving frequency f of the amplitude A of vibration of the mass is illustrated in Figure 9.27. This graph is often called a **resonance curve**.

> ★ Resonance occurs when the natural frequency of vibration of an object is equal to the driving frequency, giving a maximum amplitude of vibration.

Figure 9.27 *Resonance curve.*

Figure 9.28 *Effect of damping on the resonance curve.*

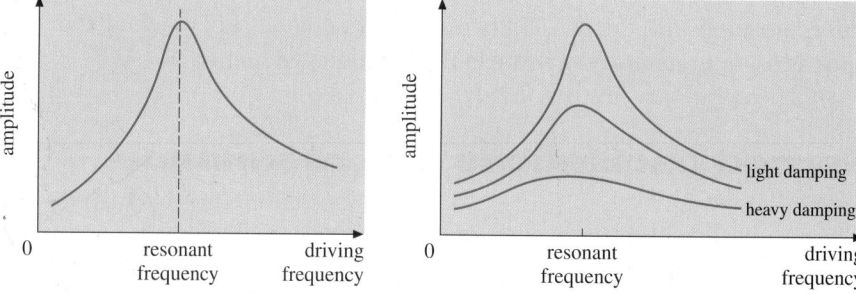

The effect of damping on the amplitude of forced oscillations can be investigated by attaching a light but stiff card to the mass in Figure 9.26. Movement of the card gives rise to air resistance and thus damping of the oscillations. The degree of damping may be varied by changing the area of the card. The effects of damping are illustrated in Figure 9.28. It can be seen that, as the degree of damping increases,

- the amplitude of oscillation at all frequencies is reduced
- the frequency at maximum amplitude shifts gradually towards lower frequencies
- the peak becomes flatter

Barton's pendulums may also be used to demonstrate resonance and the effects of damping. The apparatus consists of a set of light pendulums, made (for example) from paper cones, and a more massive pendulum (the driver), all supported on a taut string. The arrangement is illustrated in Figure 9.29. The lighter pendulums have different lengths, but one has the same length as the driver. This has the same natural frequency as the driver and will, therefore, vibrate with the largest frequency of all the pendulums.

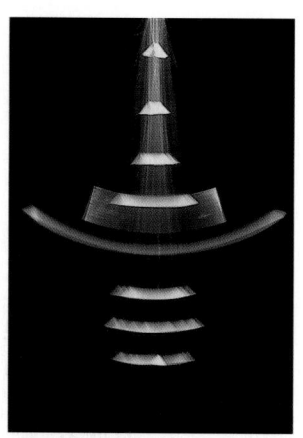

Time-exposure photographs of Barton's pendulums with light damping.

9.6 **Free and damped oscillations**

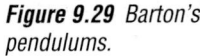
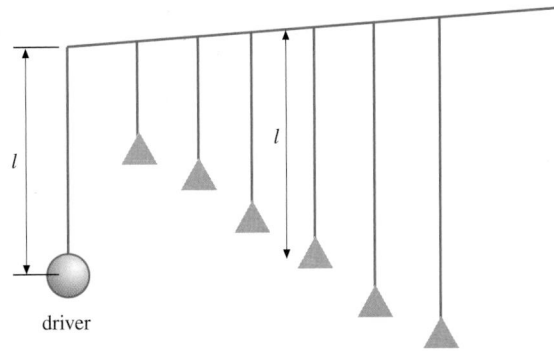

Figure 9.29 *Barton's pendulums.*

driver

Adding weights to the paper cones reduces the effect of damping. With less damping, the amplitude of the resonant pendulum is much larger.

There are many examples of resonance in everyday life. One of the simplest is that of pushing a child on a swing. We push at the same frequency as the natural frequency of oscillation of the swing and child, so that the amplitude of the motion increases.

Pushing a child on a swing.

The operation of the engine of a vehicle causes a periodic force on the parts of the vehicle, which can cause them to resonate. For example, at particular frequencies of rotation of the engine, the mirrors may resonate. To prevent excessive vibration, the mountings of the mirrors provide damping.

Musical instruments rely on resonance to amplify the sound produced. The sound from a tuning fork is louder when it is held over a tube of just the right length, so that the column of air resonates. We have met this phenomenon in Chapter 7, in connection with the resonance tube method of measuring the speed of sound in air. Stringed instruments have a hollow wooden box with a hole under the strings which acts in a similar way. To amplify all notes from all of the strings, the sounding-box has to be a complex shape so that it resonates at many different frequencies.

CHAPTER 9

The Tacoma Narrows bridge disaster.

A spectacular example of resonance that is often quoted is the failure of the first suspension bridge over the Tacoma Narrows in Washington State, USA. Wind caused the bridge to oscillate. It was used for months even though the roadway was oscillating with transverse vibrations. Approaching vehicles would appear, and then disappear, as the bridge deck vibrated up and down. One day strong winds set up twisting vibrations and the amplitude of vibration increased due to resonance, until eventually the bridge collapsed. The driver of a car that was on the bridge managed to walk to safety before the collapse, although his dog could not be persuaded to leave the car!

Section 9.6 summary

★ Free oscillations are oscillations where there are no resistive forces acting on the oscillating system.

★ The natural frequency of vibration of an object is the frequency at which the object will vibrate in the absence of resistive forces.

★ Forced oscillations occur when a periodic driving force is applied to a system which is capable of vibration.

★ Resonance occurs when the driving frequency on the system is equal to its natural frequency of vibration. The amplitude of vibration is a maximum at the resonant frequency.

★ Damping is produced by resistive forces which dissipate the energy of the vibrating system.

9.6 Free and damped oscillations

★ Light damping causes the amplitude of vibration of the oscillation to decrease gradually. Critical damping causes the displacement to be reduced to zero in the shortest time possible, without any oscillation of the object. Over-damping also causes an exponential reduction in displacement, but over a greater time than for critical damping.

Exam Questions

1 All bodies moving in a circle experience centripetal acceleration.

   (a) What does *centripetal* mean? Explain how the centripetal acceleration arises.
   (b) The Moon's orbit around the Earth may be assumed to be circular, with radius 3.84×10^5 km. The Moon moves with constant speed in the orbit, and takes 27.3 days to complete one orbit. Calculate the magnitude of the centripetal acceleration of the Moon.

2 A mountain bicycle is set up on a test rig with its front wheel clamped and its rear wheel resting on a roller. The pedals are turned so that, if the bicycle had been free, it would have been travelling at a speed of 8.3 m s^{-1}. The outside diameter of each tyre is 0.66 m.

   (a) Calculate:
      i) the angular velocity of the rear wheel of the bicycle,
      ii) how many times per second the rear wheel rotates.
   (b) A small pebble of mass 4.0 g is embedded in the tread of the rear tyre.
      i) Calculate the magnitude, and state the direction, of the force needed to keep the pebble in circular motion.
      ii) This force is equal to the maximum frictional force that the tyre can exert on the pebble. Describe and explain what happens when the rear wheel is rotated at a faster rate.

3 (a) Define simple harmonic motion.
   (b) Figure 9.30a illustrates a U-tube of uniform cross-sectional area A containing liquid of density ρ. The total length of the liquid is L. When the liquid is displaced by an amount Δx from its equilibrium position (see Figure 9.30b), it oscillates with simple harmonic motion.
      i) The weight of liquid above AB in Figure 9.30b provides the restoring force. Write down an expression for the restoring force.

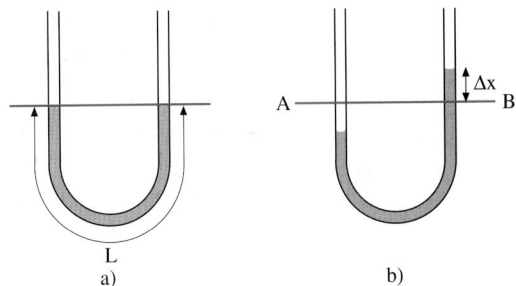

L
a)

b)

Figure 9.30

      ii) Write down an expression for the acceleration of the liquid column caused by this force.
      iii) Explain how this fulfils the condition for simple harmonic motion.
      iv) Write down an expression for the frequency of the oscillations.

4 A 'baby bouncer' consists of a harness attached to a rubber cord. A baby of mass 6.5 kg is placed gently in the harness and the cord extends by 0.40 m. The baby is then pulled down another 0.10 m and, when released, begins to move with simple harmonic motion. Calculate:

(a) the period of the motion,
(b) the maximum force on the baby.
(Acceleration of free fall $g = 9.8$ m s^{-2}.)

5 A simple model of a hydrogen molecule assumes that it consists of two oscillating hydrogen atoms connected by a spring of force constant 1.1×10^3 N m^{-1}.

(a) The mass of a hydrogen atom is 1.7×10^{-27} kg. Calculate the frequency of oscillation of the hydrogen molecule.
(b) Explain why light of wavelength 2.3 μm would be strongly absorbed by this model of the hydrogen molecule.

(Speed of light $c = 3.0 \times 10^8$ m s^{-1}.)

6 If you walk holding a full cup of coffee, you may find that, for a particular cup size and rate of taking steps, the coffee sloshes above the side of the cup, whereas if you walk more slowly, there is no risk of spillage. Explain this phenomenon, and estimate the critical combination of cup diameter and rate of taking steps.

7 As a party trick, the operatic tenor Enrico Caruso used to shatter a wine glass by singing a note of just the right frequency at full volume. If Caruso had been a physicist, what would he have understood by 'just the right frequency'? Why might he not have been successful if the glass had still contained some wine?

CHAPTER TEN
Fields

The aim of this chapter is to investigate the nature of electric and gravitational fields. We shall study the similarities and differences between these fields, and develop an appreciation of the fields in our physical world.

Electric forces hold electrons in atoms, and bind atoms together in molecules and in solids. Gravitational forces are responsible for the birth of stars in giant molecular clouds in space. Both involve fields, where forces act over a distance.

10.1 Electric charges

Some effects of static electricity are familiar in everyday life. For example, a balloon rubbed on a jumper will stick to a wall, a TV screen that has been polished attracts dust, dry hair crackles (and may actually spark!) when brushed, and you may feel a shock when you touch the metal door-handle of a car on getting out after a journey in dry weather. All these are examples of insulated objects that have gained an electric charge by friction, that is by being rubbed with other objects.

Insulators that are charged by friction will attract other objects. Some of the effects have been known for centuries. Greek scientists experimented with amber that was charged by rubbing it with fur. Our words *electron* and *electricity* come from the Greek word ηλεκτρον, which means amber. Today, electrostatics experiments are often carried out with plastic materials which are moisture-repellent and stay charged for longer.

Charging by friction can be hazardous. For example, a lorry which carries a bulk powder must be earthed before emptying its load. Otherwise, electric charge can build up on the tanker. This could lead to a spark from the tanker to earth, causing an explosion. Similarly, the pipes used for movement of highly inflammable liquids (for example, petrol), are metal-clad. An aircraft moving through the air will also become charged. To prevent the first person touching the aircraft on landing from becoming seriously injured, the tyres are made to conduct, so that the aircraft loses its charge on touchdown.

There are two kinds of electric charge. Polythene becomes negatively charged when rubbed with wool, and cellulose acetate becomes positively charged, also when rubbed with wool. To understand why this happens, we need to go back to the model of the atom. Atoms consist of a positively-charged nucleus with negatively-charged electrons orbiting it. When the polythene is rubbed with wool, friction causes some electrons to be transferred from the wool on to the polythene. So the polythene has a negative charge, and the wool is left with a positive charge. Cellulose acetate becomes positive because it loses some electrons to the wool when it is rubbed. Polythene and cellulose acetate are poorly-conducting plastics, and the charges remain static on the surface.

Putting two charged polythene rods close to each other, or two charged acetate rods close to each other, shows that similar charges repel each other. Conversely, unlike charges attract. A charged polythene rod attracts a charged acetate rod.

Figure 10.1 *Like charges repel.*

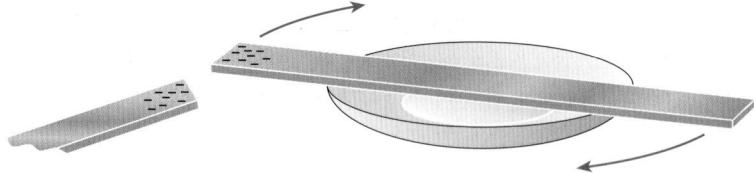

This is the basic law of force between electric charges:

> ★ like charges repel, unlike charges attract.

Charged rods will also attract uncharged objects. A charged polythene rod will pick up small pieces of paper. The presence of the negative charge on the rod causes a redistribution of the charges in the paper. Free electrons are repelled to the far side of the paper, so that the side nearest the rod is positive and is therefore attracted to the rod. The paper is said to be **charged by induction**. When the rod is removed, the free electrons will move back and cancel the positive charge.

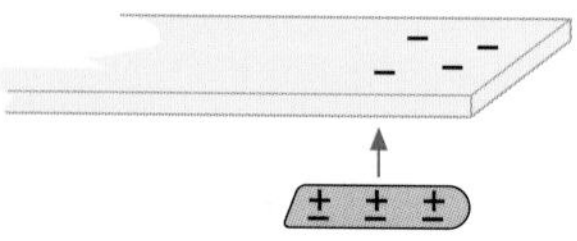

Figure 10.2 *A charged rod can induce charges in an uncharged object.*

10.1 Electric charges

Force between charges

In the late eighteenth century, the French scientist Charles Coulomb investigated the magnitude of the force between charges, and how this varies with the charges involved and the distance between them. He discovered that

> ★ the force is proportional to the product of the charges and inversely proportional to the square of the distance between them. This is known as **Coulomb's law**.

Coulomb's experiments made use of small, charged insulated spheres. Strictly speaking, the law applies to point charges, but it can be used for charged spheres provided that their radii are small compared with their separation.

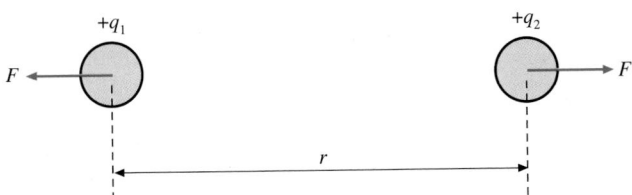

Figure 10.3

For point charges q_1 and q_2 a distance r apart (Figure 10.3), Coulomb's law gives the force F as

$$F \propto q_1 q_2 / r^2$$

giving

$$F = \frac{k q_1 q_2}{r^2}$$

where k is a constant of proportionality, the value of which depends on the insulating medium around the charges and the system of units employed. In the SI system of units, F is measured in newtons, q in coulombs and r in metres. Then the constant k is given as

$$k = \frac{1}{4 \pi \epsilon_o}$$

when the charges are in a vacuum. The quantity ϵ_o is called the **permittivity of free space** (or the permittivity of a vacuum).

The value of the permittivity of air is very close to that of a vacuum ($1.0005\epsilon_o$), so the equation can be used for the force between charges in a vacuum or in air.

The value of the permittivity of free space is given by

$$\epsilon_o = 8.85 \times 10^{-12} \, C^2 \, N^{-1} \, m^{-2}.$$

(We shall meet an alternative unit for ϵ_o later, that is $F \, m^{-1}$.) This numerical value for ϵ_o makes k equal to $8.99 \times 10^9 \, C^{-2} \, N \, m^2$.

Coulomb's law is often referred to as an **inverse square** law of force, because the variation of force with distance r between the charges is proportional to $1/r^2$. We shall meet another important inverse square law of force when we come to the gravitational force between two point masses.

Example

Calculate the force between two point charges, each of $1.0 \times 10^{-9} \, C$, which are 4.0 cm apart in a vacuum.

Using $F = kq_1q_2/r^2$, $F = 9.0 \times 10^9 \times (1.0 \times 10^{-9})^2/(4.0 \times 10^{-2})^2$
   $= \mathbf{5.6 \times 10^{-6} \, N}$.

Now it's your turn

Calculate the force between two electrons which are $1.0 \times 10^{-10} \, m$ apart in a vacuum.
(Electron charge $e = -1.6 \times 10^{-19} \, C$.)
Ans: $2.3 \times 10^{-8} \, N$

Electric fields

Electric charges exert forces on each other when they are a distance apart. The idea of an **electric field** is used to explain this action at a distance. An electric field is a region of space where a stationary charge experiences a force.

Electric fields are invisible, but they can be represented by electric lines of force, just as magnetic fields can be represented by magnetic lines of force (see Chapter 11). The direction of the electric field is defined as the direction in which a positive charge would move if it were free to do so. So the lines of force can be drawn with arrows that go from positive to negative.

For any electric field,

★ The lines of force start on a positive charge, and end on a negative charge.

★ The lines of force are smooth curves which never touch or cross.

★ The strength of the electric field is indicated by the closeness of the lines: the closer they are, the stronger the field.

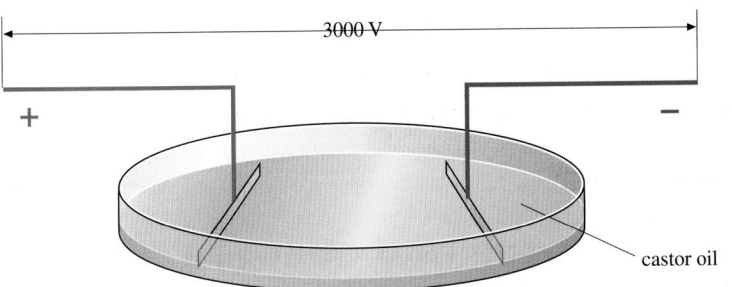

Figure 10.4 *Apparatus for investigating electric field patterns.*

The apparatus in Figure 10.4 can be used to show electric field patterns. Semolina is sprinkled on a non-conducting oil and a high voltage supply is connected to the conducting plates. The semolina becomes charged by induction and lines up along the lines of force. Some electric field patterns are illustrated in Figure 10.5. The pattern for the charged conducting sphere (Figure 10.5c) is of particular importance. Although there are no electric field lines inside the conductor, the field lines appear to come from the centre. Thus, the charge on the surface of a conducting sphere appears as if it is all concentrated at the centre. This means that small conducting spheres may be used as an approximation to point charges.

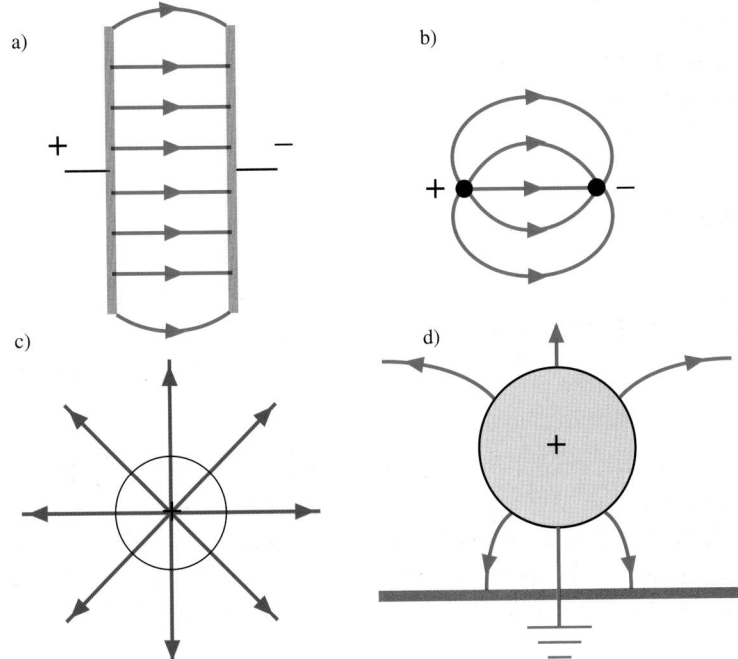

Figure 10.5 *Some electric field patterns.*

Paint spraying makes use of some of the principles of electrostatics. It can be wasteful if the paint is not sprayed where it is needed. But if the spray is given a charge, and the object to be painted is given an opposite charge, the charged droplets follow the lines of force and end up on the surface. This is illustrated in Figure 10.6.

Figure 10.6 *Paint spraying.*

Induced fields and charges

We have already stated that there can be no resultant electric field in a conductor. The reason for this is that electrons are free to move in the conductor. As soon as a charged body is placed near the conductor, the electric field of the charged body causes electrons in the conductor to move in the opposite direction to the field (because electrons have a negative charge). This is illustrated in Figure 10.7. These charges create a field in the opposite direction to the field due to the charged body. The induced charges will stop moving when the two fields are equal and opposite. Hence there is no resultant field in a conductor. This effect may be used to charge a body by induction.

Figure 10.7 *Induced charges.*

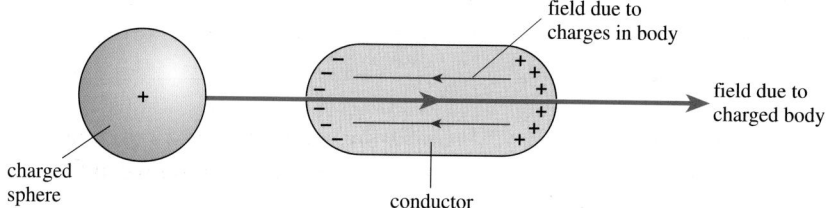

This process is illustrated in Figure 10.8. If a positively charged rod is placed near an insulated conductor, induced charges appear on the conductor, as shown in Figure 10.8a. If the conductor is now earthed, as in 10.8b, electrons move from earth to neutralise the induced positive charge. The electrons are held in position by the positively-charged rod. When the earth connection is removed, the negative charge is still held in position by the positively-charged rod, as in 10.8c. Removal of the charged rod means that the electrons will distribute themselves over the surface of the sphere, as in 10.8d. Note that if a positively-charged rod is used, the final charge on the sphere is negative, and *vice versa.*

Figure 10.8 *Charging by induction.*

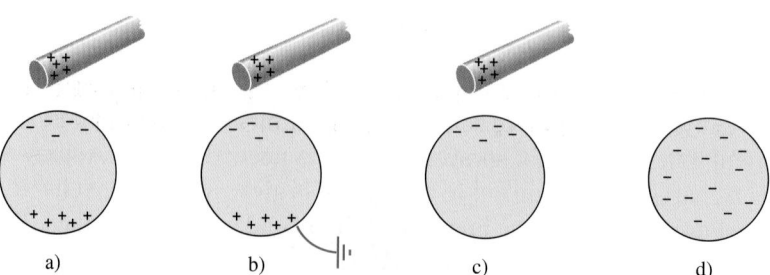

10.1 Electric charges

Electric field strength

> ★ The electric field strength at a point is defined as the force per unit charge acting on a small positive charge placed at that point.

If the force experienced by a positive test charge $+q$ placed in the field is F, the field strength E is given by

$$\star \quad E = \frac{F}{q}.$$

(Be careful not to be confused by the symbol E for field strength. E is also used for energy.) A unit of field strength can be deduced from this equation: force is measured in newtons and charge in coulombs, so the SI unit of field strength is $N\,C^{-1}$. We shall see later that there is another common SI unit for electric field strength, $V\,m^{-1}$. These two units are equivalent.

A uniform electric field

In Figure 10.5a, the electric field pattern between the parallel plates consists of parallel, equally-spaced lines, except near the edges of the plates. This means that the field between charged parallel plates (as, for example, in a capacitor) is uniform.

Figure 10.9 illustrates parallel plates a distance d apart with a potential difference V between them. A charge q in the uniform field between the plates has a force F acting on it. To move the charge towards the positive plate would require work to be done on the charge. Work is given by the product of force and distance. To move the charge from one plate to the other requires work W given by

$$W = Fd.$$

From the definition of potential difference,

$$W = Vq.$$

Thus $W = Fd = Vq$, or, rearranging,

$$F/q = V/d.$$

But F/q is the force per unit charge, and this is the definition of electric field strength. Thus, for a uniform field, the field strength E is given by

$$\star \quad E = \frac{V}{d}.$$

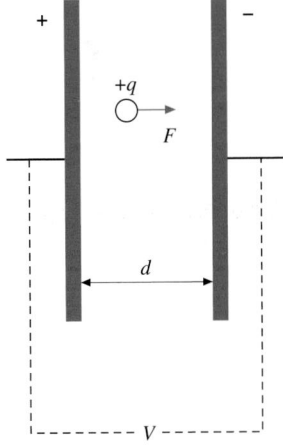

Figure 10.9 *Electric field between parallel plates.*

This equation gives an alternative unit for electric field strength (remember that we have already derived $N\,C^{-1}$ from the idea of force per unit charge). If we use the expression for a uniform field, the unit of potential difference is the volt and the unit of distance is the metre, so another unit for electric field strength is volts per metre $(V\,m^{-1})$. The two units $N\,C^{-1}$ and $V\,m^{-1}$ are equivalent.

Example

Two metal plates 5.0 cm apart have a potential difference of 1000 V between them. Calculate:
(a) the strength of the electric field between the plates,
(b) the force on a charge of 5.0 nC between the plates.

(a) From $E = V/d$, $E = 1000/5.0 \times 10^{-2} = \mathbf{2.0 \times 10^4\,V\,m^{-1}}$.
(b) From $F = EQ$, $F = 2.0 \times 10^4 \times 5.0 \times 10^{-9} = \mathbf{1.0 \times 10^{-4}\,N}$.

Now it's your turn

Two metal plates 15 mm apart have a potential difference of 750 V between them. The force on a small charged sphere placed between the plates is $1.2 \times 10^{-7}\,N$. Calculate:
(a) the strength of the electric field between the plates,
(b) the charge on the sphere.
Ans: (a) $5.0 \times 10^4\,V\,m^{-1}$; (b) $2.4 \times 10^{-12}\,C$

Field strength due to an isolated point charge

The electric field due to an isolated point charge is **radial** (see Figure 10.5c). We have to mention that the point charge is isolated. If any other object, charged or otherwise, is near it, the field could be distorted.

From Coulomb's law, the force on a test charge q a distance r from the isolated point charge Q is given by

$$F = kQq/r^2.$$

The electric field E is given by $E = F/q$. Thus, the electric field due to the isolated point charge is

$$\bigstar \quad E = \frac{kQ}{r^2}.$$

This is an inverse square law field of force because the force varies in proportion to $1/r^2$.

10.1 Electric charges

Example

In a simplified model of the hydrogen atom, the electron is at a distance of 5.3×10^{-11} m from the proton. The proton charge is $+1.6 \times 10^{-19}$ C. Calculate the electric field strength of the proton at this distance.

Assuming that the field is radial,
$E = kQ/r^2 = 9.0 \times 10^9 \times 1.6 \times 10^{-19}/(5.3 \times 10^{-11})^2 = \mathbf{5.1 \times 10^{11} \ N \ C^{-1}}$.

Now it's your turn

A Van de Graaff generator has a dome of radius 15 cm. The dome carries a charge of 2.5×10^{-6} C. It can be assumed that this charge acts as if it were all concentrated at the centre of the spherical dome. Calculate the electric field strength at the dome's surface.
Ans: $1.0 \times 10^6 \ N \ C^{-1}$

Section 10.1 summary

★ Insulators may be charged by friction.

★ Like charges repel; unlike charges attract each other.

★ When charged objects are placed near conductors, they cause a redistribution of charge in the conductor, thereby inducing charges.

★ The force between two point charges is proportional to the product of the charges and inversely proportional to the square of the distance between them. This is Coulomb's law:
$F = kq_1q_2/r^2$, where $k = 1/4\pi\epsilon_0$.

★ ϵ_0 is the permittivity of free space:
$\epsilon_0 = 8.85 \times 10^{-12} \ C^2 \ N^{-1} \ m^{-2} (F \ m^{-1})$.

★ An electric field is a region of space where a stationary charge experiences a force.

★ Electric field strength is the force per unit charge on a positive test charge: $E = F/q$.

★ The electric field between parallel plates is uniform. The field strength is given by $E = V/d$.

★ The electric field strength at a point in the field of an isolated point charge is given by $E = kQ/r^2$.

Section 10.1 questions

1 A glass rod rubbed with silk becomes positively charged. Explain what has happened to the glass. Explain also why the charged rod is able to attract small pieces of paper.

2 **(a)** Explain what is meant by an electric field.
   (b) Sketch the electric field patterns
      i) between two negatively-charged particles,
      ii) between a positive charge and a negatively-charged metal plate.

3 In a simplified model, a uranium nucleus is a sphere of radius 8.0×10^{-15} m. The nucleus contains 92 protons (and rather more neutrons). The charge on a proton is 1.6×10^{-19} C. It can be assumed that the charge of these protons acts as if it were all concentrated at the centre of the nucleus. The nucleus releases an α-particle containing two protons (and two neutrons) at the surface of the nucleus. Calculate:
   (a) the electric field strength at the surface of the nucleus,
   (b) the electric force on the α-particle.

10.2 Capacitance

Figure 10.10 *Charged spherical conductor.*

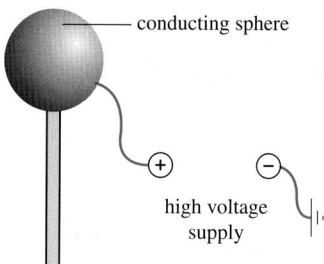

conducting sphere

high voltage supply

Consider an isolated spherical conductor connected to a high voltage supply (Figure 10.10). If it is found that, as the potential of the sphere is increased, then the charge stored on the sphere also increases. The graph showing the variation of charge on the conductor with potential is shown in Figure 10.11. It can be seen that charge q is related to potential V by

$$q \propto V.$$

Hence,

★ $q = CV,$

where C is a constant which depends on the size of the conductor. C is known as the capacitance of the conductor.

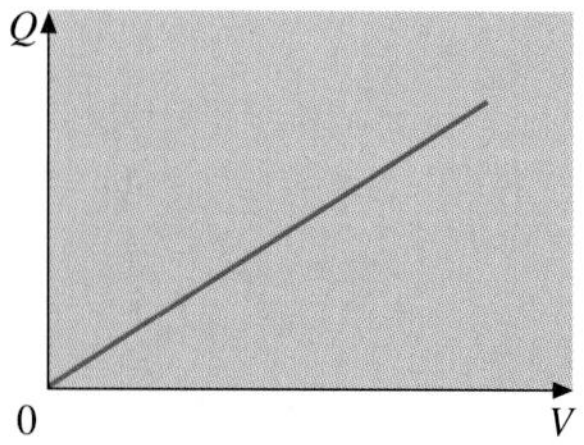

Figure 10.11 *Relation between charge and potential.*

★ Capacitance is the ratio of charge to potential for a conductor.

Another chance for confusion! The letter C is used as an abbreviation for the unit of charge, the coulomb; as an italic letter C it is used as the symbol for capacitance.

The unit of capacitance is the farad (symbol F). One farad is one coulomb per volt. The farad is an inconveniently large unit. In electronic circuits and laboratory experiments, the range of useful values of capacitance is from about 10^{-12} F (1 picofarad, or 1 pF) to 10^{-3} F (1 millifarad, or 1 mF). (See Chapter 1 for a list of decimal multiples and submultiples for use with units.)

Circuit components which store charge and therefore have capacitance are called **capacitors**.

Example

Calculate the charge stored on an isolated conductor of capacitance 470 μF when it is at a potential of 20 V.

Using $C = q/V$, we have $q = CV = 470 \times 10^{-6} \times 20 = \mathbf{9.4 \times 10^{-3}}$ **C**.

Now it's your turn

The charge on a certain isolated conductor is 2.5×10^{-2} C when its potential is 25 V. Calculate its capacitance.

Ans: 1.0 mF

Energy stored in a capacitor

When charging a capacitor, work is done by the battery to move charge on to the capacitor. Energy is transferred from the power supply and is stored as **electric potential energy** in the capacitor.

Camera flash units use a capacitor to store energy. The capacitor takes a few seconds to charge when connected to the battery in the camera. Then the energy is discharged very rapidly when the capacitor is connected to the flash-bulb to give a short but intense flash.

Since $q = CV$, the charge stored on a capacitor is directly proportional to its potential (see Figure 10.11).

From the definition of potential (see Chapter 6), the work done (and therefore the energy transferred) is the product of the potential and the charge. That is,

$$W \text{ (and } E_p) = Vq.$$

CHAPTER 10

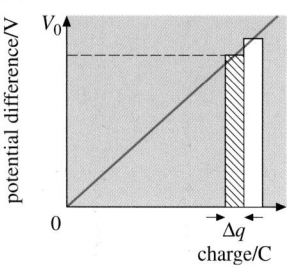

Figure 10.12 *Graph of potential against charge for a capacitor.*

However, while more and more charge is transferred to the capacitor, the potential difference is increasing. Suppose the potential is V_0 when the charge stored is q_0. When a further small amount of charge Δq is supplied, the energy transferred is given by

$$\Delta E_\mathrm{p} = V_0 \Delta q,$$

which is equal to the area of the shaded strip in Figure 10.12. Similarly, the energy transferred when a further charge Δq is transferred is given by the area of the next strip, and so on. If the amount of charge Δq is very small, the strips become very thin and their combined areas are just equal to the area between the line and the horizontal axis. Thus,

> the energy transferred from the battery when a capacitor is charged is given by the area under the graph line.

Because the graph is a straight line through the origin, this area is just the area of the right-angled triangle formed by the line and the q-axis. Thus,

$$E_\mathrm{p} = \tfrac{1}{2}qV.$$

This is the expression for the energy transferred from the battery in charging the capacitor. This is electric potential energy, and it is released when the capacitor is discharged.

Since $C = q/V$, this expression can be written in different forms:

$$\bigstar \quad E_\mathrm{p} = \tfrac{1}{2}qV = \tfrac{1}{2}CV^2 = \frac{q^2}{2C}.$$

Example

Calculate the energy stored by a 470 µF capacitor at a potential of 20 V.

Using $E_\mathrm{p} = \tfrac{1}{2}CV^2$, $E_\mathrm{p} = \tfrac{1}{2} \times 470 \times 10^{-6} \times (20)^2 = \mathbf{0.094\ J}$.

Now it's your turn

A camera flash-lamp uses a 5000 µF capacitor which is charged by a 9 V battery. Calculate the energy transferred when the capacitor is fully discharged through the lamp.

Ans: 0.203 J

10.2 Capacitance

Capacitors

The simplest capacitor in an electric circuit consists of two metal plates, with an air gap between them which acts as an insulator. This is called a parallel-plate capacitor. Figure 10.13 shows the circuit symbol for a capacitor. When the plates are connected to a battery, the battery transfers electrons from the plate connected to the positive terminal of the battery to the plate connected to the negative terminal. Thus the plates carry equal but opposite charges.

The **capacitance** of a capacitor is defined as the charge stored per unit potential difference between the plates.

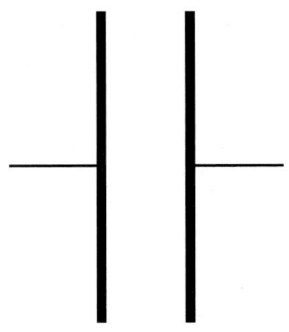

Figure 10.13 *Circuit symbol for a capacitor.*

The capacitance of an air-filled capacitor can be increased by putting an insulating material, such as mica or waxed paper, between the plates. The material between the plates is called the **dielectric**. In a type of capacitor known as an **electrolytic** capacitor the dielectric is deposited by an electrochemical reaction. These capacitors *must* be connected with the correct polarity for their plates, or they will be damaged. The circuit symbol for an electrolytic capacitor is shown in Figure 10.14. Electrolytic capacitors are available with capacitances up to about 1 mF.

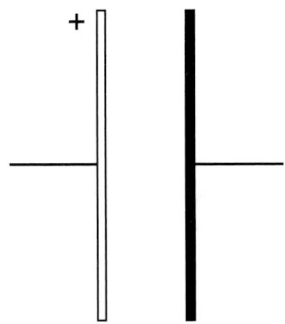

Figure 10.14 *Circuit symbol for an electrolytic capacitor.*

Factors affecting capacitance

As stated previously, the material used as a dielectric affects the capacitance of a capacitor. The other factors determining the capacitance are the area of the plates and the distance between them. Experiment shows that

 the capacitance is directly proportional to the area A of the plates, and inversely proportional to the distance d between them.

Putting these two factors together gives

$$C \propto \frac{A}{d}.$$

For a capacitor with air or a vacuum between the plates, the constant of proportionality is the permittivity of free space. Thus

 $$C = \frac{\epsilon_0 A}{d}.$$

Since C is measured in farads, A in square metres and d in metres, we can see that an alternative unit for ϵ_0 is farads per metre, $\mathrm{F\,m^{-1}}$ (see Section 10.1). We introduce a quantity called the **relative permittivity** ϵ_r of a dielectric to account for the fact that the use of a dielectric increases the capacitance.

★ The relative permittivity is defined as the capacitance of a parallel-plate capacitor with the dielectric between the plates divided by the capacitance of the same capacitor with a vacuum between the plates.

ϵ_r is a ratio, and has no units. Some values of relative permittivity are given in Figure 10.15.

Figure 10.15

Material	relative permittivity E_r
air	1.0005
polyethylene (polythene)	2.3
sulphur	4
paraffin oil	4.7
mica	6
barium titanate	1200

Including the relative permittivity factor, the full expression for the capacitance of a parallel-plate capacitor is

$$★ \quad C = \frac{\epsilon_0 \epsilon_r A}{d}.$$

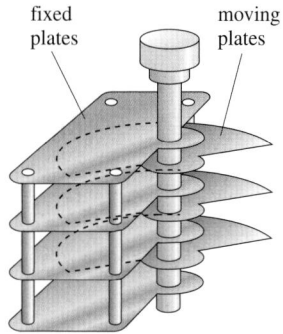

fixed plates   moving plates

Figure 10.16 *A variable capacitor.*

Variable capacitors (Figure 10.16) have one set of plates mounted on a spindle, so that the area of overlap can be changed. Varying the capacitance of a tuning circuit is one way of tuning in to different radio or TV stations.

Example

A parallel-plate, air-filled capacitor has square plates of side 30 cm a distance 1.0 mm apart. Calculate the capacitance of the capacitor.

Using $C = \epsilon_0 \epsilon_r A/d$, $C = 8.85 \times 10^{-12} \times 1 \times (30 \times 10^{-2})^2 / 1.0 \times 10^{-3}$
$= \mathbf{8.0 \times 10^{-10}}$ **F**.

Now it's your turn

A capacitor consists of two metal discs of diameter 15 cm separated by a sheet of polythene 0.25 mm thick. The relative permittivity of polythene is 2.3. Calculate the capacitance of the capacitor.
Ans: 1.4×10^{-9} F

Capacitors in series and parallel

Figure 10.17 *Capacitors in series.*

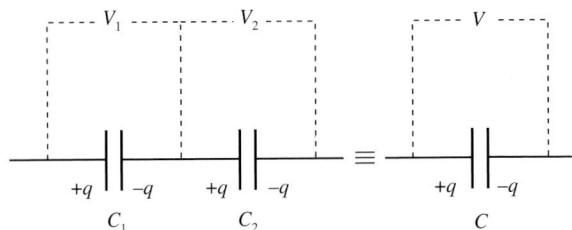

In Figure 10.17, the two capacitors of capacitance C_1 and C_2 are connected in series. We shall show that the combined capacitance C is given by

$$\star \quad \frac{1}{C} = \frac{1}{C_1} + \frac{1}{C_2}.$$

If the voltage across the equivalent capacitor is V and the charge stored on each plate is q, then $V = q/C$. The potential difference across the combination is the sum of the potential differences across the individual capacitors, $V = V_1 + V_2$, and each capacitor has charge q on each plate. A charge of $+q$ induced on one plate of one capacitor will induce a charge of $-q$ on the plate of the second capacitor.

Since $V_1 = q/C_1$ and $V_2 = q/C_2$, then

$$q/C = q/C_1 + q/C_2.$$

Dividing each side of the equation by q, we have

$$1/C = 1/C_1 + 1/C_2.$$

A similar result applies for any number of capacitors connected in series.

> ★ The reciprocal of the combined capacitance equals the sum of the reciprocals of the individual capacitances.

Note that:

- for two identical capacitors in series, the combined capacitance is equal to half of the value of each one

- for capacitors in series, the combined capacitance is always less than the value of the smallest individual capacitance

Figure 10.18 *Capacitors in parallel.*

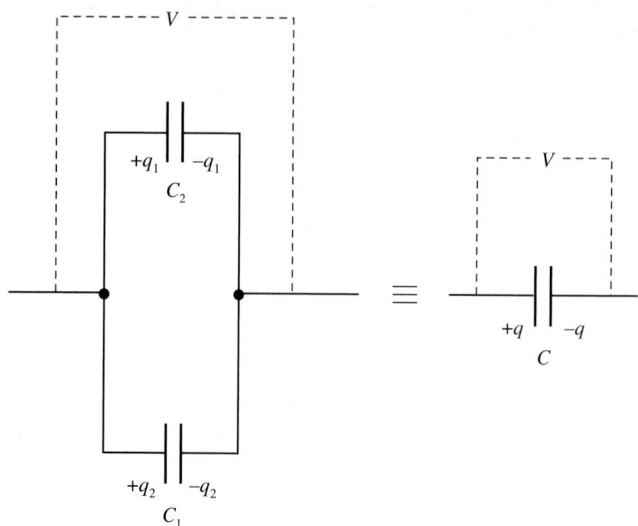

In Figure 10.18, the two capacitors of capacitance C_1 and C_2 are connected in parallel. We shall show that the combined capacitance C is given by

$$\bigstar \quad C = C_1 + C_2.$$

If the voltage across the equivalent capacitor is V and the charge stored on each plate is q, then $q = CV$. The total charge stored is the sum of the charges on the individual capacitors, $q = q_1 + q_2$, and there is the same potential difference V across each capacitor since they are connected in parallel.

Since $q_1 = C_1V$ and $q_2 = C_2V$, then

$$CV = C_1V + C_2V.$$

Dividing each side of the equation by V, we have

$$C = C_1 + C_2.$$

The same result applies for any number of capacitors connected in parallel.

> $\bigstar$ The combined capacitance equals the sum of all the individual capacitances.

Note that the equation for capacitors in *series* is similar to the equation for resistors in *parallel*, and vice versa (see Chapter 6).

10.2 Capacitance

Example

1   A 100 µF capacitor in parallel with a 50 µF capacitor is connected to a 12 V supply. Calculate:
   (a) the total capacitance,
   (b) the potential difference across each capacitor,
   (c) the charge stored on each capacitor.

   (a) Using the equation for capacitors in parallel,
       $C = C_1 + C_2 = 100 + 50 = $ **150 µF**.
   (b) The potential difference across each capacitor is the same as the potential difference across the supply. This is **12 V**.
   (c) Using $Q = CV$, the charge stored on the 100 µF capacitor is $100 \times 10^{-6} \times 12 = $ **1.2×10^{-3} C**. The charge stored on the 50 µF capacitor is $50 \times 10^{-6} \times 12 = $ **6.0×10^{-4} C**.

2   A 100 µF capacitor in series with a 50 µF capacitor is connected to a 12 V supply. Calculate:
   (a) the combined capacitance,
   (b) the charge stored by each capacitor,
   (c) the potential difference across each capacitor.

   (a) Using the equation for capacitors in series,
       $1/C = 1/C_1 + 1/C_2 = 1/100 \times 10^{-6} + 1/50 \times 10^{-6} = 3 \times 10^4$. Thus $C = 3.3 \times 10^{-5} = $ **33 µF**.
   (b) The charge stored by each capacitor is the same as the charge stored by the combination, so
       $q = CV = 33 \times 10^{-6} \times 12 = $ **4.0×10^{-4} C**.
   (c) Using $V = q/C$, the potential difference across the 100 µF capacitor is $4.0 \times 10^{-4}/100 \times 10^{-6} = $ **4.0 V**. The potential difference across the 50 µF capacitor is $4.0 \times 10^{-4}/50 \times 10^{-6} = $ **8.0 V**. Note that the two potential differences add up to the supply voltage.

Now it's your turn

(a) A 470 µF capacitor is connected to a 20 V supply. Calculate the charge stored on the capacitor.
(b) The capacitor in (a) is then connected to an uncharged 470 µF capacitor.
   i)   Explain why the capacitors are in parallel, rather than series, and why the total charge stored by the combination must be the same as the answer to (a).
   ii) Calculate the capacitance of the combination.
   iii) Calculate the potential difference across each capacitor.
   iv) Calculate the charge stored on each capacitor.
   Ans: (a) 9.4×10^{-3} C; (b) 940 µF, 10 V, 4.7×10^{-3} C

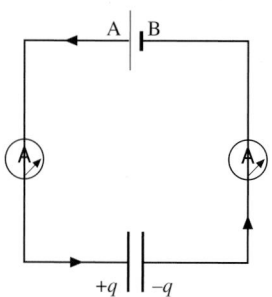

Figure 10.19 *Charging a capacitor.*

Charging capacitors

The circuit in Figure 10.19 can be used to investigate what happens when a capacitor is connected to a battery. Start with the lead to battery terminal A connected, but that to B disconnected. When connection is first made to terminal B, both ammeters flick to the right and then return to zero, indicating a momentary pulse of current. If the process is repeated, nothing further happens. The capacitor has become charged.

During the initial connection, the ammeters have similar deflections. This tells us that the same charge has moved on to the left-hand plate that has been removed from the right-hand plate. The momentary current has caused the left-hand plate to acquire a charge of $+q$, while the right-hand plate has a charge of $-q$. The capacitor is now said to store a charge of q (although if we add up the charges on the left- and right-hand plates taking account of their sign, the net charge is zero). There can be no steady current because of the gap between the capacitor plates. The current stops when the potential difference across the capacitor is the same as the e.m.f. of the battery.

Discharging capacitors

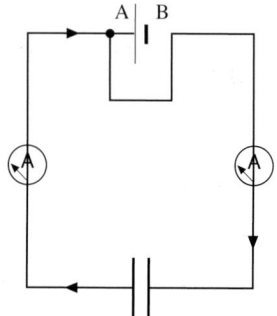

Figure 10.20 *Discharging a capacitor.*

When the battery lead to terminal B is disconnected and joined to point A (Figure 10.20) so that the battery is no longer in circuit, both ammeters give similar momentary deflections to the left. This time a current in the opposite direction has moved the charge of $+q$ from the left-hand plate to cancel the charge of $-q$ on the right-hand plate. The capacitor has become discharged.

Remember that in metal wires the current is carried by free electrons. These move in the opposite direction to that of the conventional current (see Chapter 6). When the capacitor is charged, electrons move from the negative terminal of the battery to the right-hand plate of the capacitor, and from the left-hand plate to the positive terminal of the battery. When the capacitor is discharged, electrons flow from the negative right-hand plate of the capacitor to the positive left-hand plate.

The experiment described using the circuit in Figure 10.20 showed that there is a momentary current when a capacitor discharges. A resistor connected in series with the capacitor will reduce the current, so that the capacitor discharges more slowly.

The circuit shown in Figure 10.21 can be used to investigate more precisely how a capacitor discharges. When the two-way switch is connected to point A, the capacitor will charge up until the potential difference between its plates is equal to the e.m.f. V of the supply. When the switch is moved to B, the capacitor will discharge through the resistor. When the switch makes contact with B, the current can be recorded at regular intervals of time as the capacitor discharges.

10.2 Capacitance

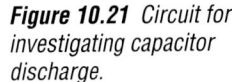
Figure 10.21 *Circuit for investigating capacitor discharge.*

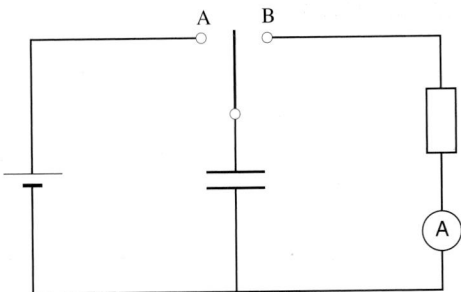

A graph of the discharge current against time is shown in Figure 10.22. The current is seen to change rapidly at first, and then more slowly. More detailed analysis shows that the curve is **exponential**. We shall meet exponential changes again when we deal with the decay of radioactive substances (Chapter 12).

All exponential decay curves have an equation of the form

$$\bigstar \quad x = x_0 e^{-kt},$$

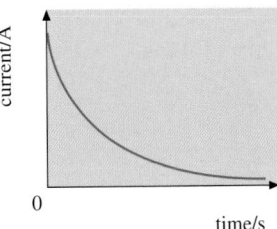

Figure 10.22 *Graph of current against time for a capacitor discharge.*

where x is the quantity which is decaying (and x_0 is the value of x at $t = 0$), e, to three decimal places is the number 2.718 (the root of natural logarithms) and k is a constant characteristic of the decay. A large value of k means that the decay is rapid, and a small value means a slow decay.

The solution for the discharge of a capacitor of capacitance C through a resistor of resistance R is of the form

$$\bigstar \quad q = q_0 e^{-t/CR}.$$

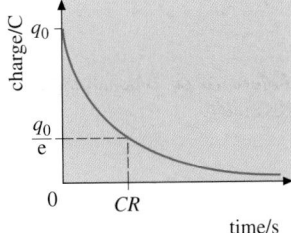

Figure 10.23 *Graph of charge against time for capacitor discharge.*

This is the equation of the graph in Figure 10.23. Here q_0 is the charge on the plates at time $t = 0$, and q is the charge at time t.

The graphs of Figures 10.22 and 10.23 have exactly the same shape, and thus the equation for the discharge of a capacitor may be written

$$\bigstar \quad I = I_0 e^{-t/CR}.$$

Furthermore, since for a capacitor, q is proportional to V, then the solution for the discharge can be written as

$$\bigstar \quad V = V_0 e^{-t/CR}.$$

10.2 Capacitance

Time constant

As time progresses, the exponential curve in Figure 10.23 gets closer and closer to the time axis, but never actually meets it. Thus, it is not possible to quote a time for the capacitor to discharge completely.

However, the quantity CR in the decay equation may be used to give an indication of whether the decay is fast or slow, as shown in Figure 10.24.

Figure 10.24 Decay curves for large and small time constants.

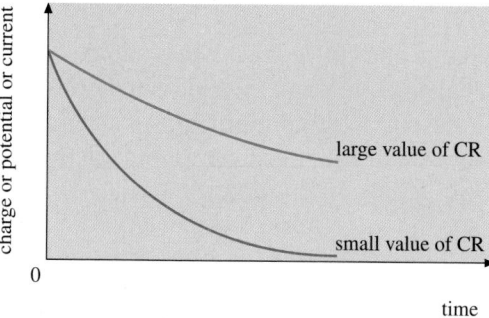

CR is called the **time constant** of the capacitor-resistor circuit.

CR has the units of time, and is measured in seconds. Its symbol is τ (Greek letter tau).

We can easily show that CR has units of time. From $C = q/V$ and $R = V/I$, then $CR = q/I$. Since charge q is in coulombs and current I is in amperes, and one ampere is equal to one coulomb per second, q/I is in seconds.

To find the charge q on the capacitor plates after a time $t = CR$, we substitute in the exponential decay equation.

$$q = q_0 e^{-CR/CR} = q_0 e^{-1} = q_0/e = q_0/2.718$$

Thus

the time constant is the time for the charge to have decreased to $1/e$ (or $1/2.718$) of its initial charge.

In one time constant the charge stored by the capacitor drops to roughly one-third of its initial value. During the next time constant it will drop by the same ratio, to about one-ninth of the value at the beginning of the decay.

Example

A 500 µF capacitor is connected to a 10 V supply, and is then discharged through a 100 kΩ resistor. Calculate:
(a) the initial charge stored by the capacitor,
(b) the initial discharge current,
(c) the value of the time constant,
(d) the charge on the plates after 100 s,
(e) the time at which the remaining charge is 2.5×10^{-3} C.

(a) From $Q = CV$, we have $Q = 500 \times 10^{-6} \times 10 = \mathbf{5.0 \times 10^{-3}}$ **C**.
(b) From $I = V/R$, we have $I = 10/(100 \times 10^{3}) = \mathbf{1.0 \times 10^{-4}}$ **A**.
(c) From $\tau = CR$, we have $\tau = 500 \times 10^{-6} \times 100 \times 10^{3} = \mathbf{50\ s}$.
(d) After 50 s, the charge on the plates is
$Q_0/e = 5 \times 10^{-3}/2.718 = 1.8 \times 10^{-3}$ C; after another 50 s, the charge is
$1.8 \times 10^{-3}/2.718 = \mathbf{6.8 \times 10^{-4}}$ **C**.
(e) Using $Q = Q_0 e^{-t/CR}$, $2.5 \times 10^{-3} = 5.0 \times 10^{-3} e^{-t/50}$, or $0.50 = e^{-t/50}$.
Taking natural logarithms of both sides, $-0.693 = -t/50$, or $t = \mathbf{35\ s}$.

Now it's your turn

A 5.0 µF capacitor is charged from a 12 V battery, and is then discharged through a 0.50 MΩ resistor. Calculate:
(a) the initial charge on the capacitor,
(b) the charge on the capacitor 2.0 s after the discharge starts,
(c) the potential difference across the capacitor at this time.
Ans: (a) 6.0×10^{-5} C; (b) 2.7×10^{-5} C; (c) 5.4 V

Section 10.2 summary

★ A capacitor stores charge. Its capacitance C is given by $C = Q/V$, where Q is the charge on the capacitor when there is a potential difference V between its plates.

★ The unit of capacitance, the farad (F), is one coulomb per volt.

★ The equivalent capacitance C of capacitors connected in series is given by $1/C = 1/C_1 + 1/C_2 + \ldots$

★ The equivalent capacitance C of capacitors connected in parallel is given by $C = C_1 + C_2 + \ldots$

★ The energy stored in a charged capacitor is given by $E = \frac{1}{2}QV = \frac{1}{2}CV^2 = \frac{1}{2}Q^2/C$.

★ When a charged capacitor discharges, the charge on the plates decays exponentially. The equation for the decay is $Q = Q_0 e^{-t/CR}$.

★ The time constant of the circuit is given by $\tau = CR$. The time constant is the time for the charge to decay to $1/e$ of its initial value.

Section 10.2 questions

1   A parallel-plate capacitor has rectangular plates of side 200 mm by 30 mm. The plates are separated by an air gap 0.50 mm thick.
   (a) Calculate the capacitance of the capacitor.
   (b) The capacitor is connected to a 12 V battery. Calculate:
      i)   the charge on each plate,
      ii) the electric field between the plates.

2   Figure 10.25 shows an arrangement of capacitors.
   (a) Calculate the capacitance of this arrangement.
   (b) The 4 µF capacitor is disconnected. Calculate the new capacitance.

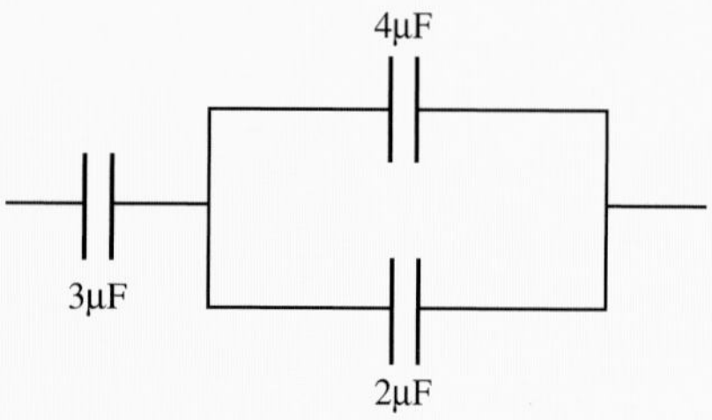

Figure 10.25

3   A 10 µF capacitor is charged from a 9.0 V battery.
   (a) Calculate:
      i)   the electric potential energy stored by the capacitor,
      ii) the charge stored by the capacitor.
   (b) The charged capacitor is discharged through a 150 kΩ resistor. Calculate:
      i)   the initial discharge current,
      ii) the time constant,
      iii) the time taken for the current to fall to 3.6×10^{-5} A.

10.3 Gravitational fields

Force between masses

We are familiar with the fact that the Earth's force of gravity is responsible for our weight, the force which pulls us towards the Earth. Isaac Newton concluded that the Earth's force of gravity is also responsible for keeping the Moon in orbit. His calculations showed that the Earth's force of gravity extends into space, but weakens with distance according to an inverse square law. That is, the Earth's force of gravity varies inversely with the square of the distance from the centre of the Earth. If you go two Earth radii from the Earth's centre, the force is a quarter of the force on the Earth's surface. This is part of Newton's **law of gravitation**.

★ Newton's law states that two point masses attract each other with a force that is proportional to the product of their masses and inversely proportional to the square of their separation.

Hence, if F is the force of attraction between two bodies of mass m_1 and m_2 respectively with distance r between their centres, then

$$F \propto m_1 m_2 / r^2$$

or

★ $$F = \frac{G m_1 m_2}{r^2},$$

where the constant of proportionality G is called the gravitational constant.

★ The value of G is 6.67×10^{-11} N m^2 kg^{-2}.

Notice that this equation has a similar form to that for Coulomb's law between two charges. Both Newton's law and Coulomb's law are inverse square laws of force. However, there are important differences.

- The electric force acts on charges, whereas the gravitational force acts on masses.

- The electric force can be attractive or repulsive, depending on the signs of the interacting charges, whereas two masses always attract each other.

It is possible to measure the gravitational constant G in a school laboratory, but the force of gravity between laboratory-sized masses is so small that it is not easy to obtain an accurate result.

Gravitational field strength

★ The gravitational field strength at a point is defined as the force per unit mass acting on a mass placed at that point.

This continues the analogy between electric and gravitational fields. Remember that electric field strength is defined as the force per unit charge.

The force per unit mass is also a measure of the acceleration of free fall, by Newton's second law of motion $F = ma$ (see Chapter 4). Thus, the gravitational field strength is given by $F/m = g$, where F is in newtons and m is in kilograms. This means that the gravitational field strength at the Earth's surface is about 9.8 N kg^{-1}. The unit N kg^{-1} is equivalent to the unit of acceleration, m s^{-2}. (Remember that we had two equivalent units for electric field strength, N C^{-1} and V m^{-1}. Although there is a clear *analogy* between N kg^{-1} and N C^{-1}, there is *no direct link* between m s^{-2} and V m$^{-1.}$)

Field strength due to an isolated point mass

We have already met the idea of a point mass in talking about the motion of a particle (Chapter 2). Of course, all the masses we come across in the laboratory have a finite size. But for calculations involving gravitational forces, it is fortunate that a spherical mass behaves as if it were a point mass at the centre of the sphere, with all the mass of the sphere concentrated at that point. This is very similar to the idea that the charge on a conducting sphere can be considered to be concentrated at the centre of the sphere.

The gravitational field outside a spherical mass is a radial one, just as the field associated with a point charge is radial. From Newton's law, the attractive force on a mass m caused by another mass M, with a distance of r between their centres, is given by

$$F = \frac{GMm}{r^2}.$$

This means that the force per unit mass or gravitational field strength g is given by

$$\bigstar \quad g = \frac{F}{m} = \frac{GM}{r^2}.$$

The field strengths due to masses that you find in a laboratory are tiny. For example, the field strength one metre away from an isolated mass of one kilogram is only 7×10^{-11} N kg^{-1}. But field strengths due to the masses of objects such as the Earth or Moon are much larger. We already know that the field strength due to the Earth at the surface of the Earth is about 10 N kg^{-1}. We can use this information to deduce information about the Earth, for example the mass of the Earth. Look at the example that follows!

Example

The radius of the Earth is 6.4×10^6 m and the gravitational field strength at its surface is 9.8 N kg^{-1}.
(a) Assuming that the field is radial, calculate the mass of the Earth.
(b) The radius of the Moon's orbit about the Earth is 3.8×10^8 m. Calculate the strength of the Earth's gravitational field at this distance.

10.3 Gravitational fields

(c) The mass of the Moon is 7.4×10^{22} kg. Calculate the gravitational attraction between the Earth and the Moon. (Gravitational constant $G = 6.7 \times 10^{-11}$ N m^2 kg^{-2}.)

(a) Using $g = GM/r^2$, we have
$M = gr^2/G = 9.8 \times (6.4 \times 10^6)^2/6.7 \times 10^{-11} = \mathbf{6.0 \times 10^{24}\ kg}$.

(b) Using $g = GM/r^2$, we have
$g = 6.7 \times 10^{-11} \times 6.0 \times 10^{24}/(3.8 \times 10^8)^2 = \mathbf{2.8 \times 10^{-3}\ N\ kg^{-1}}$.

(c) Using $F = GMm/r^2$, we have
$F = 6.7 \times 10^{-11} \times 6.0 \times 10^{24} \times 7.4 \times 10^{22}/(3.8 \times 10^8)^2 = \mathbf{2.1 \times 10^{20}\ N}$.

Now it's your turn

The mass of Jupiter is 1.9×10^{27} kg and its radius is 7.1×10^7 m. Calculate the gravitational field strength at the surface of Jupiter. (Gravitational constant $G = 6.7 \times 10^{-11}$ N m^2 kg^{-2}.)

Ans: 25 N kg^{-1}

Mass and weight

In Chapter 4, mass was said to be a measure of the inertia of a body to changes in velocity. Unless the body is travelling at speeds close to that of light, its mass is constant.

★ In a gravitational field, by definition, there is a force acting on the mass equal to the product of mass and gravitational field strength. This force is called the **weight**.

For an object of mass m in a gravitational field of strength g, the weight W is given by

$$weight = mass \times gravitational\ field\ strength$$

or

★ $W = mg$.

Although mass is invariant, weight depends on gravitational field strength. For example, a person of mass 60 kg has a weight of 600 N on Earth, but only 100 N on the Moon, although their mass is still 60 kg.

Planet and satellite orbits

Most planets in the Solar system have orbits which are nearly circular. We now bring together the idea of a gravitational force and that of a centripetal force (see Chapter 9) to derive a relation between the period and the radius of the orbit of a planet describing a circular path about the Sun, or a satellite moving round the Earth or another planet.

Consider a planet of mass m in circular orbit about the Sun, of mass M. If the radius of the orbit is r, the gravitational force F_{grav} between Sun and planet is

$$F_{grav} = GMm/r^2,$$

by Newton's law of gravitation. It is this force that provides the centripetal force as the planet moves in its orbit. The centripetal force F_{circ} is given by

$$F_{circ} = mv^2/r,$$

where v is the linear speed of the planet. As has just been stated,

$$F_{grav} = F_{circ}$$

Thus,

$$GMm/r^2 = mv^2/r.$$

The period T of the planet in its orbit is the time required for the planet to travel a distance $2\pi r$. It is moving at speed v, so

$$v = 2\pi r/T.$$

Putting this into the equation above, we have

$$GMm/r^2 = m(4\pi^2 r^2/T^2)/r,$$

or, simplifying,

$$T^2 = (4\pi^2/GM)r^3.$$

Another way of writing this is

$$\bigstar \quad T^2/r^3 = 4\pi^2/GM.$$

Look at the right-hand side of this equation. The quantities π and G are constants. If we are considering the relation between T and r for planets in the Solar system, then M is the same for each planet because it is the mass of the Sun. This equation shows that,

> $\bigstar$ for planets or satellites describing circular orbits about the same central body, the square of the period is proportional to the cube of the radius of the orbit.

10.3 Gravitational fields

This relation is known as **Kepler's third law of planetary motion**. Johannes Kepler (1571–1630) analysed data collected by Tycho Brahe (1546–1601) on planetary observations. He showed that the observations fitted a law of the form $T^2 \propto r^3$. Fifty years later Newton showed that an inverse square law of gravitation, the idea of centripetal acceleration, and the second law of motion, gave an expression of exactly the same form. Newton cited Kepler's law in support of his law of gravitation. In fact, the orbits of the planets are not circular, but elliptical, a fact recognised by Kepler. The derivation is simpler for the case of a circular orbit.

Figure 10.26 collects information about T and r for planets of the solar system. The last column shows that the value of T^2/r^3 is indeed a constant. Moreover, the value of T^2/r^3 agrees very well with the value of $4\pi^2/GM$, $2.97 \times 10^{-25} \text{ yr}^2 \text{ m}^{-3}$.

planet	T/(Earth years)	r/km	$T^2/r^3/(\text{yr}^2 \text{ m}^{-3})$
Mercury	0.241	57.9×10^6	2.99×10^{-25}
Venus	0.615	108.0×10^6	3.00×10^{-25}
Earth	1.00	150.0×10^6	2.96×10^{-25}
Mars	1.88	228.0×10^6	2.98×10^{-25}
Jupiter	11.9	778.0×10^6	3.01×10^{-25}
Saturn	29.5	1430.0×10^6	2.98×10^{-25}
Uranus	84.0	2870.0×10^6	2.98×10^{-25}
Neptune	165	4500.0×10^6	2.99×10^{-25}
Pluto	248	5900.0×10^6	2.99×10^{-25}
			(average 2.99×10^{-25})

Figure 10.26

Satellites are widely used in telecommunication. Many communications satellites are placed in what is called **geostationary** orbit. That is, they are in equatorial orbits with exactly the same period of rotation as the Earth, and move in the same direction as the Earth, so that they are always above the same point on the equator. Details of the orbit of such a satellite are worked out in the example which follows.

Example

For a geostationary satellite, calculate:
(a) the height above the Earth's surface,
(b) the speed in orbit.
(Radius of Earth $= 6.38 \times 10^6$ m; mass of Earth $= 5.98 \times 10^{24}$ kg.)

(a) The period of the satellite is 24 hours = 8.64×10^4 s. Using $T^2/r^3 = 4\pi^2/GM$, we have

$r^3 = 6.67 \times 10^{-11} \times 5.98 \times 10^{24} \times (8.64 \times 10^4)^2/4\pi^2$, giving $r^3 = 7.54 \times 10^{22}$ m^3. Taking the cube root, the radius r of the orbit is 4.23×10^7 m. The distance above the Earth's surface is $(4.23 \times 10^7 - 6.4 \times 10^6) = \mathbf{3.59 \times 10^7}$ **m**.

(b) Since $v = 2\pi r/T$, the speed is given by $v = 2\pi \times 4.23 \times 10^7/8.64 \times 10^4$ = **3070 m s^{-1}**.

Now it's your turn

The radius of the Moon's orbit is 3.84×10^8 m, and its period is 27.4 days. Use Kepler's law to calculate the period of the orbit of a satellite orbiting the Earth just above the Earth's surface. (Radius of Earth = 6.38×10^6 m.)

Ans: 1.43 hours

Weightlessness

Suppose that you are carrying out an experiment involving the use of a newton balance in a lift. When the lift is stationary, an object of mass 10 kg, suspended from the balance, will give a weight reading of $10g$ N. If the lift accelerates upwards with acceleration $0.1g$, the reading on the balance increases to $11g$ N. If the lift accelerates downwards with acceleration $0.1g$, the apparent weight of the object decreases to $9g$ N. If, by an unfortunate accident, the lift cable were to break and there were no safety restraints, the lift would accelerate downwards with acceleration g. The reading on the Newton balance would be zero. If you were to drop the pencil with which you are recording the balance readings, it would not fall to the floor of the lift, but remain stationary with respect to you. Both you and the pencil are in free fall. You are experiencing **weightlessness**. Figure 10.27 illustrates your predicament. It might be more correct to refer to this situation as *apparent* weightlessness, as you can only be truly weightless in the absence of a gravitational field, that is at an infinite distance from the Earth or any other attracting object.

Figure 10.27 *Lift experiment on weightlessness.*

A similar situation arises in a satellite orbiting the Earth. The force of gravity, which provides the centripetal force, is causing the body to fall out of its expected straight-line path. People and objects inside the satellite are experiencing a free fall situation and apparent weightlessness.

Section 10.3 summary

★ The attractive force between two point masses is proportional to the product of the masses and inversely proportional to the square of the distance between them. This is Newton's law of gravitation: $F = Gm_1m_2/r^2$.

★ A gravitational field is a region round a mass where another mass feels a force. Gravitational field strength g is the force per unit mass: $g = F/m$. g is also the acceleration of free fall.

★ For a planet or satellite in circular orbit about a body, the square of the period is proportional to the cube of the radius of the orbit. This is Kepler's third law of planetary motion: $T^2/r^3 = 4\pi^2/GM$.

Section 10.3 questions

1 The Earth's radius is about 6.4×10^6 m. The acceleration of free fall at the Earth's surface is 9.8 m s^{-2}. The gravitational constant G is 6.7×10^{-11} N m^2 kg^{-2}. Use this information to estimate the mean density of the Earth.

2 The weight of a passenger in an aircraft on the runway is W. His weight when the aircraft is flying at an altitude of 16 km above the Earth's surface is W_a. The percentage change F in his weight is given by $F = 100(W_a - W)/W$ %. Taking the radius of the Earth as 6.378×10^3 km, calculate F.

3 The times for Mars and Jupiter to orbit the Sun are 687 days and 4330 days respectively. The radius of the orbit of Mars is 228×10^6 km. Calculate the radius of the orbit of Jupiter.

Exam Questions

1 Two metal plates 10 cm apart are connected to a 1000 V supply. A small plastic ball with a conducting surface is suspended between the plates on an insulating thread. When the ball is made to touch the positive plate, it gains a charge of 5.0×10^{-9} C.

(a) Calculate the strength of the electric field between the plates.

(b) Explain why the ball shuttles to and fro between the plates.

(c) Calculate the force on the ball when it is in the field.

(d) Explain why the ball may start to shuttle to and fro if it is not charged, but only moves like this if it starts off slightly slightly closer to one of the plates.

2 In a simplified model of the hydrogen atom, the electron is separated from the proton by a distance r of 5.3×10^{-11} m. Use the following data to calculate the ratio of the electric force to the gravitational force at this distance. What would be the ratio if the charges were separated to a distance $2r$?

Electron charge $e = -1.6 \times 10^{-19}$ C
Electron mass $m_e = 9.1 \times 10^{-31}$ kg
Proton mass $m_p = 1.7 \times 10^{-27}$ kg
Permittivity of free space $\epsilon_0 = 8.9 \times 10^{-12}$ F m^{-1}
Gravitational constant $G = 6.7 \times 10^{-11}$ N m^2 kg^{-2}

3 In Figure 10.28, there are two similar table-tennis balls, each of mass 2.5 g. The balls have equal charges and are suspended by insulating thread.

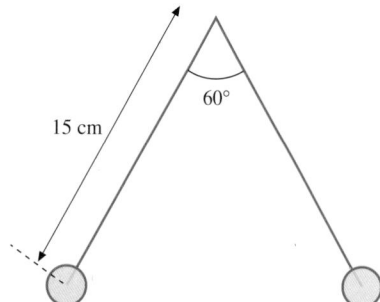

Figure 10.28

(a) Calculate the tension in one of the threads.

(b) Show that the electrostatic force between the two balls is 1.4×10^{-2} N.

(c) Calculate the charge on each ball.

4 A capacitor is made from two strips of metal foil 50 mm wide and 50 m long, with paper of thickness 0.10 mm between them. The relative permittivity of the paper is 2.0.

(a) Calculate the capacitance of the capacitor.

(b) The strips are then rolled up into a cylinder, to make the capacitor easier to handle. However, a second sheet of paper must be applied on top of one of the metal sheets. Explain why this is necessary, and why the capacitance of the rolled-up sheets is greater than that in (a).

5 Figure 10.29 illustrates the apparatus used by Cavendish in 1798 to find a value for the gravitational constant G. In a school experiment using similar apparatus, two lead spheres are attached to a light horizontal beam which is suspended by a wire. When a flask of mercury is brought close to each sphere, the gravitational attraction causes the beam to twist through a small angle. From measurements of the twisting (torsional) oscillations of the beam, a value can be found for the force producing a measured deflection. G can then be calculated if the large and small masses are known.

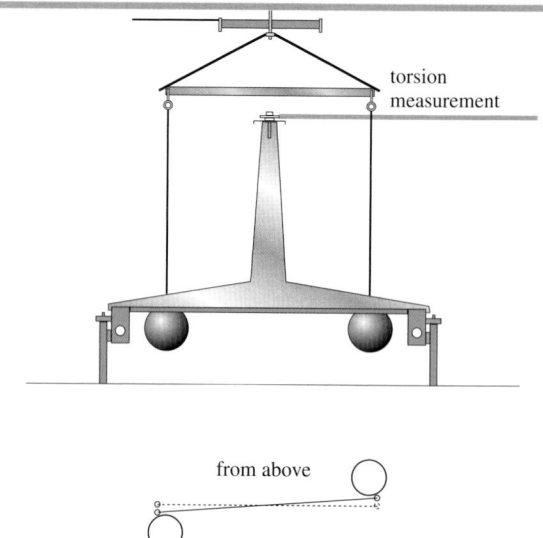

Figure 10.29 Cavendish's experiment for G.

(a) In such an experiment, one lead sphere has mass 6.22×10^{-3} kg and the mass of the mercury flask is 0.713 kg. Calculate the force between them when they are 72.0 mm apart.

(b) Comment on the size of the force.

6 Two point charges of $+2.4\,\mu C$ and $-2.9\,\mu C$ are placed at points A and B respectively in a vacuum, 0.15 m apart (Figure 10.30). It is required to find a point P at which the resultant electric field due to these two charges is zero.

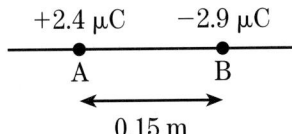

$+2.4\,\mu C$ $-2.9\,\mu C$

A   B

0.15 m

Figure 10.30

(a) Explain why P cannot lie off the line joining the charges (extended if necessary).
(b) Decide whether P can lie to the left of A, between A and B, to the right of B, or in more than one of these regions. Give reasons for your answer.
(c) Hence deduce a location for P.

7 The gravitational field strength at the surface of the Moon is $1.62\,\text{N kg}^{-1}$. The radius of the Moon is 1740 km.

(a) Show that the mass of the Moon is $7.35 \times 10^{22}\,\text{kg}$.
(b) The Moon rotates about its axis (as well as moving in orbit about the Earth). In the future, scientists may wish to put a satellite into an orbit about the Moon such that the satellite remains stationary above one point on the Moon's surface.
    i) Explain why this orbit must be an equatorial orbit.
    ii) The period of rotation of the Moon about its axis is 27.4 days. Calculate the radius of the required orbit.

(Gravitational constant $G = 6.67 \times 10^{-11}\,\text{N m}^2\,\text{kg}^{-2}$.)

CHAPTER ELEVEN
Electromagnetism

Electric motors depend on electromagnetic effects for their operation – from the largest to the smallest. The photographs show electric motors from a high-speed electric train and from a 'nano-technology' robot as it travels round a human body.

A study of electricity and of electromagnetic effects is one of the key areas for all students of physics. Not only are electromagnetic effects vital to modern living but they also give some insight into the pioneering research carried out by scientists such as Volta, Ampere, Ohm and Faraday. Their discoveries were exceptional because they were attempting to understand the invisible effects of electric currents and magnetic fields.

11.1 Magnetic fields

Some of the properties of magnets have been known for many centuries. The ancient Greeks discovered an iron ore called lodestone which, when hung from a thread, will come to rest always pointing in the same direction. This was the basis of the magnetic compass which has been in use since about 1500 BC as a means of navigation.

The magnetic compass is dependent on the fact that a freely suspended magnet will come to rest pointing north-south. The ends of the magnet are said to be poles. The pole pointing to the north is referred to as the north-seeking pole (the north pole) and the other, the south-seeking pole (the south pole). It is now known that a compass behaves in this way because the Earth is itself a magnet.

Magnets exert forces on each other. These forces of either attraction or repulsion are the basis of many children's toys such as 'fridge magnets'. The effects of the forces may be summarised in the **law of magnets**.

> ★ Like poles repel
> Unlike poles attract.

The law of magnets implies that around any magnet, there is a region where a magnetic pole will experience a force. This region is known as a **magnetic field**. Magnetic fields are not visible but they may be represented by lines of magnetic force or **magnetic field lines**. A simple way of imaging magnetic field lines is to think of one such line as the direction in which a free magnetic north pole would move if placed in the field. Magnetic fields may be plotted using a small compass (a plotting compass) or by the use of iron filings and a compass.

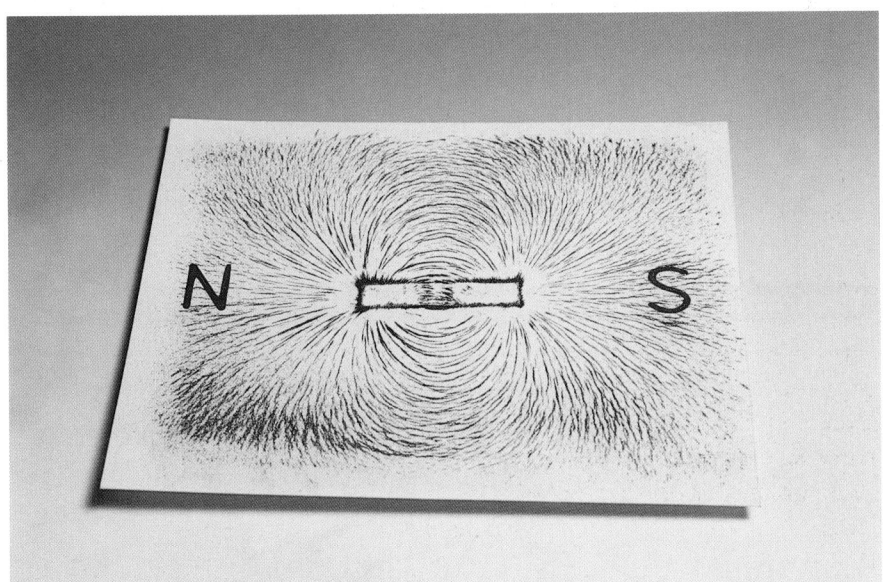

The iron filings line up in the direction of the magnetic field of the bar magnet which is under the sheet of paper. The plotting compass gives the direction of the field.

The magnetic field of a bar magnet is shown in Figure 11.1. Effects due to the Earth's magnetic field have not been included since the Earth's field is relatively weak and would be of importance only at some distance from the magnet. It is important to realise that, although the magnetic field has been drawn in two dimensions, the actual magnetic field is three-dimensional.

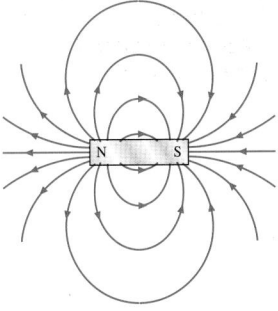

Figure 11.1 *Magnetic field pattern of a bar magnet.*

For any magnetic field,

- the magnetic field lines are smooth curves,
- the magnetic field lines never touch or cross,
- the strength of the magnetic field is indicated by the distance between the lines – closer lines mean a stronger field.

It can be seen that these properties are very similar to those for electric field lines (Chapter 10).

Figure 11.2 illustrates the magnetic field pattern between the north pole of one magnet and the south pole of another. This pattern is similar to that produced between the poles of a horseshoe magnet.

Figure 11.2 *Magnetic field pattern between a north and south pole.*

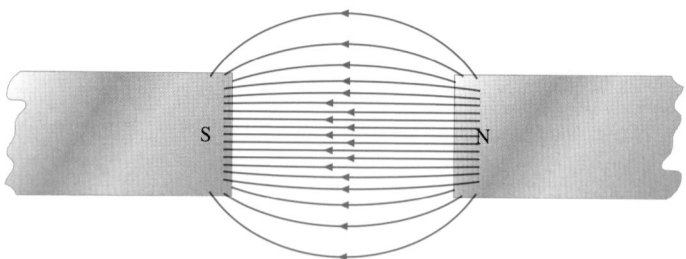

Figure 11.3 shows the magnetic field between the north poles of two magnets. The magnetic field due to one magnet opposes that due to the other. The field lines cannot cross and consequently there is a point X, known as a **neutral point**, where there is no resultant magnetic field because the two fields are equal in magnitude but opposite in direction.

Figure 11.3 *Magnetic field pattern between two south poles.*

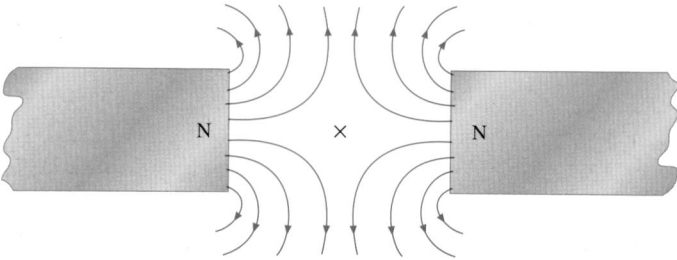

Example

A circular magnet is made with its north pole at the centre and separated from the south pole by an air gap, as shown in Figure 11.4.

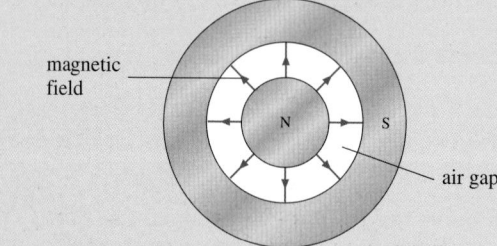

Figure 11.4

Draw the magnetic field lines in the gap.

Now it's your turn

1   Draw a diagram to illustrate the magnetic field between the south poles of two magnets.

Ans: See Figure 11.3 but with reversed direction

2   Two bar magnets are placed on a horizontal surface as illustrated in Figure 11.5.

Figure 11.5

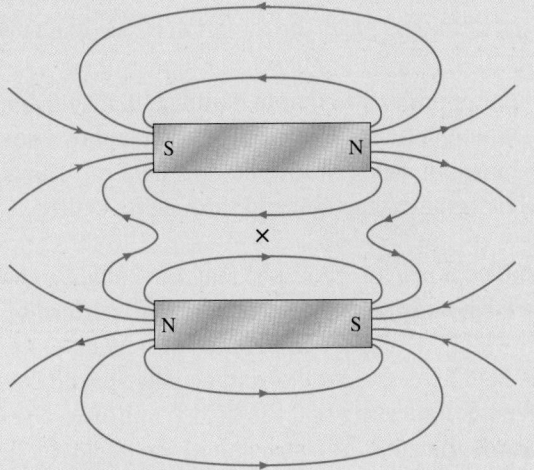

Draw the two magnets, and on your diagram, mark the position of any neutral points.

Magnetic effect of an electric current

The earliest discovery of the magnetic effect of an electric current was made in 1820 when Oersted, a Danish physicist, noticed that a compass was deflected when brought near to a wire carrying an electric current. It is now known that all electric currents produce magnetic fields. The size and shape of the magnetic field depends on the size of the current and the shape (configuration) of the conductor through which the current is travelling.

The magnetic field due to a long straight wire may be plotted using the apparatus illustrated in Figure 11.6. Note that the current must be quite large (about 5 A). Iron filings are sprinkled on to the horizontal board and a plotting compass is used to determine the direction of the field.

Hans Christian Oersted

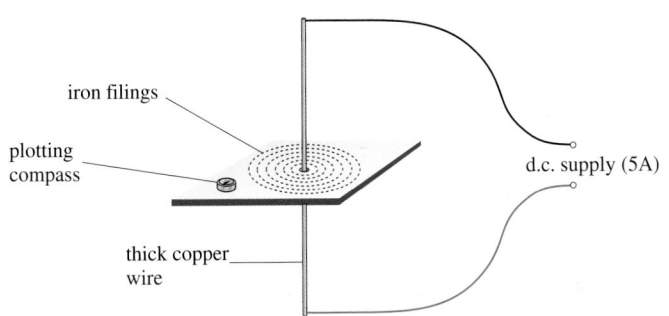

Figure 11.6 *Apparatus to plot the magnetic field due to a long wire.*

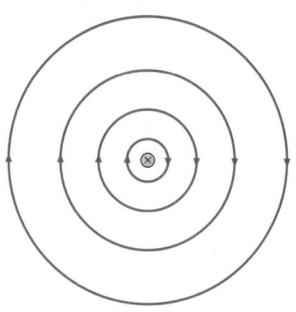

Figure 11.7 *Magnetic field pattern due to a long straight wire.*

Figure 11.7 shows the field pattern due to a long straight wire. The lines are concentric circles centred on the middle of the wire. The separation of the lines increases with distance from the wire, indicating that the field is decreasing in strength. The direction of the magnetic field may be found using the **right-hand rule** (see Figure 11.8).

> ★ Imagine holding the conductor in the right hand with the thumb pointing in the direction of the current. The direction of the fingers gives the direction of the magnetic field.

Similar apparatus to that in Figure 11.6 may be used to investigate the shapes of the magnetic field due to a flat coil and to a solenoid. Figure 11.9 illustrates the magnetic field pattern due to a flat coil. The field has been drawn in a plane normal to the coil and through its centre.

A solenoid is a long coil and may be thought to be made up of many flat coils placed side-by-side. The magnetic field pattern of a long solenoid (that is, a coil which is long in comparison with its diameter) is shown in Figure 11.10. The field lines are parallel and equally spaced over the centre section of the solenoid, indicating that the field is uniform. The field lines spread out towards the ends. The strength of the magnetic field at each end is one half that at the centre. The direction of the magnetic field in a flat coil and in a solenoid may be found using the **right-hand grip rule**, as illustrated in Figure 11.11.

> ★ Grasp the coil or solenoid in the right hand with the fingers pointing in the direction of the current. The thumb gives the direction of the magnetic field.

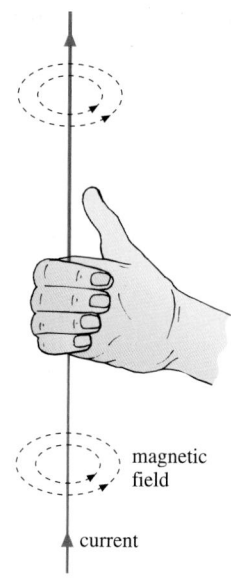

Figure 11.8 *The right-hand rule.*

Figure 11.10 *Magnetic field pattern due to a solenoid.*

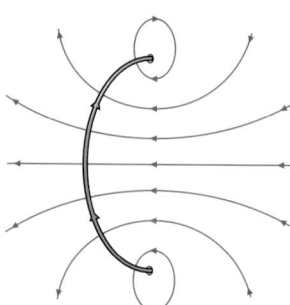

Figure 11.9 *Magnetic field pattern due to a flat coil.*

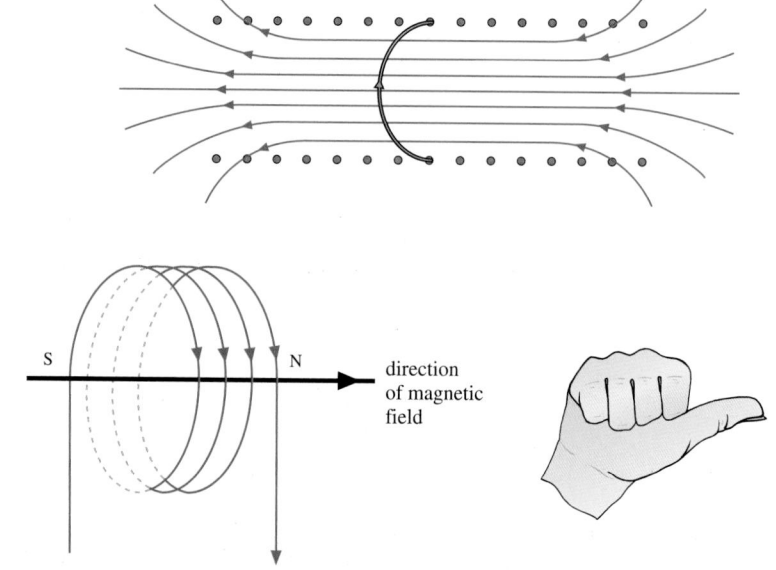

Figure 11.11 *The right-hand grip rule.*

11.1 Magnetic fields

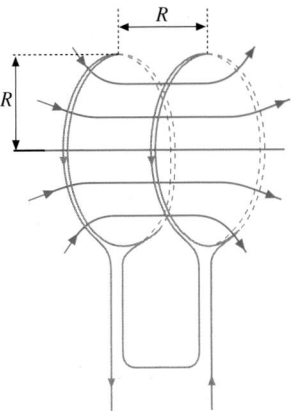

The magnetic north end of the coil or solenoid is the end from which the lines of magnetic force are emerging. Note the similarities and, more importantly, the differences between this rule and the right-hand rule for the long straight wire.

Uniform magnetic fields are of importance in the study of charged particles. A uniform field is produced in a solenoid but this field is inside the solenoid and consequently, it may be difficult to make observations and to take measurements. This problem is overcome by using **Helmholtz coils**. These are two identical flat coils, with the same current flowing in each. The coils are positioned so that their planes are parallel and separated by a distance equal to the radius of either coil. The coils and their resultant magnetic field are illustrated in Figure 11.12.

Figure 11.12 Magnetic field in Helmholtz coils.

Example

Two long straight wires, of circular cross-section, are each carrying the same current directly away from you, down into the page. Draw the magnetic field due to the two current-carrying wires.

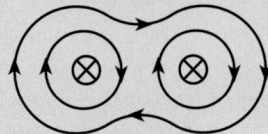

Figure 11.13

Now it's your turn

1 Draw magnetic field patterns, one in each case to represent
   (a) a uniform field,
   (b) a field which is decreasing in strength in the direction of the field,
   (c) a field which is increasing in strength along the direction of the field.

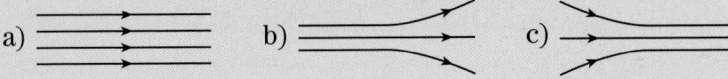

Figure 11.13a

   Ans: see Figure 11.13a

2 Draw a diagram of the magnetic field due to two long straight wires when the currents in the two wires are in opposite directions.

Figure 11.13b

   Ans: see Figure 11.13b

CHAPTER 11

Electromagnets and their uses

The strength of the magnetic field due to a flat coil or a solenoid may be increased by winding the coil on a bar of soft iron. The bar is said to be the **core** of the coil. The iron is referred to as being 'soft' because it can be magnetised and demagnetised easily. With such a core, the strength of the magnetic field may be increased by up to 1000 times. If the iron is alloyed with cobalt and nickel, the field may be 10^4 times stronger. Magnets such as these are called **electromagnets**. Electromagnets have many uses because, unlike a permanent magnet, the magnetic field can be switched off by removing the current in the coil.

Magnetic imaging of body structures requires the use of a very large uniform magnetic field.

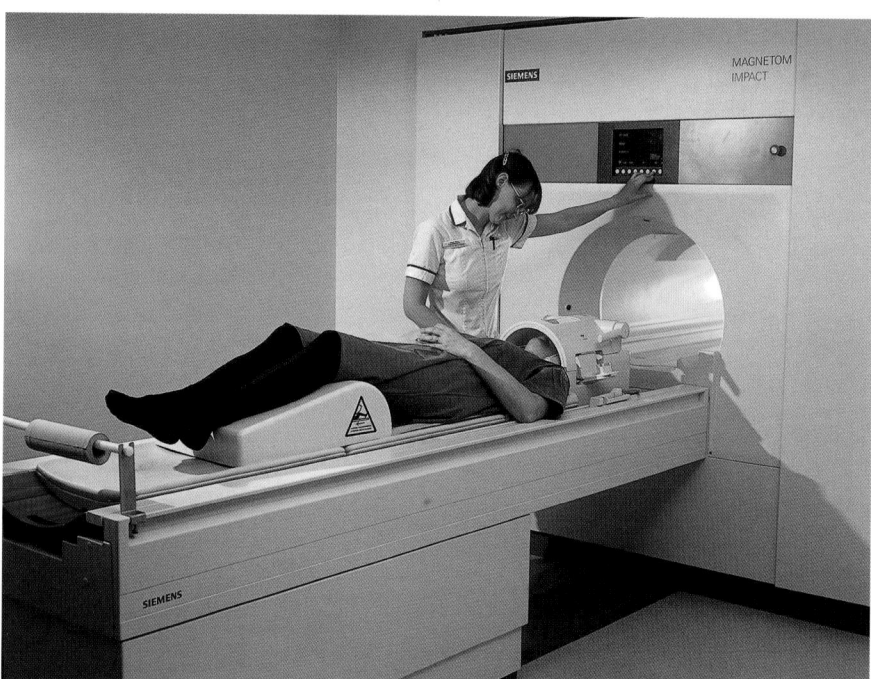

Section 11.1 summary

★ A magnetic field is a region of space where a magnet will experience forces.

★ Magnetic field lines never touch or cross.

★ The separation of magnetic field lines indicates the strength of the magnetic field.

★ An electric current gives rise to a magnetic field, the strength and direction of which depends on the size of the current and the shape of the current-carrying conductor.

★ The direction of the field due to a straight wire is given by the right-hand rule.

★ The direction of the field in a solenoid is given by the right-hand grip rule.

11.2 Forces and fields

In Section 11.1, it was seen that the plotting of lines of magnetic force gives the direction and shape of the magnetic field. Also, the distance between the lines indicates the strength of the field. However, the strength of the magnetic field was not defined. Physics is the science of measurement and consequently, the topic would not be complete without defining and measuring magnetic field strength. Magnetic field strength is defined through a study of the **motor effect**.

The motor effect

The interaction of the magnetic fields produced by two magnets causes forces of attraction or repulsion between the two. If a conductor is placed in a magnetic field and a current is passed through the conductor, the magnetic fields of the current-carrying conductor and the magnet may interact, causing forces between them. This is illustrated in Figure 11.14.

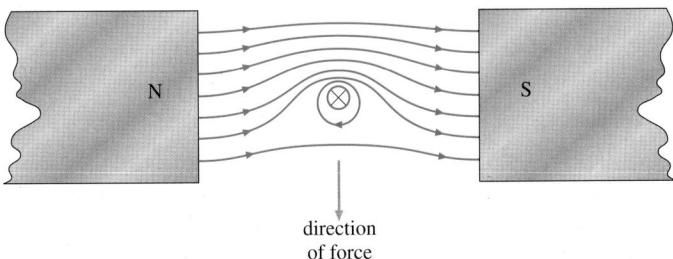

direction
of force

Figure 11.14 *The interacting magnetic fields of a current-carrying conductor and a bar magnet.*

The existence of the force may be demonstrated with the apparatus shown in Figure 11.15. The strip of aluminium foil is held loosely between the poles of a horseshoe magnet so that the foil is at right angles to the magnetic field. When the current is switched on, the foil jumps and becomes taut, showing that a force is acting on it. The direction of the force, known as the **electromagnetic force**, depends on the direction of the magnetic field and of the current. This phenomenon, when a current-carrying conductor is at an angle to a magnetic field, is called the **motor effect**.

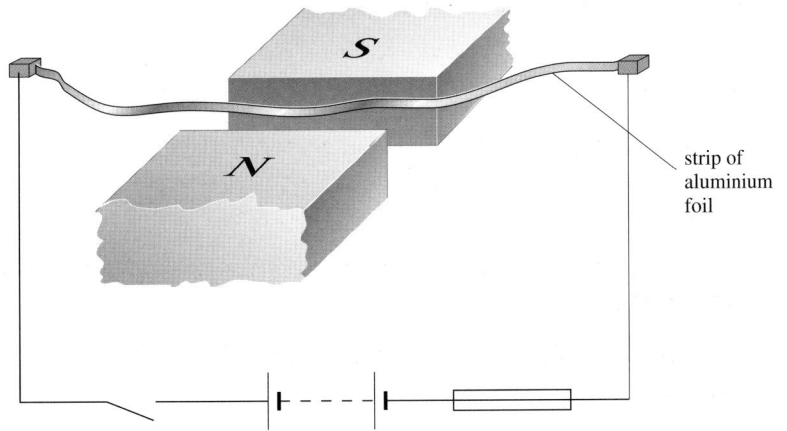

strip of
aluminium
foil

Figure 11.15 *The motor effect acting on a piece of aluminium.*

The direction of the force relative to the direction of the current and the magnetic field may be predicted using **Fleming's left-hand rule**. This is illustrated in Figure 11.16.

Figure 11.16 *Fleming's left-hand rule.*

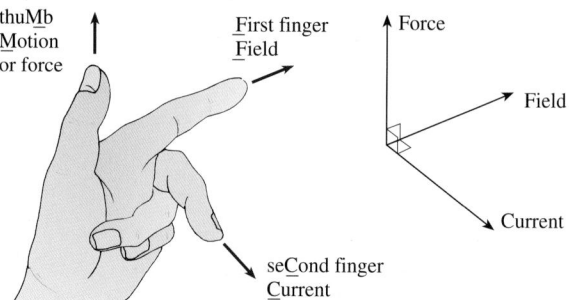

> ★ If the first two fingers and thumb of the left hand are held at right angles to one another with the **F**irst finger in the direction of the **F**ield and the se**C**ond finger in the direction of the **C**urrent, then the thu**M**b gives the direction of the force or **M**otion.

Note that, if the conductor is held fixed, motion will not be seen but nevertheless, there will be an electromagnetic force.

Figure 11.17 *Apparatus to investigate the magnitude of the electromagnetic force.*

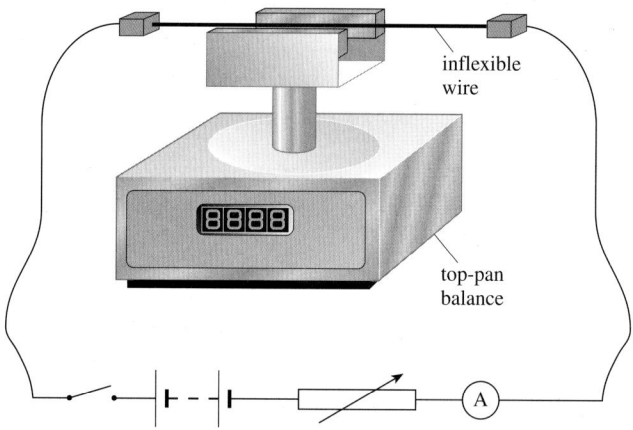

The magnitude of the electromagnetic force may be investigated using the apparatus shown in Figure 11.17. A horseshoe magnet is made by placing two flat magnadur magnets on a U-shaped iron core. A length of inflexible wire is firmly fixed between the poles of the magnet. When the current is switched on, the reading on the top-pan balance changes. This change in reading is a measure of the electromagnetic force. Variation of the current leads to the conclusion that the electromagnetic force F is proportional to the current I. By using magnadur magnets of different lengths (but of similar strengths), the force F is found to be proportional to the length of wire L in the magnetic field. Finally, by varying the angle θ between the wire and the direction of the magnetic field, the force F is found to be proportional to sin θ. Hence the expression

$$F \propto IL \sin \theta,$$

11.2 Forces and fields

is derived.

The expression can be rewritten as

$$\bigstar \quad F = BIL \sin \theta,$$

where B is a constant.

The constant B depends on the strength of the magnet and, if stronger magnets are used, the constant has a greater value. The equation can, therefore, be used as the defining equation for magnetic field strength.

The equation can be rewritten as

$$\bigstar \quad \frac{F}{L} = BI \sin \theta.$$

So

> $\bigstar$ for a long straight conductor carrying unit current at right angles to a uniform magnetic field, the magnetic field strength is numerically equal to the force per unit length of the conductor.

In the SI system of units, **magnetic field strength** or **magnetic flux density** is measured in **tesla** (**T**). An alternative name for this unit is **weber per square metre** (**Wb m^{-2}**).

> $\bigstar$ One tesla is the uniform magnetic flux density which, acting normally to a long straight wire carrying a current of 1 ampere, causes a force per unit length of 1 N m^{-1} on the conductor.

Since force is measured in newtons, length in metres and current in amperes, it can be derived from the defining equation for the tesla that the tesla may also be expressed as N m^{-1} A^{-1}. The unit involves force which is a vector quantity and thus magnetic flux density is also a vector.

When using the equation $F = BIL \sin \theta$, it is sometimes helpful to think of the term $B \sin \theta$ as being the component of the magnetic flux density which is at right angles to the conductor, (see Figure 11.18).

The tesla is a large unit for the measurement of flux density. A strong magnet may have a flux density between its poles of a few teslas. The magnetic flux density due to the Earth in the UK is about 44 µT at an angle of 66° to the horizontal.

Figure 11.18 B *sin* θ *is the component of the magnetic field which is normal to the conductor.*

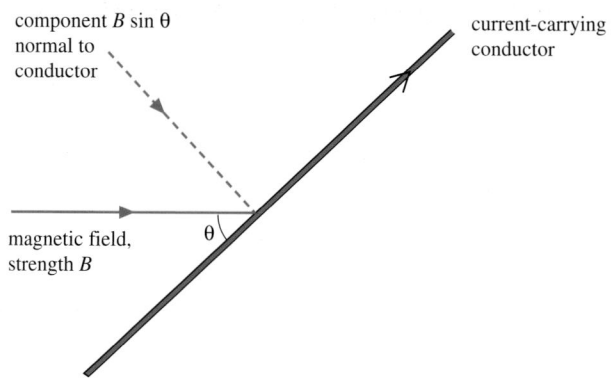

component $B \sin \theta$ normal to conductor

current-carrying conductor

magnetic field, strength B

θ

Example

The horizontal component of the Earth's magnetic flux density is 1.8×10^{-5} T. The current in a horizontal cable is 150 A. Calculate, for this cable:

(a) the maximum force per unit length,

(b) the minimum force per unit length.

In each case, state the angle between the cable and the magnetic field.

(a) force per unit length $= \dfrac{F}{L} = BI \sin \theta$.

Force per unit length is a maximum when $\theta = 90°$ and $\sin \theta = 1$.

Force per unit length $= 1.8 \times 10^{-5} \times 150 = 2.7 \times 10^{-3}$ N m^{-1}

Maximum force per unit length $= \mathbf{2.7 \times 10^{-3}}$ **N m^{-1}** when the cable is at right angles to the field.

(b) Force per unit length is a minimum when $\theta = 0$ and $\sin \theta = 0$.

Minimum force per unit length $= \mathbf{0}$ when the cable is along the direction of the field.

Now it's your turn

1 The effective length of the filament in a light bulb is 3.1 cm and, for normal brightness, the current in the filament is 0.25 A. Calculate the maximum electromagnetic force on the filament when in the Earth's field of flux density 44 µT.

Ans: 3.4×10^{-7} N

2 A straight conductor carrying a current of 6.5 A is situated in a uniform magnetic field of flux density 4.3 mT.

Calculate the electromagnetic force per unit length of the conductor when the angle between the conductor and the field is:

(a) 90°,

(b) 45°.

Ans: (a) 2.8×10^{-2} N m^{-1}; (b) 2.0×10^{-2} N m^{-1}

Force between parallel conductors

A current-carrying conductor has a magnetic field around it. If a second current-carrying conductor is placed parallel to the first, this second conductor will be in the magnetic field of the first and, by the motor effect, will experience a force.

By similar reasoning, the first conductor will also experience a force. By Newton's third law (see Chapter 4), these two forces will be equal in magnitude and opposite in direction. The effect can be demonstrated using the apparatus in Figure 11.19. It can be seen that, if the currents are in the same direction, the pieces of foil move towards one another (the pinch effect). If the currents are in opposite directions, the pieces of foil move apart. An explanation for the effect can be found by reference to Figure 11.20.

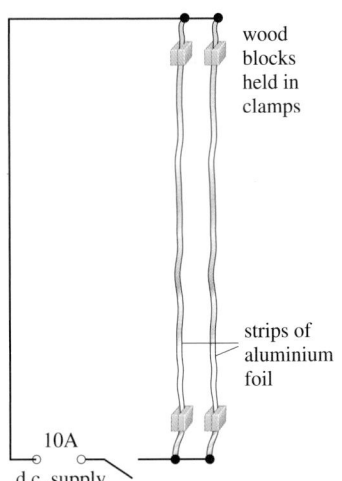

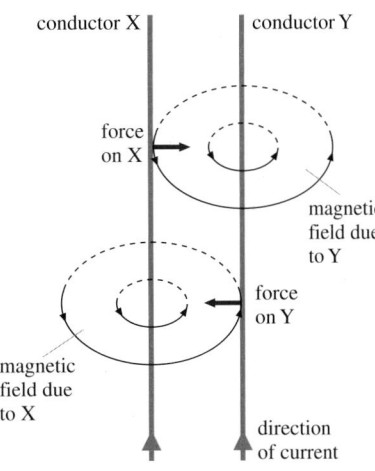

Figure 11.19 Apparatus to demonstrate the force between parallel current-carrying conductors.

Figure 11.20 Diagram to illustrate the force between parallel current-carrying conductors.

The current in conductor X causes a magnetic field and the field lines are concentric circles. These field lines will be at right angles to conductor Y and so, using Fleming's left-hand rule, there will be a force on Y in the direction of X. Using similar reasoning, the force on X due to the magnetic field of Y is towards Y. Reversing the direction of the current in one conductor will reverse the directions of the two forces and thus, when the currents are in opposite directions, the conductors tend to move apart.

The force per unit length on each of the conductors depends on the magnitude of the current in each conductor and also their separation. Since force can be expressed in terms of base units, then it is possible to use the effect to define the ampere in terms of SI units. The definition may be given as follows.

Consider two long straight parallel conductors of negligible area of cross-section, situated one metre apart in a vacuum. Then, if the current in each conductor is 1 ampere, the force per metre length acting on each conductor is 2.0×10^{-7} N.

It may not be necessary to learn the definition of the ampere, but it is important to realise how the ampere is defined in terms of SI units and that the definition does not involve the property of a substance.

Torque on a coil in a magnetic field

Consider a rectangular loop PQRS of wire carrying a current I such that its plane makes an angle φ with a uniform magnetic field of flux density B, as illustrated in Figure 11.21. The loop is free to rotate about the axis XY.

Figure 11.21

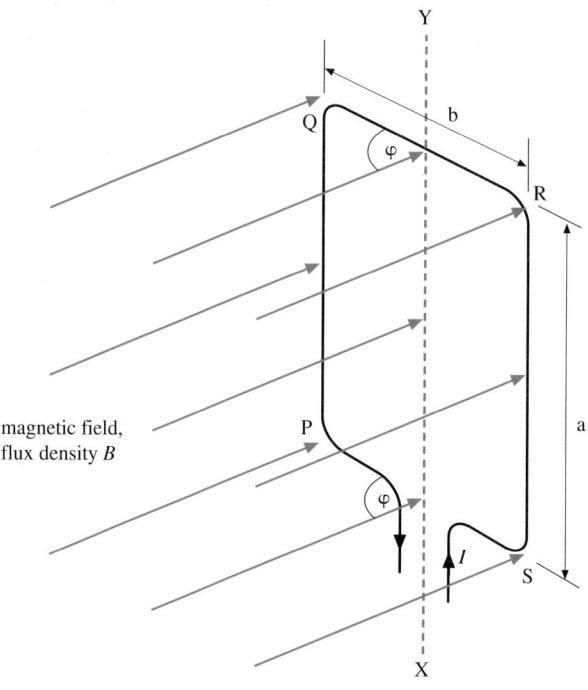

magnetic field, flux density B

All four sides of the loop are at angles to the magnetic field and, by Fleming's left hand rule, will experience electromagnetic forces. However, the forces acting on sides PS and QR will be along the axis of rotation of the loop and will, therefore, not cause any rotation. The forces on PQ and SR will tend to cause rotation. A plan view of the loop, indicating the directions of the currents, magnetic field and the forces on sides PQ and SR is given in Figure 11.22.

Figure 11.22

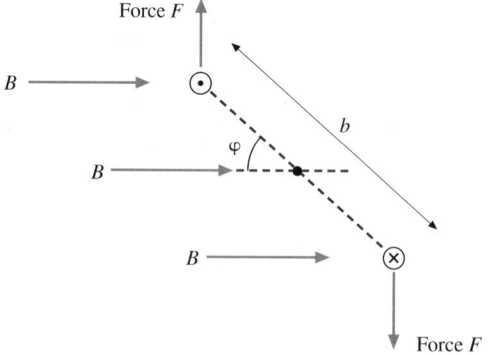

Consider the loop having sides PQ and SR of length a and sides PS and QR of length b. Then, force on PQ = force on SR = BIa.

These two forces act in opposite directions and cause a turning effect about the axis of rotation.

11.2 Forces and fields

Two forces, equal in magnitude and opposite in direction, but not acting along the same line form what is known as a **couple**. The turning effect of a couple is known as the **torque** of the couple and is equal to the product of one of the forces and the perpendicular distance between the forces (see Chapter 3).

$$torque\ on\ the\ loop = force \times perpendicular\ distance\ between\ the\ forces$$
$$= BIa \times b\cos\varphi.$$

The cross-sectional area A of the loop is equal to ab. If the loop is replaced by a coil having the same cross-sectional area but with N turns of wire, then

$$torque\ on\ coil = BIAN\cos\varphi.$$

The torque produced on a current-carrying coil in a magnetic field is used as a practical means of measuring current in the **moving-coil meter**. The coil is made to rotate against a hair spring so that the larger the current in the coil, the greater the torque and hence the greater the deflection. In order that the scale of the meter may be linear, the coil moves in a radial magnetic field (see Section 11.1). This means that the magnetic field is always in the plane of the coil and, referring to the equation for the torque on the coil $\varphi = 0$ and $\cos\varphi = 1$.

A moving-coil meter.

A coil in a uniform magnetic field will come to rest in the position where the torque on the coil is zero. This will occur when the plane of the coil is normal to the direction of the magnetic field. In this position, $\varphi = 90°$ and $\cos\varphi = 0$.

Figure 11.23 *A split-ring commutator.*

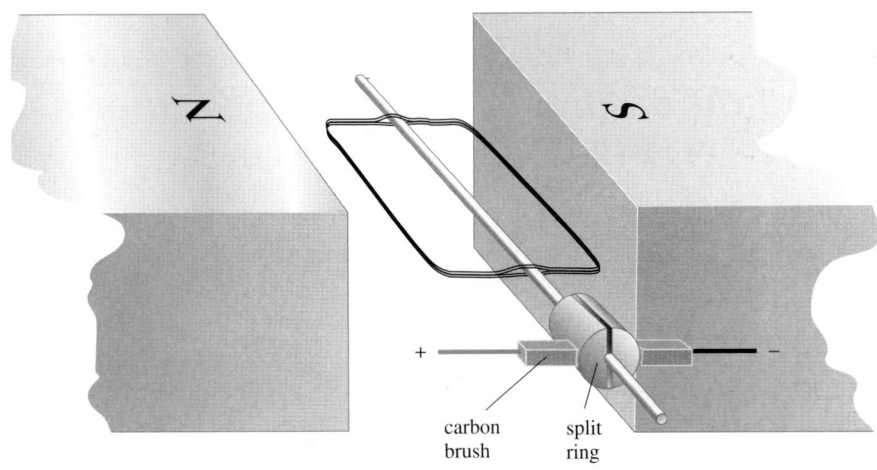

carbon brush   split ring

In order for the coil to rotate continuously, the current in the coil must be reversed when it is normal to the field. In the **d.c. motor**, this is achieved by means of a split-ring commutator, as illustrated in Figure 11.23. In a practical motor, there are several coils wound on a soft-iron core so that the torque produced by the motor is more uniform. The soft-iron core increases the magnetic flux density (field strength).

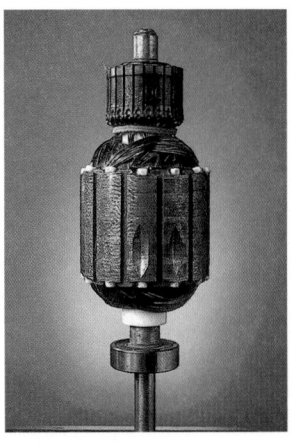
The coils of a d.c. motor.

Example

A square coil has sides of length 3.5 cm. The coil consists of 250 turns of insulated wire carrying a current of 0.15 A. The plane of the coil is at an angle φ to a uniform magnetic field of flux density 32 mT. Calculate the torque on the coil for angle $\varphi = 30°$.

$$\text{Torque} = BANI \cos \varphi$$
$$= 32 \times 10^{-3} \times (3.5 \times 10^{-2})^2 \times 250 \times 0.15 \times \cos 30$$
$$= \mathbf{1.3 \times 10^{-3} \, N \, m}$$

Now it's your turn

1   A square coil has sides of length 2.5 cm and is situated with its plane at an angle φ to a uniform magnetic field of flux density 65 mT. The coil has 50 turns of insulated wire carrying a current of 25 mA. Calculate the torque on the coil for angle φ equal to:
  (a) 0,
  (b) 50°,
  (c) 90°.
  Ans: (a) 5.1×10^{-5} N m; (b) 3.3×10^{-5} N m; (c) zero

2   A coil of area 1.5×10^{-3} m^2 has 60 turns of insulated wire carrying a current of 0.12 A. When the plane of the coil is at an angle of 45° to a uniform magnetic field, the torque on the coil is 8.7×10^{-6} N m. Calculate the magnitude of the flux density of the magnetic field.
  Ans: 1.1×10^{-3} T

Force on a moving charged particle in a magnetic field

An electric current is charge in motion. Since charge is always associated with particles, then the current in a conductor is a movement of charged particles. If a current-carrying conductor is placed in a magnetic field, it may experience a force depending on the angle between the field and the conductor. The force arises from the force on the individual moving charged particles in the conductor. It has been shown that a conductor of length L carrying a current I at an angle θ to a uniform magnetic field of flux density B experiences a force F given by

$$F = BIL \sin \theta.$$

If there are n charged particles in a length L of the conductor, each carrying a charge q, which pass a point in the conductor in time t, then the current in the conductor is given by

$$I = \frac{nq}{t},$$

and the speed of charged particles is given by $v = \dfrac{L}{t}$

Hence,

$$F = B\frac{nq}{t}L \sin \theta$$

and

$$F = Bnqv \sin \theta.$$

This force is the force on n charged particles. Therefore,

> ★ the force on a particle of charge q moving at speed v at an angle θ to a uniform magnetic field of flux density B is given by
>
> $$F = Bqv \sin \theta.$$

The direction of the force will be given by Fleming's left-hand rule. However, it must be remembered that the second finger is used to indicate the direction of the *conventional* current. If the particles are positively charged, then the second finger is placed in the same direction as the velocity. However, if the particles are negatively charged (e.g. electrons), the finger must point in the opposite direction to the velocity.

Consider a positively charged particle of mass m carrying charge q and moving with velocity v as shown in Figure 11.24. The particle enters a uniform magnetic field of flux density B which is normal to the direction of motion of the particle. As the particle enters the field, it will experience a force normal to its direction. This force will not change the speed of the particle but it will change its direction of motion. As the particle moves through the field, the force will remain constant since the speed has not changed and it will always be normal to the direction of motion. The particle will, therefore, move in an arc of a circle (see Chapter 8).

Figure 11.24

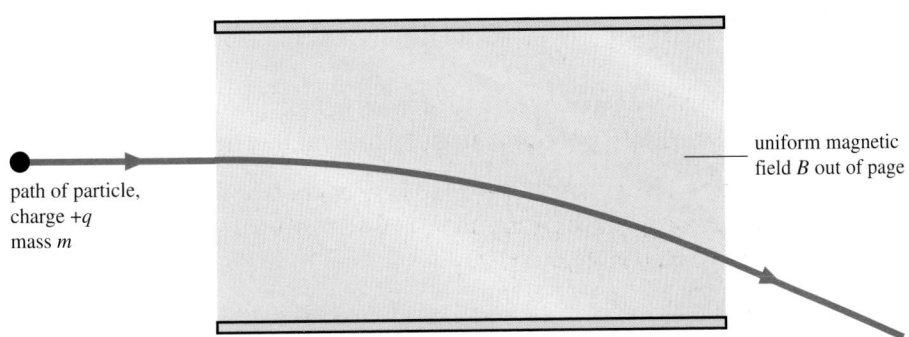

path of particle,
charge $+q$
mass m

uniform magnetic
field B out of page

The force $F = Bqv \sin \theta$ (in this case, $\sin \theta = 1$), provides the centripetal force for the circular motion. Hence,

$$\text{centripetal force} = mv^2/r = Bqv$$

Rearranging,

$$\text{radius of path} = r = \frac{mv}{Bq}.$$

The importance of this equation is that, if the speed of the particle and the radius of its path are known, then the specific charge, i.e. the ratio of charge to mass can be found. Then, if the charge on the particle is known, its mass may be calculated. The technique is also used in nuclear research to try to identify some of the fundamental particles. The tracks of charged particles are made visible in a bubble chamber. Analysing these tracks gives information as to the sign of the charge on the particle and its specific charge.

Tracks of particles produced in a bubble chamber.

Specific charge of the electron

How do you find the mass of an electron? This cannot be done directly but the ratio of charge to mass q/m can be found and, knowing the charge q on the electron, the mass m can then be calculated.

We have seen that a particle of mass m and charge q moving with speed v at right angles to a uniform magnetic field of flux density B experiences a force F_B given by

$$F_B = Bqv.$$

The direction of this force is given by Fleming's left-hand rule and is always normal to the velocity. The fact that we have a constant force which is normal to the direction of motion is the condition necessary for circular motion. The force F_B provides the centripetal force. The motion can be represented by the equation

$$Bqv = mv^2/r.$$

Rearranging the terms,

$$q/m = v/Br.$$

The ratio q/m, sometimes referred to as the specific charge, may be determined using a fine-beam tube, as shown. The path of electrons is made visible by having low-pressure gas in the tube and thus the radius of the orbit may be measured. By accelerating the electrons through a known potential difference, their speed on entry into the region of the magnetic field may be calculated (see Section 6.1). The magnetic field is provided by a pair of current-carrying coils (Helmholtz coils, page 279).

Teltron fine-beam tube.

It is of interest to rotate the tube slightly, so that the velocity of the electrons is not normal to the magnetic field. In this case, the path of the electrons is seen to be a helix (rather like the coils of a spring). The component of the velocity normal to the field gives rise to circular motion. However, there is a component of velocity along the direction of the field. There is no force on the electrons resulting from this component of velocity. Consequently, the electrons execute circular motion *and* move in a direction normal to the plane of the circle. The circles are 'pulled out' into a helix (see Figure 11.25). The helical path is an important aspect of the focusing of electron beams by magnetic fields in an electron microscope.

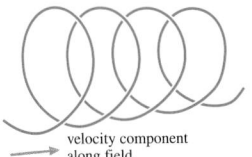

velocity component
along field

Figure 11.25 *Electrons moving in a helix.*

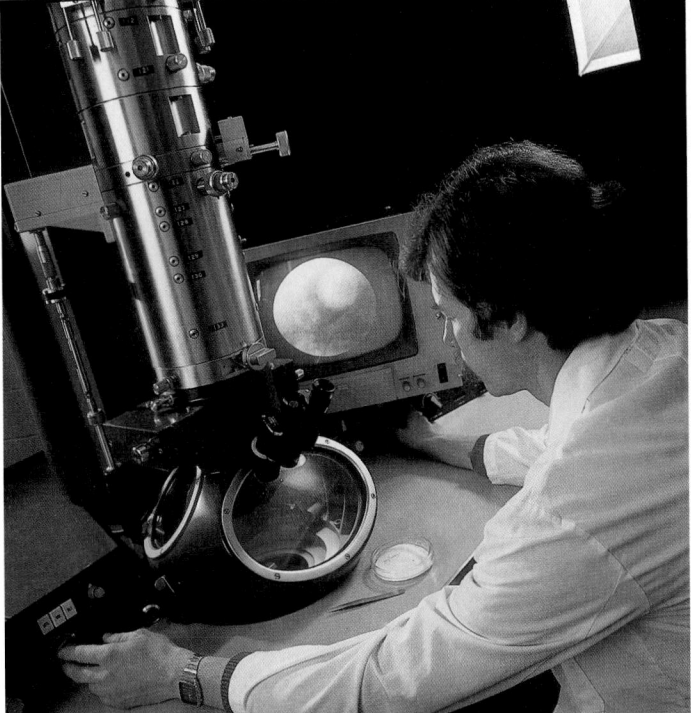

Electron microscope.

Example

Electrons are accelerated to a speed of 8.4×10^6 m s^{-1}. They then pass into a region of uniform magnetic flux of flux density 0.50 mT. The path of the electrons in the field is a circle with a radius 9.6 cm. Calculate:

(a) the specific charge of the electron,

(b) the mass of the electron, assuming the charge on the electron is 1.6×10^{-19} C.

(a) $q/m = v/Br$
$$= 8.4 \times 10^6/(0.50 \times 10^{-3} \times 0.096)$$
$$= \mathbf{1.8 \times 10^{11} \ C \ kg^{-1}}$$

(b) $q/m = 1.8 \times 10^{11} = 1.6 \times 10^{-19}/m$
$$m = \mathbf{9.1 \times 10^{-31} \ kg}$$

Now it's your turn

Electrons are accelerated through a potential difference of 220 V. They then pass into a region of uniform magnetic flux of flux density 0.54 mT. The path of the electrons is normal to the magnetic field. Given that the charge on the electron is 1.6×10^{-19} C and its mass is 9.1×10^{-31} kg, calculate:

(a) the speed of the accelerated electron,

(b) the radius of the circular path in the magnetic field.

Ans: (a) 8.8×10^6 m s^{-1}; (b) 9.3 cm

Velocity selection of charged particles

We have seen that when particles of mass m and charge $+q$ enter an electric field of field strength E, there is a force F_E on the particle given by

$$F_E = qE.$$

If the velocity of the particles before entry into the field is v at right angles to the field lines (see Figure 11.26), the particles will follow a parabolic path as they pass through the field.

Figure 11.26

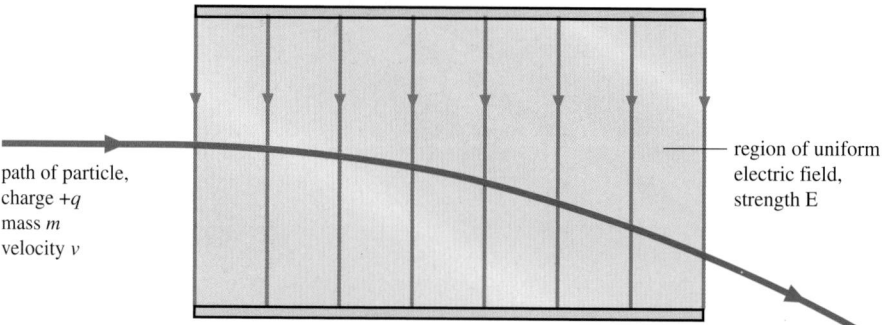

path of particle,
charge $+q$
mass m
velocity v

region of uniform electric field, strength E

Now suppose that a magnetic field acts in the same region as the electric field. If this field acts downwards into the plane of the page, then, by Fleming's left-hand rule (see page 282), a force will act on the charged particle in the direction opposite to the force due to the electric field. The magnitude F_B of this force is given by

$$F_B = Bqv,$$

where B is the flux density of the magnetic field (see Section 11.2).

If the magnitude of one of the two fields is adjusted, then a situation can arise where the two forces, F_E and F_B are equal in magnitude but opposite in direction. Thus

$$Bqv = qE$$

and

$$v = E/B.$$

For the value of the velocity given by E/B, the particles will not be deflected, as shown in Figure 11.27. Particles with any other velocities will be deflected. If a parallel beam of particles enters the field then all the particles passing undeviated through the slit will have the same velocity. Note that the mass does not come into the equation for the speed and so, particles with a different mass (but the same charge) will all pass undeviated through the region of the fields if they satisfy the condition $v = E/B$.

The arrangement shown in Figure 11.27 is known as a velocity selector. Velocity selectors are very important in the study of ions. Frequently, the production of the ions gives rise to different speeds but to carry out investigations on the ions, the ions must have one speed only.

Figure 11.27

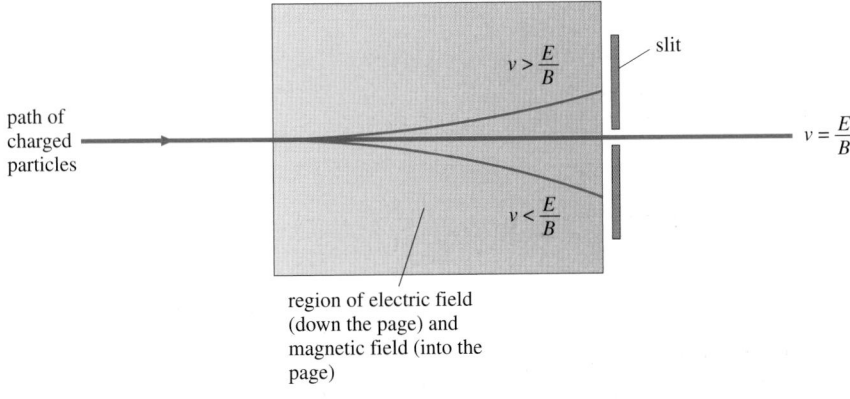

path of charged particles

slit

$v > \dfrac{E}{B}$

$v < \dfrac{E}{B}$

$v = \dfrac{E}{B}$

region of electric field (down the page) and magnetic field (into the page)

Example

It is required to select charged ions which have a speed of 4.2×10^6 m s^{-1}. The electric field strength in the velocity selector is 3.2×10^4 V m^{-1}. Calculate the magnetic field strength required.

$$v = E/B$$
$$B = 3.2 \times 10^4/4.2 \times 10^6$$
$$= \mathbf{7.6 \times 10^{-3}\ T}$$

Now it's your turn

Singly charged ions pass undeviated through a velocity selector. The electric field strength in the selector is 3.6×10^4 V m^{-1} and the magnetic flux density is 8.5 mT. Calculate the selected speed of the ions.

Ans: 4.2×10^6 m s^{-1}

The Hall effect

Consider a thin slice of a conductor which is normal to a magnetic field, as illustrated in Figure 11.28. When there is a current in the conductor in the direction shown, charge carriers (electrons in a metal) will be moving at right-angles to the magnetic field. They will experience a force which will tend to make them move to one side of the conductor. A potential difference known as the Hall voltage V_H will develop across the conductor. The Hall voltage does not increase indefinitely but instead, reaches a constant value when the force on the charge carrier due to the magnetic field is equal to the force due to the electric field set up as a result of the Hall voltage.

The size of the Hall voltage depends on the material of the conductor, the current in the sample and the magnetic flux density. If the current is kept constant, then the Hall voltage across a sample will be proportional to the magnetic flux density. The Hall effect provides a means by which flux densities may be measured using a Hall probe. In general, the material used

Figure 11.28 The Hall effect.

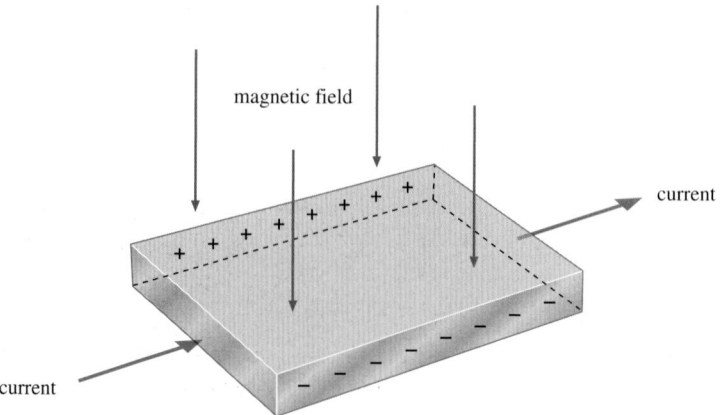

for a Hall probe is a semiconductor. This gives a Hall voltage many thousands of times greater than that for a metal under the same conditions.

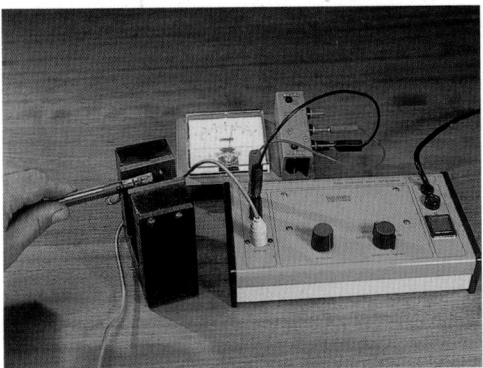

The Hall probe.

Example

A charged particle has mass 6.7×10^{-27} kg and charge $+3.2 \times 10^{-19}$ C. It is travelling at speed 2.5×10^8 m s^{-1} when it enters a region of space where there is a uniform magnetic field of flux density 1.6 T at right angles to its direction of motion. Calculate:

(a) the force on the particle due to the magnetic field,
(b) the radius of its orbit in the field.

(a) Force $= BQv \sin \theta$
$\qquad = 1.6 \times 3.2 \times 10^{-19} \times 2.5 \times 10^8 \times 1$
$\qquad = \mathbf{1.3 \times 10^{-10}}$ **N**.

(b) Centripetal force is provided by the electromagnetic force
$$mv^2/r = BQv$$
$$6.7 \times 10^{-27} \times (2.5 \times 10^8)^2/r = 1.3 \times 10^{-13}$$
$$\boldsymbol{r = 3.2}\ \textbf{m}$$

Now it's your turn

1 An electron of mass 9.1×10^{-31} kg and charge -1.6×10^{-19} C is travelling at a speed of 7.0×10^7 m s^{-1} when it enters a region of space where there is a uniform magnetic field of flux density 4.5 mT at right angles to its direction of motion. Calculate:
   (a) the force on the electron due to the magnetic field,
   (b) the radius of its orbit in the field.
   Ans: (a) 5.0×10^{-14} N; (b) 8.8 cm

2 An electron and an α-particle travelling at the same speed both enter the same region of uniform magnetic flux which is at right angles to their direction of motion. State and explain any differences between the two paths in the field.
   Ans: Deflections in opposite directions because the particles are oppositely charged. Electron has smaller radius of path because q/m is larger.

Section 11.2 summary

★ There is a force on a current-carrying conductor whenever it is at an angle to a magnetic field.

★ The direction of the force is given by Fleming's left-hand rule – place the first two fingers and thumb of the left hand at right angles to each other, first finger in the direction of the magnetic field, second finger in the direction of the current, then the thumb gives the direction of the force.

★ Magnetic flux density (field strength) is measured in tesla (T). $1\text{ T} = 1\text{ Wb m}^{-2}$.

★ The magnitude of the force F on a conductor of length L carrying a current I at an angle θ to a magnetic field of flux density B is given by the expression

$$F = BIL \sin \theta$$

★ The force between parallel current-carrying conductors provides a means by which the ampere may be defined in terms of SI units.

★ A current-carrying coil may experience a torque when it is placed in a magnetic field.

★ The force F on a particle with charge Q moving at speed v at an angle θ to a magnetic field of flux density B is given by the expression

$$F = BQv \sin \theta$$

★ The path of a charged particle, moving at constant speed in a plane at right angles to a uniform magnetic field, is circular.

Section 11.2 questions

1 A stiff straight wire has a mass per unit length of 45 g m^{-1}. The wire is laid on a horizontal bench and a student passes a current through it to try to make it lift off the bench. The horizontal component of the Earth's magnetic field is 18 µT in a direction from south to north and the acceleration of free fall is 10 m s^{-2}.

   (a) i) State the direction in which the wire should be laid on the bench.

       ii) Calculate the minimum current required.

   (b) Suggest whether the student is likely to be successful with his experiment.

2 The magnetic flux density B at a distance r from a long straight wire carrying a current I is given by the expression

$$B = 2.0 \times 10^{-7} \times \frac{I}{r},$$

where r is in metres and I is in amps.

Section 11.2 questions

(a) Calculate:
  i) the magnetic flux density at a point distance 2.0 cm from a wire carrying a current of 15 A.
  ii) the force per unit length on a second wire, also carrying a current of 15 A, which is parallel to, and 2.0 cm from, the first wire.

(b) Suggest why the force between two wires is demonstrated in the laboratory using aluminium foil rather than copper wires.

3 A small horseshoe magnet is placed on a balance and a stiff wire is clamped in the space between its poles. The length of wire between the poles is 5.0 cm. When a current of 3.5 A is passed through the wire, the reading on the balance increases by 0.027 N.

(a) Calculate the magnetic flux density between the poles of the magnet.
(b) State three assumptions which you have made in your calculation.

4 The diagram shows the track of a particle in a bubble chamber as it passes through a thin sheet of metal foil. A uniform magnetic field was applied into the plane of the paper. State with a reason:
(a) in which direction the particle was moving,
(b) whether the particle is positively or negatively charged.

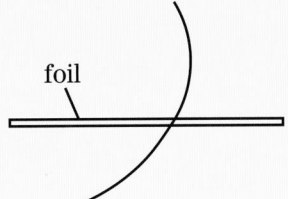

11.3 Electromagnetic induction

Magnetic flux

Figures 11.7, 11.9 and 11.10 illustrate the pattern of the magnetic field in the region of various conductors. However, the patterns are not complete. In fact, the magnetic field lines should be continuous, as illustrated in Figure 11.29. Early experimenters thought that there was a flow of something along these lines and this gave rise to the idea of a magnetic flux, since 'flux' means 'flow'. Magnetic flux density (field strength) varies and this is shown by the spacing of the lines.

> Magnetic flux density may be considered as the number of lines of magnetic force per unit area of an area at right angles to the lines.
> **Magnetic flux** indicates the total number of lines and this is given the symbol ϕ.

Magnetic flux is the product of the magnetic flux density and the area normal to the lines of flux.

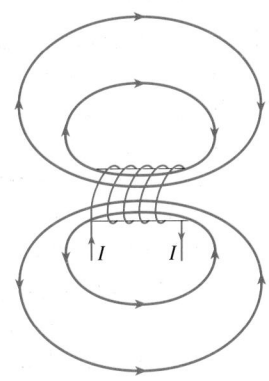

Figure 11.29 *Continuous magnetic field lines.*

★ For a uniform magnetic field of flux density B which makes an angle θ with an area A, the magnetic flux ϕ is given by the expression
$\phi = BA \sin \theta$.

The unit of magnetic flux is the weber (Wb). One weber is equal to one tesla metre-squared i.e. $T\ m^2$.

Section 11.3 summary

★ Magnetic field is a general term used for a region of space where a current-carrying conductor experiences an electromagnetic force.

★ Magnetic flux density (magnetic field strength) measures the strength of a magnetic field and has a magnitude and direction at every point in the field.

★ Magnetic flux is the product of flux density and area normal to the flux.

The concept of magnetic flux is used frequently when studying electromagnetic induction.

Electromagnetic induction

The link between electric current and magnetic field was discovered by Oersted (1819). In 1831, Henry in the United States and Faraday in England demonstrated that an e.m.f. could be induced by a magnetic field. The effect was called **electromagnetic induction**.

Figure 11.30 Apparatus to demonstrate electromagnetic induction.

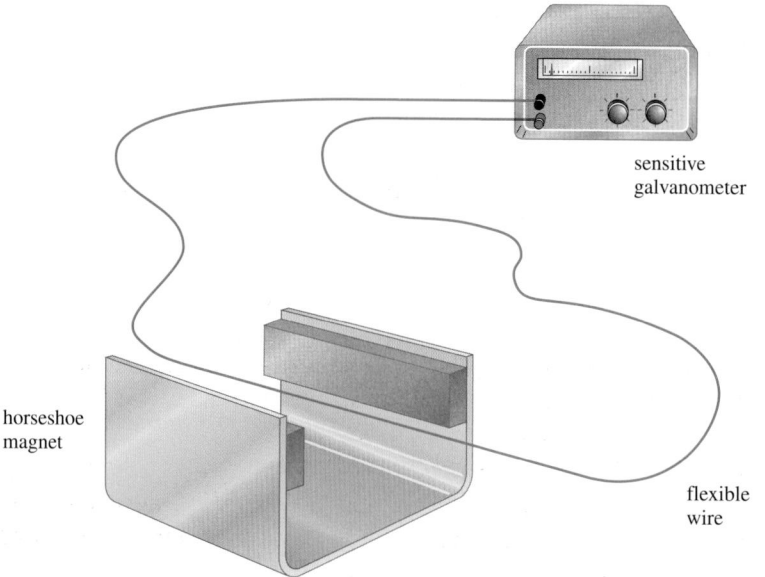

sensitive
galvanometer

horseshoe
magnet

flexible
wire

Electromagnetic induction is now easy to demonstrate in the laboratory because sensitive meters are available. Figure 11.30 illustrates apparatus which may be used for this purpose. The galvanometer detects very small currents but it is important to realise that what is being detected are small electromotive forces (e.m.f.s). The current arises because there is a complete circuit incorporating an e.m.f. The following observations can be made.

- An e.m.f. is induced when
  i)   the wire is moved through the magnetic field, across the face of the pole-pieces,
  ii) the magnet is moved so that the wire passes across the face of the pole-pieces.
- An e.m.f. is *not* induced when
  i)   the wire is held stationary between the pole-pieces,
  ii) the magnet is moved so that the pole-pieces move along the length of the wire,
  iii) the wire moves lengthways so that it does not change its position between the poles of the magnet.

These observations lead to the conclusion that an e.m.f. is induced whenever lines of magnetic flux are cut. The cutting may be caused by a movement of either the wire or of the magnet. The magnitude of the e.m.f. is also observed to vary.

- The magnitude of the e.m.f.
  i)   increases as the speed at which the wire is moved increases,
  ii) increases as the speed at which the magnet is moved increases,
  iii) increases if the wire is made into a loop with several turns (see Figure 11.31),
  iv) increases as the number of turns on the loop increases.

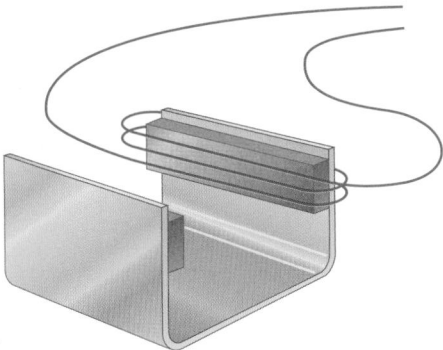

Figure 11.31 *Wire wound to form a loop of several turns.*

It can be concluded that the magnitude of the induced e.m.f. depends on the rate at which magnetic flux lines are cut. The rate may be changed by varying the rate at which the flux lines are cut by a single wire or by using different numbers of turns of wire. The two factors are taken into account by using the term **magnetic flux linkage**. Change in magnetic flux linkage $\Delta(N\phi)$ is equal to the product of the change in magnetic flux $\Delta\phi$ and the number of turns N of a conductor involved in the change in flux.

11.3 Electromagnetic induction

> ★ *Change in magnetic flux linkage* $\Delta(N\phi) = N\Delta\phi$

The experimental observations are summarised in **Faraday's law of electromagnetic induction**.

> ★ The e.m.f. induced is proportional to the rate of change of magnetic flux linkage.

The experimental observations made with the apparatus of Figures 11.30 and 11.31 have been concerned with the magnitude of the e.m.f. However, it is noticed that the direction of the induced e.m.f. changes and that the direction is dependent on the direction in which the magnetic flux lines are being cut. The direction of the induced e.m.f. or current in a wire moving through a magnetic field at right angles to the fields may be determined using **Fleming's right-hand** rule.

> If the first two fingers and thumb of the right hand are held at right angles to one another, the **F**irst finger in the direction of the magnetic **F**ield and the thu**M**b in the direction of **M**otion, then the se**C**ond finger gives the direction of the induced e.m.f. or **C**urrent.

The rule is illustrated in Figure 11.32.

Figure 11.32 *Fleming's right-hand rule.*

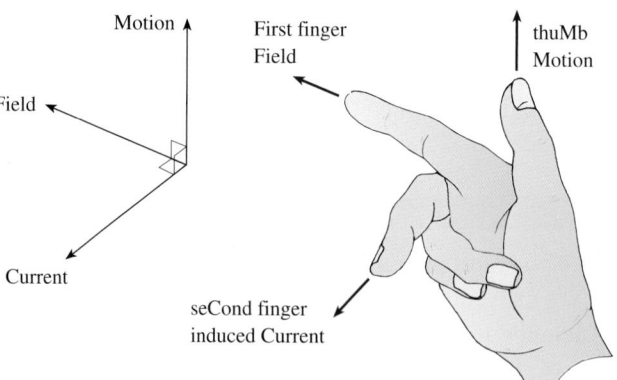

An explanation for the direction of the induced e.m.f. can be found by reference to the motor effect and conservation of energy.

Figure 11.33 shows a wire being moved downwards through a magnetic field. Since the wire is in the form of a continuous loop, the induced e.m.f. gives rise to a current, and the direction of this current can be found using Fleming's right-hand rule. This current is at right angles to the magnetic flux and, by the motor effect, there will be a force on the wire. Using Fleming's left-hand rule, the force is upwards when the wire is moving downwards. Reversing the direction of motion of the wire causes a current in the opposite

Figure 11.33

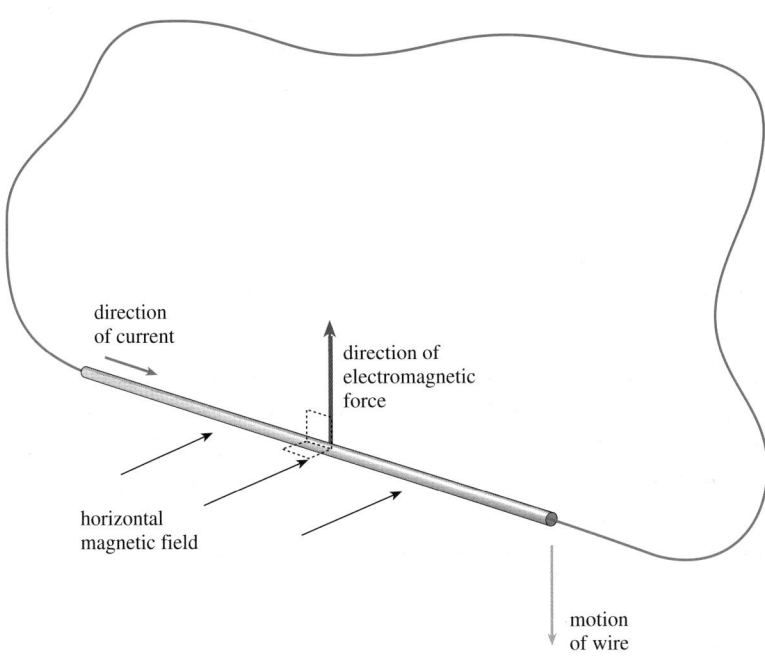

direction and hence the electromagnetic force would once again oppose the motion. This conclusion is not surprising when conservation of energy is considered. An electric current is a form of energy and this energy must have been converted from some other form. Movement of the wire against the electromagnetic force means that work has been done in overcoming this force and it is this work which is seen as electrical energy. Anyone who has ridden a bicycle with a dynamo will realise that work has to be done to light the lamp!

This application of conservation of energy is summarised in **Lenz's law**.

> ★ The direction of the induced e.m.f. is such as to cause effects to oppose the change producing it.

Faraday's law of electromagnetic induction and Lenz's law may be summarised using the equation

$$\bigstar \; e = -\frac{\mathrm{d}(N\phi)}{\mathrm{d}t},$$

where e is the e.m.f. induced by a rate of change of flux linkage of $\mathrm{d}(N\phi)/\mathrm{d}t$. The minus sign indicates that the induced e.m.f. causes effects to oppose the change producing it.

The conversion of mechanical energy to electrical energy may be shown by spinning a metal disc in a magnetic field, as illustrated in Figure 11.34. The disc is seen to slow down much more rapidly with the magnet in place than when it has been removed. As the disc spins, it cuts through the flux lines of the magnet. This cutting becomes more obvious if a radius of the disc is considered. As the radius rotates, it will cut flux. An e.m.f. will be induced in the disc but, because the rate of cutting of flux varies from one part of the disc to another, the e.m.f. will have different magnitudes in different regions. The disc is metal and therefore electrons will move between regions with different e.m.f. values. Currents are induced in the disc. Since these currents vary in magnitude and direction there are called **eddy currents**. The eddy currents cause heating in the disc and the dissipation of the energy of rotation of the disc is referred to as **eddy current damping**. If the permanent magnet in Figure 11.34 is replaced by an electromagnet, the spinning disc will be slowed down whenever there is a current in the electromagnet. This is the principle behind electromagnetic braking. The advantage over conventional brakes is that there is no physical contact with the spinning disc. This makes such brakes very useful for trains travelling at high speeds. However, the disadvantage is that, as the disc slows down, the induced eddy currents will be smaller and therefore the braking will be less efficient. This system would be useless as the parking brake on a car!

Figure 11.34 *Aparatus to demonstrate eddy current damping.*

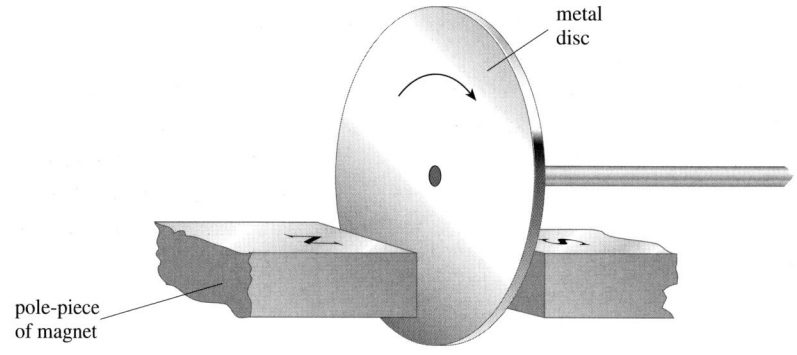

metal disc

pole-piece of magnet

Example

The uniform flux between the poles of a magnet is 0.080 T. A small coil of area of cross-section 6.5 cm^2 has 250 turns and is placed with its plane normal to the magnetic field. The coil is withdrawn from the field in a time of 0.26 s.
Determine:
(a) the magnetic flux through the coil when it is between the poles of the magnet,
(b) the change in magnetic flux linkage when the coil is removed from the field,
(c) the average e.m.f. induced in the coil whilst it is being withdrawn.

(a) *magnetic flux* $\phi = BA \sin \theta$
$$= 0.080 \times 6.5 \times 10^{-4}$$
$$= \mathbf{5.2 \times 10^{-5} \ Wb}$$

11.3 Electromagnetic induction

(b) *change in flux linkage* $= (N\phi)_{\text{FINAL}} - (N\phi)_{\text{INITIAL}}$
$$= 0 - (250 \times 5.2 \times 10^{-5})$$
$$= -1.3 \times 10^{-2} \text{ Wb}$$

(the sign indicates that the flux linkage is decreasing)

(c) induced e.m.f. $= \dfrac{\text{change in flux linkage}}{\text{time taken}}$

$$= (1.3 \times 10^{-2})/0.26$$
$$= 0.050 \text{ V}$$

Now it's your turn

1 An aircraft has a wingspan of 17 m and is flying horizontally in a northerly direction at a speed of 94 m s^{-1}. The vertical component of the Earth's magnetic field is 40 µT in a downward direction.

   (a) Calculate:

      i)   the area swept out per second by the wings,

      ii) the magnetic flux cut per second by the wings,

      iii) the e.m.f. induced between the wingtips.

   (b) State which wing-tip will be at the higher potential.

   Ans: (a) 1600 m^2, 0.064 Wb, 0.0641 V; (b) wing pointing west

2 A current-carrying solenoid produces a uniform magnetic flux of density 4.6×10^{-2} T along its axis. A small circular coil of radius 1.2 cm has 350 turns of wire and is placed on the axis of the solenoid with its plane normal to the axis. Calculate the average e.m.f. induced in the coil when the current in the solenoid is switched off in a time of 85 ms.

   Ans: 86 mV

3 A metal disc is made to spin at 15 revolutions per second about an axis through its centre normal to the plane of the disc. The disc has radius 24 cm and spins in a uniform magnetic field of flux density 0.15 T, parallel to the axis of rotation.

   Calculate:

   i)   the area swept out each second by a radius of the disc,

   ii) the flux swept out each second by a radius of the disc,

   iii) the e.m.f. induced in the disc.

   Ans: 2.7 m^2, 0.41 Wb, 0.41 V

e.m.f. induced between two coils

A current-carrying solenoid or coil is known to have a magnetic field. If the current in the coil is switched off, there will be a change in flux in the region around the coil. Consider the apparatus illustrated in Figure 11.35.

Figure 11.35

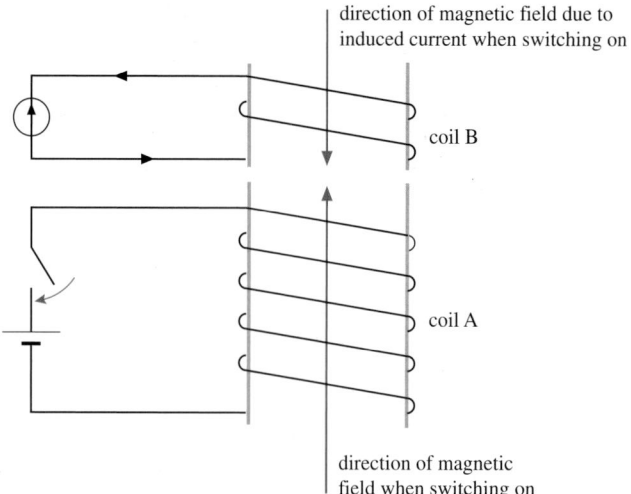

direction of magnetic field due to
induced current when switching on

coil B

coil A

direction of magnetic
field when switching on

As the current in the coil A is being switched on, the magnetic field in this coil grows. The magnetic field links with the turns on coil B and, as a result, there is a change in flux linkage in coil B and an e.m.f. is induced in this coil. Coil B forms part of a complete circuit and hence there is a current in coil B. The direction of this current can be determined using Lenz's law.

The change which brought about the induction of a current was a *growth* in the magnetic flux in coil A. The induced current in coil B will give rise to a magnetic field in coil B and this field will, by Lenz's law, try to oppose the growth of the field in coil A. Consequently, since the field in coil A is vertically upwards (the right-hand grip rule), the field in coil B will be vertically downwards and the induced current will be in an anticlockwise direction through the meter. When the current in coil A is switched off, the magnetic field in coil A will *decay*. The magnetic field in coil B due to the induced current must try to prevent this decay and hence it will be vertically upwards. The induced current has changed in direction.

The magnitude of the induced e.m.f. can be increased by inserting a soft iron core into the coil (but be careful not to damage the meter as any induced e.m.f. will be very much greater) or by increasing the number of turns on the coils or by switching a larger current in coil A.

It is important to realise that an e.m.f. is induced only when the magnetic flux in coil A is changing. That is, the current in coil A is changing. A steady current in coil A will not give rise to an induced e.m.f. An e.m.f. may be induced continuously in coil B if an alternating current is provided for coil A. This is the principle of the **transformer**.

A simple transformer is illustrated in Figure 11.36. Two coils of insulated wire are wound on to a laminated soft-iron core. An alternating e.m.f. is applied across one coil (the primary coil) and an e.m.f. is induced in the secondary coil. The ratio of the output e.m.f. to the applied e.m.f. is dependent on the ratio of the number of turns on the two coils. The applied e.m.f. and the e.m.f. across the secondary coil have the same frequency but are out of phase by π rad (180°).

Figure 11.36

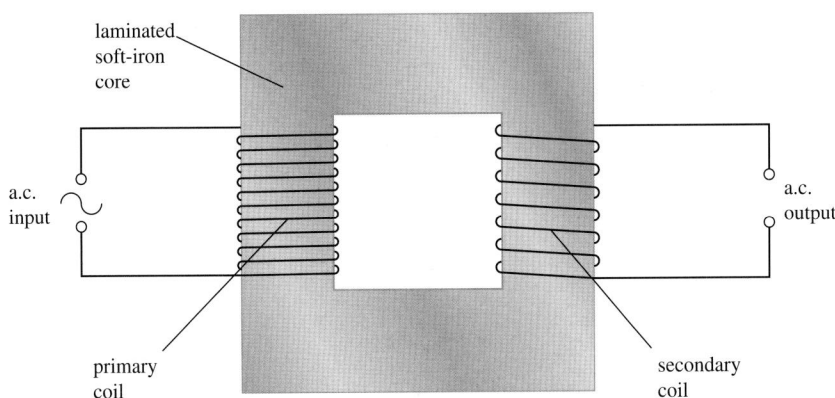

laminated
soft-iron
core

a.c.
input

a.c.
output

primary
coil

secondary
coil

Consider a transformer with N_p turns and N_s turns on the primary and secondary coils respectively, with currents in the coils of I_p and I_s when the e.m.f.s across the coils are V_p and V_s.

For a transformer which is 100% efficient, (an **ideal** transformer),

$$input\ power = output\ power$$
$$V_p I_p = V_s I_s,$$

and

$$\bigstar \quad \frac{N_s}{N_p} = \frac{V_s}{V_p} = \frac{I_p}{I_s}$$

When V_s is greater than V_p, there are more turns on the secondary coil than the primary and the transformer is said to be a **step-up transformer**. Conversely, if V_s is less than V_p, the secondary coil has fewer turns than the primary and the transformer is said to be a **step-down transformer**.

In practice, the transformer will not be 100% efficient due to power losses. Some sources of these losses are:

- loss of magnetic flux between the primary and secondary coils
- resistive heating in the primary and secondary coils
- heating of the core due to eddy currents
- heating of the core due to repeated magnetisation and demagnetisation

The use of soft iron reduces heating due to magnetisation and demagnetisation of the core. However, since iron (an electrical conductor) is used as the material for the core so that the magnetic flux linkage is large, eddy currents cannot be prevented. Laminating the core, i.e. building up the core from thin strips of soft iron which are electrically insulated from one another, reduces eddy current losses.

Transformers are produced in many shapes and sizes.

Example

A transformer is to be used with an alternating supply of 240 V to power a lamp rated as 12 V, 3 A. The secondary coil of the transformer has 80 turns. Assuming that the transformer is ideal, calculate, for the primary coil:

(a) the number of turns,
(b) the current.

(a) For an ideal transformer,

$$V_s/V_p = N_s/N_p$$
$$12/240 = 80/N_p$$
$$N_p = \textbf{1600 turns}$$

(b) For an ideal transformer,

$$V_s/V_p = I_p/I_s$$
$$12/240 = I_p/3$$
$$I_p = \textbf{0.15 A}$$

Now it's your turn

1   The primary coil of an ideal transformer has 1200 turns and is connected to a 240 V alternating supply. The transformer is to be used to step down the voltage to 9.0 V.
    (a) Calculate the number of turns on the secondary coil.
    (b) An appliance, rated as 9.0 V, 1.5 A is connected to the secondary coil of the transformer. Calculate the current in the primary coil.
    Ans: (a) 45; (b) 56 mA

2   An ideal transformer is to be used to step up a voltage from 220 V to supply a current of 15 mA at 6.6 kV. The primary coil has 250 turns. Calculate:
    (a) the ratio of the number of turns on the secondary coil to that on the primary coil,
    (b) the number of turns on the secondary coil,
    (c) the current in the primary coil.
    Ans: (a) 30; (b) 7500; (c) 450 mA

Section 11.3 summary

★   The direction of the induced current in a conductor moving through a magnetic field is given by Fleming's right-hand rule. That is:

If the first two fingers and thumb of the right hand are held at right angles to each other, the first finger in the direction of the magnetic field and the thumb in the direction of motion, then the second finger gives the direction of the induced e.m.f. or current.

★ Faraday's law of electromagnetic induction states that the e.m.f. induced is proportional to the rate of change of magnetic flux linkage.

★ Lenz's law states that the direction of the induced e.m.f. is such as to cause effects to oppose the change producing it.

★ Faraday's law of electromagnetic induction and Lenz's law may be summarised using the equation

$$e = -\frac{d(N\phi)}{dt}$$

where e is the e.m.f. induced by a rate of change of flux linkage of $d(N\phi)/dt$. The sign indicates the relative direction of the e.m.f. and the change in flux linkage.

Section 11.3 questions

1 A coil is constructed by winding 400 turns of wire on to a cylindrical iron core. The mean radius of the coil is 3.0 cm. It is found that the flux density B in the core due to a current I in the coil is given by the expression

$$B = 2.2\, I,$$

where B is in tesla.

(a) Calculate, for a current of 0.64 A in the coil:
   i) the magnetic flux density in the core,
   ii) the magnetic flux in the core,
   iii) the flux linkage of the coil.

(b) The current in the coil is switched off in a time of 0.011 s. Calculate the e.m.f. induced and state where this e.m.f. will be observed.

(c) Hence suggest why, when switching off a large electromagnet, the current is reduced gradually rather than switched off suddenly.

2 A flat coil contains 250 turns of insulated wire and has a mean radius of 1.5 cm. The coil is placed in a region of uniform magnetic flux of flux density 85 mT such that there is an angle θ between the plane of the coil and the flux lines. The coil is withdrawn from the magnetic field in a time of 0.30 s. Calculate the average e.m.f. induced in the coil for an angle θ of:
   (a) zero,
   (b) 90°,
   (c) 35°.

3 An ideal transformer is to be used to step down a 220 V alternating supply to 7.5 V.

(a) Calculate the ratio of the number of turns on the secondary coil to that on the primary coil,

(b) The number of turns on the primary coil given that there are 140 turns on the secondary coil.

(c) Suggest why the wire forming the secondary coil would, in this case, be thicker than that for the primary coil.

Exam Questions

1 A long straight current-carrying wire is held in a north-south direction vertically above a small compass. The vertical distance between the compass and the wire is 4.6 cm.
The horizontal component of the Earth's field is 18 µT and the magnetic flux density B due to a current I in a long straight wire is given by the expression

$$B = 2.0 \times 10^{-7} \times \frac{I}{r}$$

where B is measured in tesla and r is the distance from the wire in metres.

(a) State the direction in which the compass points when there is no current in the wire.
(b) A steady current of 6.0 A is switched on in the wire.
  i) State the direction of the magnetic field of the wire in the region of the compass.
  ii) Calculate the magnetic flux density due to the current in the wire at the position of the compass.
  iii) Determine the angle through which the compass will rotate when the current in the wire is switched on.
(c) The compass is replaced by a long straight wire so that this wire is parallel to and 4.5 cm from the first wire. This second wire carries a current of 7.7 A. Calculate the force per unit length between the two wires.

2 A square coil of wire of side 2.5 cm is made to rotate about an axis such that a radial magnetic field is always in the plane of the coil. The coil has 250 turns and the magnetic field has flux density 8.4 mT in the region of the wire of the coil.

(a) Suggest how a radial magnetic field may be produced.
(b) The coil carries a current of 12 mA. Calculate the torque on the coil due to this current.
(c) The coil is controlled by a spiral spring which allows the coil to rotate through an angle of 20° for every additional 1.0×10^{-6} N m of torque applied to the coil.
  i) Calculate the angle through which the coil rotates before coming to rest when a current of 12 mA is passed through it.
  ii) Calculate the current in the coil for an angle of rotation of 48°.

3 The magnetic flux density B at the centre of a long solenoid carrying a current I is given by the expression

$$B = 2.5\,I$$

where B is in millitesla and I is in amps. The cross-section of the solenoid has a diameter of 2.5 cm. A length of wire is wound tightly round the centre of the solenoid to form a small coil of 50 turns, as illustrated below.

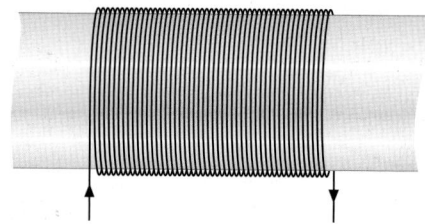

Figure 11.37

(a) Calculate, for a current of 1.8 A in the solenoid:
  i) the flux density in the solenoid,
  ii) the magnetic flux linkage of the small coil.
(b) The current is reversed in direction in 0.031 s. Calculate the average e.m.f. induced in the small coil.

4 Two coils A and B are placed as shown in Figure 11.38.

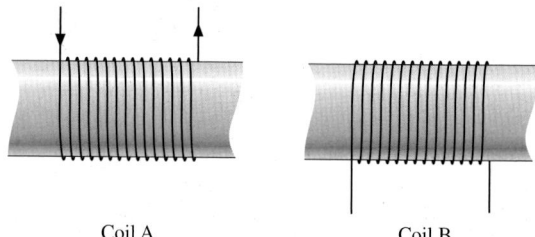

Coil A   Coil B

Figure 11.38

The current in coil A is switched on and then off so that the variation with time t of the current I is as shown in Figure 11.39.

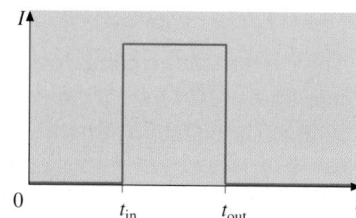

Figure 11.39

Sketch a graph to show the variation with time t of the e.m.f. E induced in coil B.

CHAPTER TWELVE
Nuclear physics

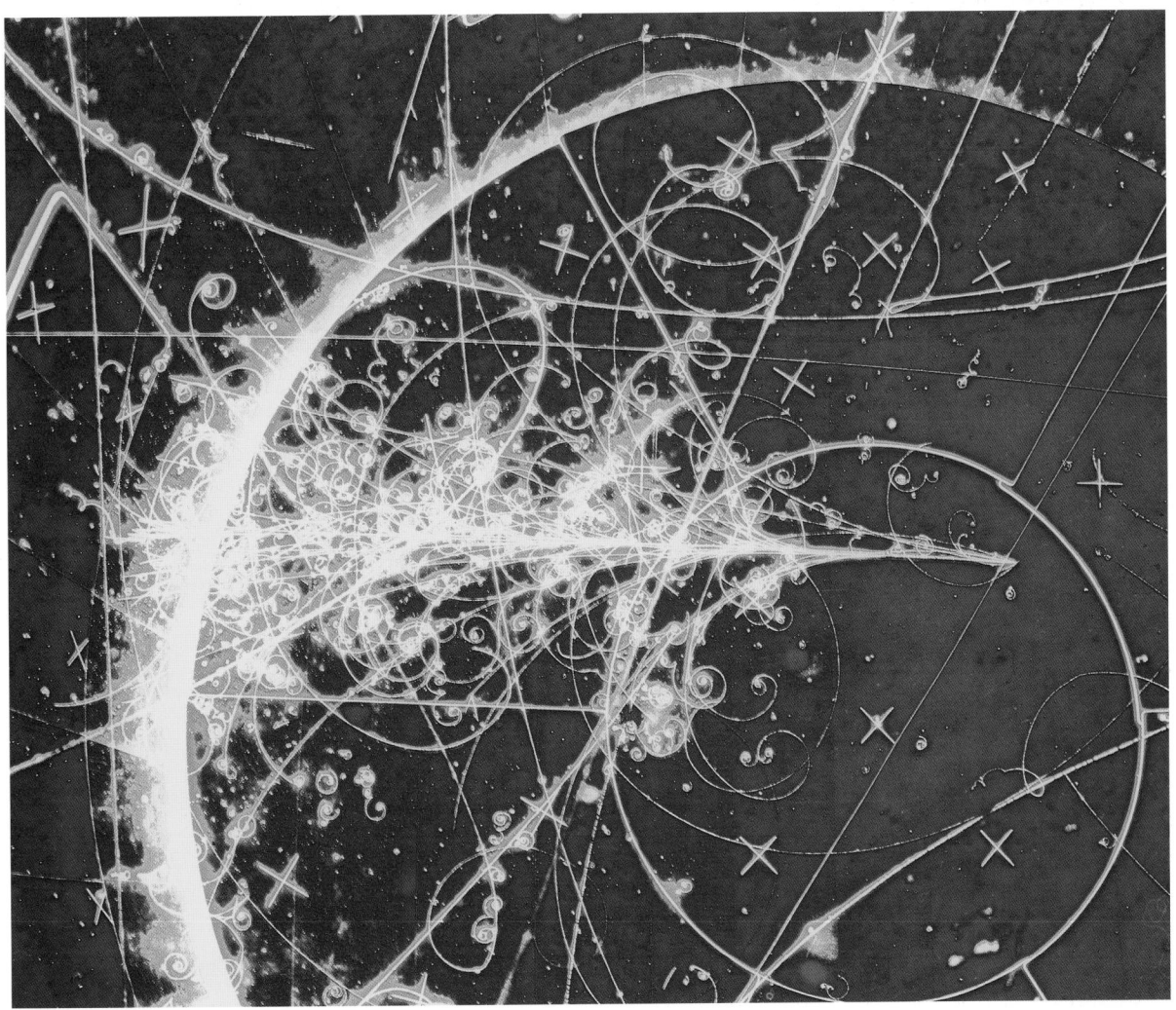

In the early 1900s Rutherford showed that the atom consisted of a very small nucleus containing the majority of the atom's mass, surrounded by orbiting electrons. In this chapter we shall look at some of the physics of the nucleus. We start by looking at the structure of the nucleus. The decay of unstable nuclei leads to radioactive emissions. We shall investigate the properties of these emissions. Nuclei may also be transformed by particle impact. We shall look at the processes of fission and fusion, with particular reference to energy changes.

Tracks made by sub-atomic particles in a bubble chamber.

12.1 Atomic structure

> ★ The atoms of all elements are made up of three particles called **protons**, **neutrons**, and **electrons**. The protons and neutrons are at the centre or nucleus of the atom. The electrons orbit the nucleus.

We shall see later that the diameter of the nucleus is only about 1/10 000 of the diameter of an atom.

Figure 12.1 illustrates very simple models of a helium atom and a lithium atom.

Figure 12.1 *Structures of a) a helium atom and b) a lithium atom.*

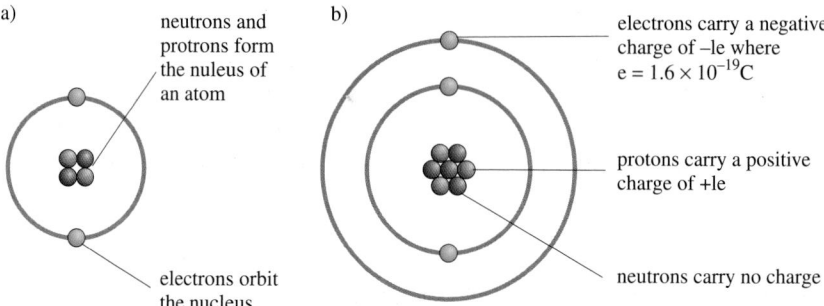

a)
neutrons and protrons form the nuleus of an atom

electrons orbit the nucleus

b)
electrons carry a negative charge of –1e where e = 1.6 × 10⁻¹⁹C

protons carry a positive charge of +1e

neutrons carry no charge

The protons and neutrons both have a mass of about one atomic mass unit (1 u). (Remember that $1\ u = 1.67 \times 10^{-22}$ kg.) By comparison, the mass of an electron is very small, about 1/2000 of 1 u. The vast majority of the mass of the atom is therefore in the nucleus.

The basic properties of the proton, neutron and electron are summarised in Figure 12.2.

Figure 12.2

	approximate mass	charge	position
proton	u	$+e$	in nucleus
neutron	u	0	in nucleus
electron	u/2000	$-e$	orbiting nucleus

Atoms and ions

Atoms are uncharged because they contain equal numbers of protons and electrons. If an atom loses one or more electrons so that it does not contain an equal number of protons and electrons, it becomes charged and is called an **ion**.

For example, if a sodium atom loses one of its electrons it becomes a positive sodium ion.

$$\text{Na} \quad \rightarrow \quad \text{Na}^+ \quad + \quad \text{e}^-$$

sodium atom $\qquad$ sodium ion $\qquad$ electron

If an atom gains an electron, it becomes a negative ion.

Proton number and nucleon number

★ The number of protons in the nucleus of an atom is called the **proton number** (or **atomic number**).

★ The number of protons together with the number of neutrons in the nucleus is called the **nucleon number** (or **mass number**).

This is because a **nucleon** is the name given to either a proton or a neutron. The difference between the nucleon number A and the proton number Z gives the number of neutrons in the nucleus. Because atoms are electrically neutral, the proton number is equal to the number of electrons orbiting the nucleus.

Nuclear representation

The nucleus of one form of sodium contains 11 protons and 12 neutrons. Therefore its proton number Z is 11 and the nucleon number A is $11 + 12 = 23$. This nucleus can be shown as $^{23}_{11}\text{Na}$. In general, a nucleus can be represented as

$$^{\text{nucleon number}}_{\text{proton number}}\text{X}$$

where X is the chemical symbol.

Example

1   An oxygen nucleus is represented by $^{16}_{8}\text{O}$. Describe its atomic structure.

   The nucleus has a proton number of 8 and a nucleon number of 16. Thus, its nucleus contains *8 protons* and $16 - 8 = 8$ *neutrons*. There are also *8 electrons* (equal to the number of protons) orbiting the nucleus.

Now it's your turn

Write down the proton number and the nucleon number for the potassium nucleus $^{40}_{19}\text{K}$. Deduce the number of neutrons in the nucleus.

   Ans: 19; 40; 21

CHAPTER 12

Isotopes

Sometimes atoms of the same element have different numbers of neutrons in their nuclei. The most abundant form of chlorine contains 17 protons and 18 neutrons in its nucleus, giving it a nucleon number of $17 + 18 = 35$. This is often called chlorine-35. Another form of chlorine contains 17 protons and 20 neutrons in the nucleus, giving it a nucleon number of 37. This is chlorine-37. Chlorine-35 and chlorine-37 are said to be **isotopes** of chlorine.

> Isotopes are different forms of nuclei of the same element which have the same numbers of protons but different numbers of neutrons in their nuclei.

Some elements have many isotopes, but others have very few. For hydrogen, the most common isotope is hydrogen-1. Its nucleus is a single proton. Hydrogen-2 is called deuterium; its nucleus contains one proton and one neutron. Hydrogen-3, with one proton and two neutrons, is called tritium.

Note that the term isotope is used to describe nuclei with the same proton number (that is, nuclei of the same element) but with different nucleon numbers. You may also come across the term **nuclide**.

> ★ A nuclide is one type of nucleus with a particular nucleon number and a particular proton number.

Section 12.1 summary

★ An atom consists of a nucleus containing protons and neutrons surrounded by orbiting electrons.

★ Most of the mass of an atom is contained in its nucleus.

★ Neutral atoms contain equal numbers of protons and electrons.

★ Atoms which have gained or lost electrons are charged, and are called ions.

★ The nucleon number A of a nucleus is the number of nucleons (protons and neutrons) in the nucleus.

★ The proton number Z of a nucleus is the number of protons in the nucleus hence the number of neutrons in the nucleus is $A - Z$.

★ A nucleus (chemical symbol X) may be represented by
$$_{\text{proton number}}^{\text{nucleon number}} \text{X}$$

★ Isotopes are different forms of the same element (that is, with the same proton number) but with a different nucleon number.

12.2 α-particles, β-particles and γ-radiation

Some elements have nuclei which are unstable. In order to become more stable, they emit particles and/or electromagnetic radiation. The nuclei are said to be **radioactive**, and the emission is called **radioactivity**. The emissions are invisible to the eye, but their tracks may be made visible in a device called a cloud chamber. The photograph shows tracks created by one type of emission, alpha-particles.

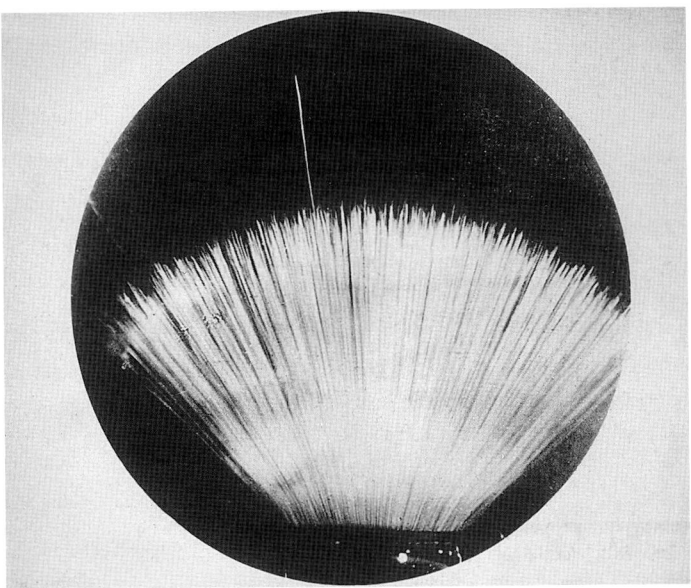

Cloud-chamber tracks of α-particles.

Examination of the nature and properties of the emitted particles or radiation shows that the emissions are of three different types. These are α-particles (alpha-particles), β-particles (beta-particles) and γ-radiation (gamma radiation). All three emissions orginate from the nucleus.

α-particles

> ★ An α-particle is identical to the nucleus of a helium atom.

Like a helium atom, an α-particle contains two protons and two neutrons, and hence carries a charge of $+2e$. α-particles travel at speeds of up to about 10^7 m s^{-1} (about 5% of the speed of light). α-particle emission is the least penetrating of the three types of emission. It can pass through very thin paper, but is unable to penetrate thin card. Its range in air is a few centimetres. Because α-particles are charged, they can be deflected by electric and magnetic fields.

As α-particles travel through matter, they interact with nearby atoms causing them to lose one or more electrons. The ionised atom and the dislodged electron are called an ion pair. The production of an ion pair requires the separation of unlike charges, and this process requires energy. α-particles have a relatively large mass and charge, and consequently they are efficient ionisers. They may produce as many as 10^5 ion pairs for every centimetre of air through which they travel. Thus they lose energy relatively quickly, and have low penetrating power.

When the nucleus of an atom emits an α-particle, it is said to undergo α-decay. The nucleus loses two protons and two neutrons in this emission.

> ★ This means that the proton number of the nucleus decreases by two, and the nucleon number by four.

Each element has a particular proton number, and therefore α-decay causes one element to change into another. (This process is sometimes called **transmutation**.) The original nuclide is called the **parent** nuclide, and the new one the **daughter** nuclide.

For example, uranium-234 (the parent nuclide) may emit an α-particle. The daughter nuclide is thorium-230. In addition, energy is released. This emission is represented by the nuclear equation

$$^{234}_{92}\text{U} \rightarrow {}^{230}_{90}\text{Th} + {}^{4}_{2}\text{He} + \text{energy}.$$

β-particles

> ★ β-particles are fast-moving electrons.

They have speeds in excess of 99% of the speed of light. These particles have half the charge and very much less mass than α-particles. Consequently, they are much less efficient than α-particles in producing ion pairs. They are thus far more penetrating than α-particles, being able to travel up to about a metre in air. They can penetrate card and sheets of aluminium up to a few millimetres thick. Their charge means that they are affected by electric and magnetic fields. However, there are important differences between the behaviour of α- and β-particles in these fields. β-particles carry a negative charge, and are thus deflected in the opposite direction to the positively-charged α-particles. Because the mass of a β-particle is much less than that of an α-particle, β-particles experience a much larger deflection when moving at the same speed as α-particles.

It was stated in Section 12.1 that the nucleus contains protons and neutrons. What, then, is the origin of β-particle emission? The β-particles certainly come from the nucleus, not from the electrons outside the nucleus. Just prior to β-emission a neutron in the nucleus forms a proton and an electron, in order to change the ratio of protons to neutrons in the nucleus. This makes the daughter nucleus more stable.

In fact, *free* neutrons are known to decay like this:

$$\,^1_0n \rightarrow \,^1_1p + \,^0_{-1}e + \text{energy}$$

The same process happens in the nucleus. In β-decay the electron is emitted from the nucleus as a β-particle. This leaves the nucleus with the same number of nucleons as before, but with one extra proton and one less neutron.

> ★ A daughter nuclide has been formed with the proton number increased by one, but with the same nucleon number.

For example, strontium-90 (the parent nuclide) may decay with the emission of a β-particle to form the daughter nuclide yttrium-90. As in the case of α-emission, energy is evolved. The decay is represented by the nuclear equation

$$\,^{90}_{38}Sr \rightarrow \,^{90}_{39}Y + \,^0_{-1}e + \text{energy}.$$

γ-radiation

> ★ γ-radiation is part of the electromagnetic spectrum with wavelengths between 10^{-11} m and 10^{-13} m.

Since γ-radiation has no charge, its ionising power is much less than that of either α- or β-particles. γ-radiation penetrates almost unlimited thicknesses of air, several metres of concrete or several centimetres of lead.

α- and β-particles are emitted by unstable nuclei which have excess energy. The emission of these particles results in changes in the ratio of protons to neutrons, but the nuclei may still have excess energy. The nucleus may return to its unexcited or ground state by emitting energy in the form of γ-radiation.

> ★ For γ-emission, no particles are emitted, and there is therefore no change to the proton number or nucleon number of the parent nuclide.

For example, when uranium-238 decays by emitting an α-particle, the resulting nucleus of thorium-234 contains excess energy (it is in an excited state) and emits a photon of γ-radiation to return to the ground state. This process is represented by the nuclear equation

$$\,^{234}_{90}Th^* \rightarrow \,^{234}_{90}Th + \,^0_0\gamma$$

12.2 α-particles, β-particles, and γ-radiation

The * next to the symbol Th on the left-hand side of the equation shows that the thorium nucleus is in an excited state.

Summary of the properties of α- and β-particles and γ-radiation

Figure 12.3 summarises the properties of the three types of radioactive emission.

Figure 12.3

property	α-particle	β-particle	γ radiation
mass	4 u	about u/2000	0
charge	$+2e$	$-e$	0
nature	helium nucleus (2 protons + 2 neutrons)	electron	short-wavelength electromagnetic waves
speed	up to 0.05 c	more than 0.99 c	c
penetrating power	few cm of air	few mm of aluminium	few cm of lead
relative ionising power	10^4	10^2	1
affects photographic film?	yes	yes	yes
deflected by electric, magnetic fields?	see Figure 12.4	see Figure 12.4	no

Figure 12.4 illustrates a hypothetical demonstration of the effect of a magnetic field on the three types of emission. The direction of the magnetic field is perpendicularly into the page. γ-radiation is uncharged, and is not deflected by the magnetic field. Because α- and β-particles have opposite charges, they are deflected in opposite directions. Note that the deviation of the α-particles is less than that of the β-particles, and that α-particles all deviate by the same amount. β-particles by contrast have a range of deflections indicating that they have a range of energies.

Figure 12.4 Deflection of α- and β-particles and γ-rays by a magnetic field.

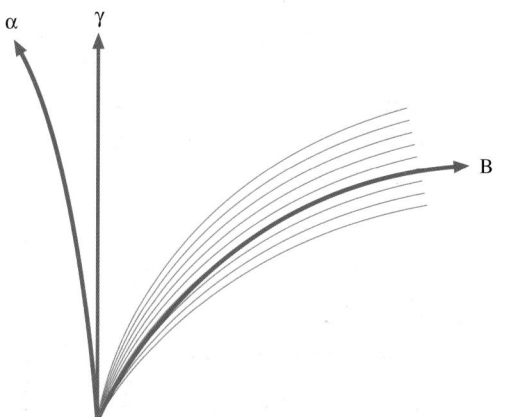

12.2 α-particles, β-particles, and γ-radiation

12.3 Radioactive decay series

The daughter nuclide of a radioactive decay may itself be unstable and so may emit radiation to give another different nuclide. This sequence of radioactive decay from parent nuclide through succeeding daughter nuclides is called a **radioactive decay series**. The series ends when a stable nuclide is reached.

Part of such a radioactive decay series, the uranium series, is shown in Figure 12.5.

decay	radiation emitted
$^{238}_{92}U \rightarrow ^{234}_{90}Th + ^{4}_{2}He + ^{0}_{0}\gamma$	α, γ
$^{234}_{90}Th \rightarrow ^{234}_{91}Pa + ^{0}_{-1}e + ^{0}_{0}\gamma$	β, γ
$^{234}_{91}Pa \rightarrow ^{234}_{92}U + ^{0}_{-1}e + ^{0}_{0}\gamma$	β, γ
$^{234}_{92}U \rightarrow ^{230}_{90}Th + ^{4}_{2}He + ^{0}_{0}\gamma$	α, γ
$^{230}_{90}Th \rightarrow ^{226}_{88}Ra + ^{4}_{2}He + ^{0}_{0}\gamma$	α, γ
$^{226}_{88}Ra \rightarrow ^{222}_{86}Rn + ^{4}_{2}He$	α
$^{222}_{86}Rn \rightarrow ^{218}_{84}Po + ^{4}_{2}He$	α

Figure 12.5

12.4 Detecting radioactivity

Some of the methods used to detect radioactive emissions are based on the ionising properties of the particles or radiation.

The cloud chamber

Figure 12.6 illustrates a cloud chamber. Alcohol in the chamber evaporates. This vapour is cooled by solid carbon dioxide ('dry ice') and, near the base of the chamber, the vapour becomes supersaturated. This means that it will tend to condense. When an α- or β-particle is emitted by a radioactive source, it passes through the supersaturated vapour, creating large numbers of ion pairs. Tiny droplets of liquid form around these ions producing tracks of

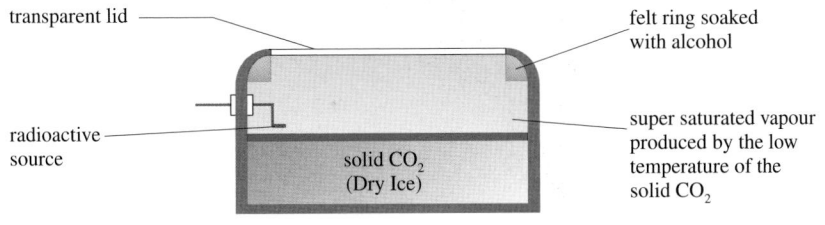

Figure 12.6

condensed vapour, showing the path that the ionising particle has taken. As shown in the photograph on page 313, the tracks created by α-particles are thick, straight, and usually of a definite length. This provides evidence for α-particles being highly ionising and having specific energies. The tracks created by β-particles, as illustrated in the following photograph, are broken and tortuous, since β-particles are less ionising and are easily deflected by air molecules. γ-radiation cannot usually be detected in a cloud chamber because its ionising power is so much smaller. The radiation penetrates and passes right through the chamber with very little ionisation.

Cloud-chamber tracks of β-particles. Please note that the long, straight track through the centre of the image is a consequence of the exposure, and is not a β-particle track.

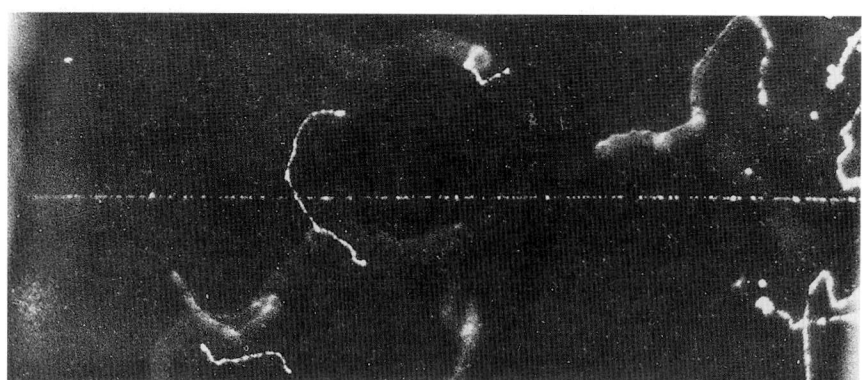

The Geiger counter

Figure 12.7 illustrates a Geiger-Müller tube with a scaler connected to it. When radiation enters the window it creates ion pairs in the gas in the tube. These charged particles, and particularly the electrons, are accelerated by the potential difference between the central wire anode and the cylindrical cathode. These accelerated particles then cause further ionisation. The result of this continuous process is described as an **avalanche effect**. That is, the entry of one particle into the tube and the production of one ion pair results in very large numbers of electrons and ions arriving at the anode and cathode respectively. This gives a pulse of charge which is amplified and counted by the scaler or ratemeter. (A scaler measures the total count of pulses in the tube during the time that the scaler is operating. A ratemeter continuously monitors the number of counts per second.) Once the pulse has been registered the charges are removed from the gas in readiness for further radiation entering the tube.

Figure 12.7 Geiger-Müller tube and scaler.

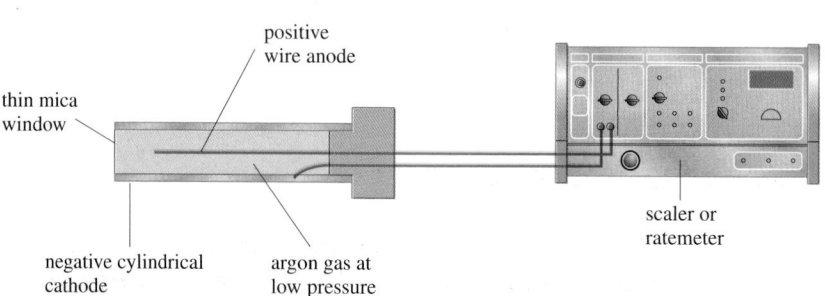

positive wire anode

thin mica window

negative cylindrical cathode

argon gas at low pressure

scaler or ratemeter

Photographic plates

When radioactive emission strikes a photographic film, the film reacts as if it had been exposed to a small amount of visible light. When the film is developed, fogging or blackening is seen. This fogging can be used to detect not only the presence of radioactivity, but also the intensity of the radiation.

Film badge dosimeter.

The photograph illustrates a film badge dosimeter. It contains a piece of photographic film which becomes fogged when exposed to radiation. Workers who are at risk from radiation wear such badges to gauge the type and dose of radiation to which they have been exposed. The radiation passes through different filters before reaching the film. Consequently the type of radiation, as well as the quantity, can be assessed.

The scintillation counter

Early workers with radioactive materials used glass screens coated with zinc sulphide to detect radiation. When radiation is incident on the zinc sulphide, it emits a tiny pulse of light called a **scintillation**. The rate at which these pulses are emitted indicates the intensity of the radiation.

The early researchers worked in darkened rooms, observing the zinc sulphide screen by eye and counting the number of flashes of light occurring in a certain time. Now a *scintillation counter* is used (Figure 12.8). Often a scintillator crystal is used instead of a zinc sulphide screen. The crystal is mounted close to a device known as a photomultiplier, a vacuum-tube device which uses the principle of photoelectric emission (see Chapter 8). Flashes of light cause the emission of photoelectrons from the negative electrode of the photomultiplier. The photoelectric current is amplified inside the tube. The output electrode is connected to a scaler or ratemeter, as with the Geiger-Müller tube.

Figure 12.8 *Scintillation counter.*

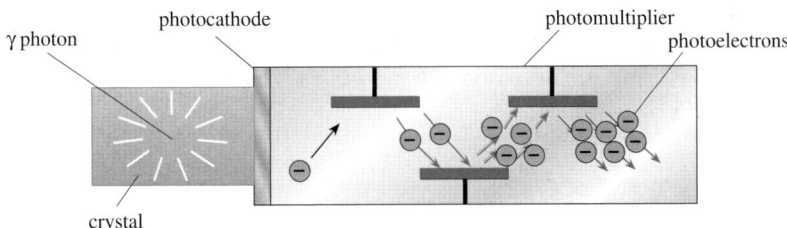

Background radiation

Radioactivity is a natural phenomenon. Rocks such as granite contain small amounts of radioactive nuclides, some foods we eat emit radiation, and even our bodies are naturally radioactive. Although the atmosphere provides life on Earth with some shielding, there is, nevertheless, some radiation from outer space (cosmic radiation). In addition to this natural radioactivity, we are exposed to radiation from man-made sources. These are found in medicine, in fallout from nuclear explosions, and in leaks from nuclear power stations. The sum of all this radiation is known as **background radiation**. Figure 12.9 indicates the relative proportions of background radiation coming from various sources.

Figure 12.9 *Sources of background radiation.*

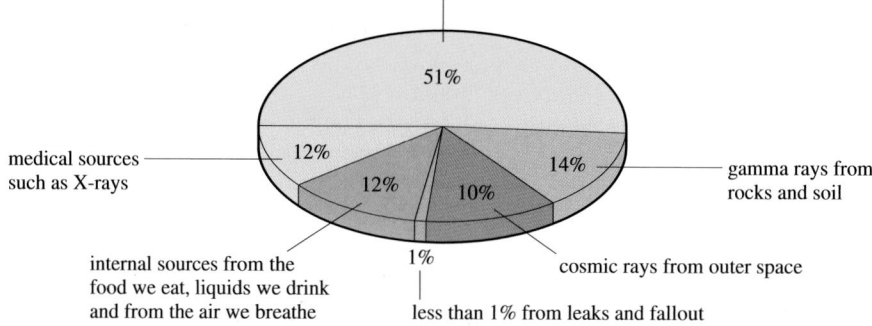

In carrying out experiments with radioactive sources, it is important to take account of background radiation.

The activity of a radioactive source

> ★ The activity of a radioactive source is the number of nuclear disintegrations produced per unit time in the source. Activity is measured in becquerels (Bq), where 1 becquerel is 1 disintegration per second.

$$\bigstar \ 1\,\text{Bq} = 1\,\text{s}^{-1}$$

In order to determine the count-rate due to the radioactive source, the background count-rate must be subtracted from the total measured count-rate. Allowance for background radiation gives the corrected count-rate.

Sections 12.1–12.4 summary

★ An α-particle is a helium nucleus (two protons + two neutrons).

★ A β-particle is a fast-moving electron.

★ γ-radiation consists of short-wavelength electromagnetic waves.

★ Nuclear notation: $^{\text{nucleon number}}_{\text{proton number}}\text{X}$, where X is the symbol for the chemical element.

★ α-particle $^4_2$He; β-particle $^0_{-1}$e; γ-radiation $^0_0$γ.

★ α-emission reduces the nucleon number of the parent nucleus by 4, and reduces the proton number by 2.

★ β-emission causes no change to the nucleon number of the parent nucleus, and increases the proton number by 1.

★ γ-emission causes no change to nucleon number or proton number of the parent nucleus.

★ Activity of a radioactive source is the number of disintigrations per unit time. The unit of activity is the becquerel (Bq): 1 becquerel = 1 disintegration per second.

12.5 The spontaneous and random nature of radioactive decay

The emission of radiation is both *spontaneous* and *random*. It is a spontaneous process because it is not affected by any external factors, such as temperature or pressure. Decay is random in that it is not possible to predict which nucleus in a sample will decay next. There is, however, a constant probability (or chance) that a nucleus will decay in any fixed period of time.

We now look at some of the consequences of the random nature of radioactive decay. If six dice are thrown simultaneously (Figure 12.10), it is likely that one of them will show a six. If twelve dice are thrown, it is likely that two of them will show a six, and so on. While it is possible to predict the likely number of sixes that will be thrown, it is impossible to say which of the dice will actually show a six. We describe this situation by saying that the throwing of a six is a **random process**.

Figure 12.10 *The dice experiment.*

In an experiment similar to the one just described, some students throw a large number of dice (say 6000). Each time a six is thrown, that die is removed. The results for the number of dice remaining after each throw are shown in Figure 12.11.

Figure 12.11

number of throws	Results of dice-throwing experiment	
	number of dice remaining	number of dice removed
0	6000	
1	5000	1000
2	4173	827
3	3477	696
4	2897	580
5	2414	483
6	2012	402
7	1677	335
8	1397	280

Figure 12.12 is a graph of the number of dice remaining against the number of throws. This kind of graph is called a **decay curve**. The rate at which dice are removed is not linear, but there is a pattern. After between 3 and 4 throws, the number of dice remaining has halved. Reading values from the graph shows that approximately 3.8 throws would be required to halve the number of dice. After another 3.8 throws the number has halved again, and so on.

Figure 12.12

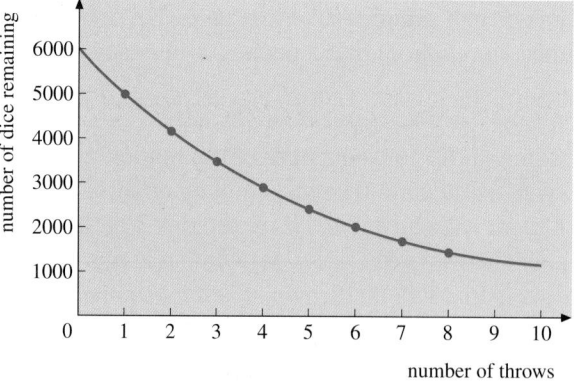

Remember that we have met this type of curve before, in talking about the decay of current in a circuit containing a capacitor and a resistor (Chapter 10).

We can apply the dice experiment to model radioactivity decay. The 6000 dice represent radioactive nuclei. To score a six represents radioactive emission. All dice scoring six are removed, because once a nucleus has undergone radioactive decay, it is no longer available for further decay. Thus, we can describe how rapidly a sample of radioactive material will decay.

A graph of the number of undecayed nuclei in a sample against time has the typical decay-curve shape shown in Figure 12.13. It is not possible to state

12.5 The spontaneous and random nature of radioactive decay

Figure 12.13

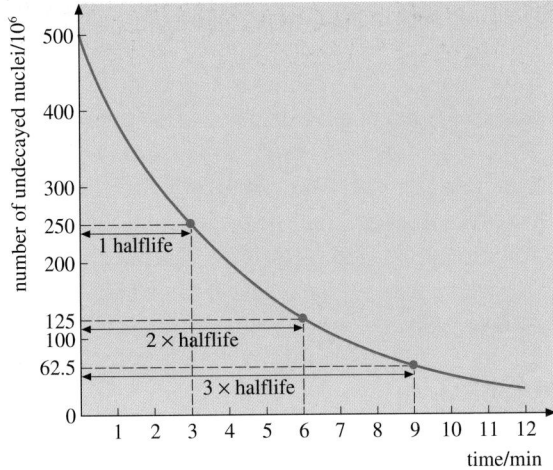

how long the entire sample will take to decay (its 'life'). However, after 3 minutes the number of undecayed nuclei in the sample has halved. After a further 3 minutes, the number of undecayed atoms has halved again. We describe this situation by saying that this radioactive nuclide has a **half-life** $t_{1/2}$ of 3 minutes.

> ★ The half-life of a radioactive nuclide is the time taken for the number of undecayed nuclei to be reduced to half its original number.

The half-lives of different nuclides have a very wide range of values. Examples of some radioactive nuclides and their half-lives are given in Figure 12.14.

radioactive nuclide	half-life
uranium-238	4.5×10^9 years
radium-226	1.6×10^3 years
radon-222	3.8 days
francium-221	4.8 minutes
astatine-217	0.03 seconds

Figure 12.14

We shall see later that half-life may also be expressed in terms of the activity of the material.

Note that half-life is similar to the time constant, which we met in Chapter 10. Both half-life and the time constant relate to the time taken for a certain fraction of the quantity concerned to decay. For the half-life, that fraction is just 1/2. For the time constant, the fraction is 1/e, or about 1/2.7.

Example

The half-life of francium-221 is 4.8 minutes. Calculate the fraction of a sample of francium-221 remaining undecayed after a time of 14.4 minutes.

The half-life of francium-221 is 4.8 min, so after 4.8 min half of the sample will remain undecayed. After two half-lives (9.6 min), $0.5 \times 0.5 = 0.25$ of the sample will remain undecayed. After three half-lives (14.4 min), $0.5 \times 0.25 = 0.125$ will remain undecayed. So the fraction remaining undecayed is **0.125** or **1/8**.

Now it's your turn

Using the half-life values given in Figure 12.14, calculate **(a)** the fraction of a sample of uranium-238 remaining undecayed after 9.0×10^9 years, **(b)** the fraction of a sample of astatine remaining undecayed after 0.30 s, **(c)** the fraction of a sample of radium-226 that has decayed after 3200 years, **(d)** the fraction of a sample of radon-222 that has decayed after 15.2 days.

Ans: (a) 1/4; (b) 1/1024; (c) 3/4; (d) 15/16

12.6 Mathematical descriptions of radioactive decay

Disintegration constant and activity

As we saw in the dice experiment, increasing the number of dice increases the number of sixes that appear with each throw. Similarly, if we investigate the decay of a sample of radioactive material we find that the greater the number of radioactive nuclei in the sample the greater the rate of decay. This can be described mathematically as $dN/dt \, \alpha \, N$.

This gives

$$\star \quad \frac{dN}{dt} = -\lambda N.$$

N is the number of undecayed atoms in the sample. dN/dt is the rate at which the number of nuclei in the sample is changing, so $-dN/dt$ represents the rate of decay. $-dN/dt$ is known as the **activity** of the source, and is measured in becquerels. λ is the constant of proportionality, known as the **disintegration constant**. It has the units s^{-1}, yr^{-1} and so on and is defined as the probability per unit time that a nucleus will undergo decay.

This is an important equation because it relates a quantity we can measure ($-dN/dt$, the rate of decay or activity) to a quantity which cannot in practice be determined (N, the number of undecayed nuclei). We shall see later that the disintegration constant λ is directly related to the half-life $t_{1/2}$, which can be obtained by experiment. This opens the way to calculating the number of undecayed nuclei in a sample. Trying to count nuclei when they are decaying is similar to counting sheep in a field while some are escaping through a gap in the hedge!

Example

Calculate the number of phosphorus-32 nuclei in a sample which has an activity of 5.0×10^6 Bq. (Disintegration constant of phosphorus-32 = $5.6 \times 10^{-7}\,\text{s}^{-1}$.)

From $dN/dt = -\lambda N$, $N = (-dN/dt)/\lambda = -5.0 \times 10^6/5.6 \times 10^{-7}$ = -8.6×10^{12}. The minus sign in this answer arises because dN/dt is the rate of decay. The quantity measured by a ratemeter is the rate of decay, and so should be negative, but it is always displayed as a positive quantity. Similarly, activities are always quoted as positive. So don't be worried about discarding the minus sign here! The number of phosphorus-32 nuclei = **8.6×10^{12}**.

Note also, in this type of calculation, that because the activity is measured in becquerel (s^{-1}), the disintegration constant λ must also be measured in consistent units. If λ had been quoted as 4.2 day^{-1}, it would have been necessary to convert to s^{-1}.

Now it's your turn

Calculate the activity of the following samples of radioactive materials:

(a) 6.7×10^{21} atoms of strontium-90. Disintegration constant of strontium-90 = $8.3 \times 10^{-10}\,\text{s}^{-1}$.

(b) 2.0 mg of uranium-238. 0.238 kg of uranium-238 contains 6.0×10^{23} atoms. Disintegration constant of uranium-238 = $5.0 \times 10^{-13}\,\text{s}^{-1}$.

Ans: (a) 5.6×10^{12} Bq; (b) 2.5×10^6 Bq

To solve the equation $dN/dt = -\lambda N$ requires mathematics beyond the scope of AS/A2 Physics. However, it is important to know the solution, in order to find the variation with time of the number of nuclei remaining in the sample. The solution is

$$\bigstar \quad N = N_0\,e^{-\lambda t},$$

where N_0 is the initial number of undecayed nuclei in the sample, and N is the number of undecayed atoms at time t. This equation is of exactly the same form as the one used to describe the decay of current in a circuit containing a capacitor and a resistor. Note that the decay constant λ is analogous to the reciprocal of the time constant, that is to $1/CR$.

Example

A sample of phosphorus-32 contains 8.6×10^{12} nuclei at time $t = 0$. The disintegration constant of phosphorus-32 is 4.8×10^{-2} day^{-1}. Calculate the number of undecayed phosphorus-32 nuclei in the sample after 10 days.

From $N = N_0 e^{-\lambda t}$, we have $N = 8.6 \times 10^{12} e^{-0.048 \times 10}$, so $N = \mathbf{5.3 \times 10^{12}}$. (Again, it is important to measure λ and t in consistent units. Here λ is in day^{-1} and t is in days, so there is no problem.)

Now it's your turn

A sample of phosphorus-32 contains 8.6×10^{12} nuclei at time $t = 0$. The disintegration constant of phosphorus-32 is 4.8×10^{-2} day^{-1}. Calculate the number of undecayed phosphorus-32 nuclei in the sample after
(a) 20 days,
(b) 40 days.
        Ans: (a) 2.3×10^{12}; (b) 1.3×10^{12}

12.7 Disintegration constant and half-life

Using the equation $N = N_0 e^{-\lambda t}$, we can derive an equation which relates the half-life to the disintegration constant.

For any radioactive nuclide, the number of undecayed nuclei after one half-life is, by the definition of half-life, equal to $N_0/2$, where N_0 is the original number of undecayed nuclei. Using the radioactive decay equation

$$N = N_0 e^{-\lambda t},$$

we have, at time $t = t_{1/2}$,

$$N_0/2 = N_0 e^{-\lambda t_{1/2}}$$

and dividing each side of the equation by N_0,

$$1/2 = e^{-\lambda t_{1/2}}$$

or

$$2 = e^{\lambda t_{1/2}}$$

Taking natural logarithms of both sides,

$$\ln 2 = \lambda t_{1/2}$$

so that

$$\bigstar \quad t_{1/2} = \frac{\ln 2}{\lambda}$$

or

$$\bigstar \quad t_{1/2} = \frac{0.693}{\lambda}$$

Example

Calculate the half-life of radium-226, which has a disintegration constant of $1.42 \times 10^{-11}\ \text{s}^{-1}$.

Using $t_{1/2} = 0.693/\lambda$, we have $t_{1/2} = 0.693/1.42 \times 10^{-11} = \mathbf{4.88 \times 10^{10}\ s}$.

Now it's your turn

1 Calculate the half-lives of the following radioactive nuclides:
   (a) bismuth-214, which has a disintegration constant of $4.3 \times 10^3\ \text{s}^{-1}$;
   (b) carbon-14, which has a disintegration constant of $4.1 \times 10^{-12}\ \text{s}^{-1}$.
   Ans: (a) $1.6 \times 10^{-4}\ \text{s}$; (b) $1.7 \times 10^{11}\ \text{s}$

2 Calculate the disintegration constants of the following radioactive nuclides:
   (a) helium-5, which has a half-life of $6.0 \times 10^{-20}\ \text{s}$;
   (b) sodium-24, which has a half-life of 15 h.
   Ans: (a) $1.2 \times 10^{19}\ \text{s}^{-1}$; (b) $4.6 \times 10^{-2}\ \text{h}^{-1}$

Sections 12.6–12.7 summary

★ Radioactive decay is a spontaneous, random process.

★ The half-life $t_{1/2}$ of a radioactive nuclide is the time taken for the number of undecayed nuclei to be reduced to half the original number.

★ The activity of dN/dt of a source is related to the number N of undecayed nuclei by the equation $dN/dt = -\lambda N$, where λ is the disintegration constant.

★ The disintegration constant is defined as the probability of decay, per unit time, of a nucleus.

★ The number N of undecayed nuclei in a radioactive sample at time t is given by the equation $N = N_0\,e^{-\lambda t}$, where N_0 is the number of undecayed nuclei at time $t = 0$.

★ The half-life $t_{1/2}$ and the decay constant λ are related by the equation $t_{1/2} = 0.693/\lambda$.

1 **(a)** You are provided with a radioactive source which could be emitting α- or β-particles, or γ-radiation, or any combination of these. Describe a simple experiment, based on their relative penetrating qualities, you might carry out to determine the nature of the radiation(s) being emitted.

(b) Explain the changes that take place to the nucleus of an atom when it emits i) an α-particle, ii) a β-particle, iii) γ-radiation.

(c) Complete the following radioactive series:

$$^a_bX \rightarrow {}^?_?Y + {}^4_2He$$
$$^?_?Y \rightarrow {}^?_?Z + {}^0_{-1}e$$
$$^?_?Z \rightarrow {}^?_?Z + {}^?_??$$

2 Figure 12.15 shows the variation with time t of the activity of a sample of a radioactive nuclide X.
The average background count during the experiment was 36 min^{-1}.

(a) Plot a graph to show the variation with time of the corrected count-rate.

(b) Use the graph to determine the half-life of the nuclide X.

3 A radioactive source with a half-life of 10 days has an initial activity of 2.0×10^{10} Bq. Calculate the activity of this source after 30 days.

4 The half-lives of two radioactive nuclides X and Y are 40 s and 70 s respectively. Initially, samples of each contain 2.0×10^7 undecayed nuclei.

(a) Calculate the initial activity of each sample.

(b) Draw, on the same axes, two lines to represent the decay of these samples. Estimate the time at which the activities will be the same. Verify your estimate by calculation.

5 Calculate the mass of caesium-137 that has an activity of 4.0×10^5 Bq. 0.137 kg of caesium-137 contains 6.0×10^{23} atoms. The half-life of caesium-137 is 30 years.

6 **(a)** The activity of a radioactive source X falls from 5.0×10^{10} Bq to 1.0×10^{10} Bq in 5.0 hr. Calculate the half-life.

(b) The activity of a certain mass of carbon-14 is 5.00×10^9 Bq. The half-life of carbon-14 is 5570 yr. Calculate the number of carbon-14 nuclei in the sample.

t/hour	0	1	2	3	4	5	6	7	8	9	10
activity/min^{-1}	854	752	688	576	544	486	448	396	362	334	284

Figure 12.15

12.8 Mass defect

At a nuclear level the masses we deal with are so small that it would be very clumsy to measure them in kilograms. Instead, we measure the masses of nuclei and nucleons in **atomic mass units** (**u**).

★ One atomic mass unit (1 u) is defined as being equal to one-twelfth of the mass of a carbon-12 atom. 1 u is equal to 1.66×10^{27} kg.

Using this scale of measurement, to six decimal places, we have:

the proton mass m_p = 1.007276 u,
the neutron mass m_n = 1.008665 u,
the electron mass m_e = 0.000549 u.

Because all atoms and nuclei are made up of protons, neutrons and electrons, we should be able to use these figures to calculate the mass of any atom or nucleus.

For example, the mass of a helium-4 nucleus, consisting of two protons and two neutrons, should be

$$(2 \times 1.007276) + (2 \times 1.008665) = 4.031882 \text{ u.}$$

However, the actual mass of a helium nucleus is 4.001508 u.

The difference between the expected mass and the actual mass of a nucleus is called the **mass defect** of the nucleus. In the case of the helium-4 nucleus, the mass defect is $4.031882 - 4.001508 = 0.030374$ u.

★ The mass defect of a nucleus is the difference between the total mass of the separate nucleons and the combined mass of the nucleus.

Example

Calculate the mass defect for a carbon-14 ($^{14}_{6}$C) nucleus. The measured mass is 14.003240 u.

The nucleus contains 6 protons and 8 neutrons, of total mass $(6 \times 1.007276) + (8 \times 1.008665) = 14.112976$ u. The mass defect is $14.112976 - 14.003240 = \mathbf{0.109736 \text{ u.}}$

Now it's your turn

1 Calculate the mass defect for a nitrogen-14 ($^{14}_{7}$N) nucleus. The measured mass is 14.003070 u.
   Ans: 0.108517 u

12.9 Mass-energy equivalence

In 1905 Albert Einstein proposed that there is an equivalence between mass and energy, that is, it is possible to interchange mass and energy. The relationship between energy E and change in mass Δm is

$$E = \Delta mc^2,$$

where c is the speed of light. E is measured in joules, Δm in kilograms and c in metres per second.

Using this relation, we can calculate that 1.0 kg of matter is equivalent to $1.0 \times (3.0 \times 10^8)^2 = 9.0 \times 10^{16}$ J.

The mass defect of the helium nucleus, calculated previously as 0.030374 u, is equivalent to $0.030374 \times 1.66 \times 10^{-27} \times (3.00 \times 10^8)^2 = 4.54 \times 10^{-12}$ J. (Note that the mass in u must be converted to kg by multiplying by 1.66×10^{-27}.)

The joule is an inconveniently large unit to use for nuclear calculations. A more convenient energy unit is the mega electron volt (MeV), as many energy changes that take place in the nucleus are of the order of several MeV. One mega electron volt is the energy gained by one electron when it is accelerated through a potential difference of one million volts.

Since electrical *energy = charge × potential difference*, and the electron charge is 1.60×10^{-19} C,

$$1 \text{ MeV} = 1.60 \times 10^{-19} \times 1.00 \times 10^6$$

or

★ $1 \text{ MeV} = 1.60 \times 10^{-13}$ J.

The energy equivalent of the mass defect of the helium nucleus is thus $4.54 \times 10^{-12}/1.60 \times 10^{-13} = 28.4$ MeV.

If mass is measured in u and energy in MeV,

★ 1 u is the equivalent of 931 MeV.

12.10 Binding energy

Within the nucleus there are strong forces which bind the protons and neutrons together. To completely separate all these nucleons requires energy. This energy is referred to as the **binding energy** of the nucleus. Stable nuclei, those which have little or no tendency to disintegrate, have large binding energies. Less stable nuclei have smaller binding energies.

Similarly, when protons and neutrons are joined together to form a nucleus, this binding energy must be released. The binding energy is the energy equivalent of the mass defect.

We have seen that the binding energy of the helium-4 nucleus is 28.4 MeV. 28.4 MeV of energy is required to separate, to infinity, the two protons and two neutrons of this nucleus.

> ★ Binding energy is the energy equivalent of the mass defect of a nucleus. It is the energy required to separate to infinity all the nucleons of a nucleus.

Example

Calculate the binding energy, in MeV, of a carbon-14 nucleus with a mass defect of 0.109736 u.

Using the equivalence 1 u = 931 MeV, 0.109736 u is equivalent to 102 MeV. Since the binding energy is the energy equivalent of the mass defect, the binding energy = **102 MeV**.

Now it's your turn

Calculate the binding energy, in MeV, of a nitrogen-14 nucleus with a mass defect of 0.108517 u.

Ans: 101 MeV

Stability of nuclei

A stable nucleus is one which has a very low probability of decay. Less stable nuclei are more likely to disintegrate. A useful measure of stability is the **binding energy per nucleon** of the nucleons in the nucleus.

> ★ Binding energy per nucleon is defined as the total energy needed to completely separate all the nucleons in a nucleus divided by the number of nucleons in the nucleus.

Figure 12.16

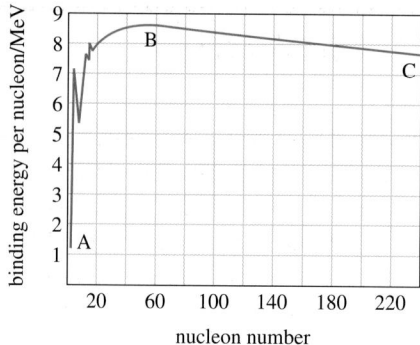

Figure 12.16 shows the variation with nucleon number of the binding energy per nucleon for different nuclides. The most stable nuclides are those with the highest binding energy per nucleon, that is, those near point B on the graph. Iron is one of these stable nuclides. Typically, very stable nuclides have binding energies per nucleon of about 8 MeV.

Light nuclei, between A and B on the graph, may combine or fuse to form larger nuclei with larger binding energies per nucleon. This process is called **nuclear fusion**. For the process to take place, conditions of very high temperature and pressure are required, such as in stars like the Sun.

Heavy nuclei, between B and C on the graph, when bombarded with neutrons, may break into two smaller nuclei, again with larger binding energy per nucleon values. This process is called **nuclear fission**.

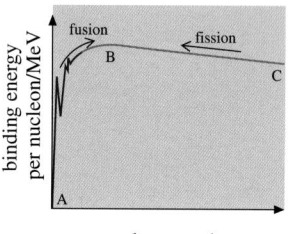

Figure 12.17

Figure 12.17 highlights nuclides which may undergo fusion or fission in order to increase their binding energy per nucleon.

When nuclear fusion or fission takes place, the nucleon numbers of the nuclei involved change. A higher binding energy per nucleon is achieved, and this is accompanied by a *release* of energy. This release of energy during fission reactions is how the present generation of nuclear power stations produce electricity.

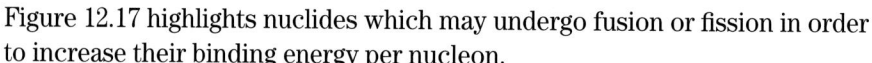

Example

The binding energy of a helium-4 nucleus is 28.4 MeV. Calculate the binding energy per nucleon.

The helium-4 nucleus has 4 nucleons. The binding energy per nucleon is thus 28.4/4 = **7.1 MeV per nucleon**.

Now it's your turn

The binding energy of a carbon-14 nucleus is 102 MeV. Calculate the binding energy per nucleon.

Ans: 7.3 MeV per nucleon

12.10 Binding energy

12.11 Nuclear fission

Within the nucleus of an atom the nucleons experience both attractive and repulsive forces. The attractive force is called the **strong nuclear force**. This acts like a 'nuclear glue' to hold the nucleons together. The repulsive forces are the electric (Coulomb-law) forces between the positively-charged protons. Gravitational forces of attraction exist, but are negligible in comparison to the other forces. Stable nuclei have much larger attractive forces than repulsive forces. Stable nuclides generally have approximately the same number of neutrons and protons in the nucleus, that is, the neutron-to-proton ratio is close to one. In heavy nuclei such as uranium and plutonium there are far more neutrons than protons, giving a neutron-to-proton ratio of more than one. For example, uranium-235 has 92 protons and 143 neutrons, giving a neutron-to-proton ratio of 1.55. This leads to a much lower binding energy per nucleon compared with iron, and such nuclides are less stable. Any further increase in the number of neutrons in such nuclei is likely to cause the nucleus to undergo **nuclear fission**.

> ★ Nuclear fission is the splitting of a heavy nucleus into two lighter nuclei of approximately the same mass.

When a uranium-235 nucleus absorbs a neutron, it becomes unstable and splits into two lighter, more stable nuclei. The most likely nuclear reaction is

$$^{235}_{92}U + ^{1}_{0}n \rightarrow ^{236}_{92}U \rightarrow ^{141}_{56}Ba + ^{92}_{36}Kr + 3^{1}_{0}n + \text{energy}$$

This process is called *induced* nuclear fission, because it is started by the capture of a neutron by the uranium nucleus.

Each of the fission reactions described by this equation results in the release of three neutrons. Other possible fission reactions release two or three neutrons. If these neutrons are absorbed by other uranium-235 nuclei, these too may become unstable and undergo fission, thereby releasing even more neutrons. The reaction is described as being a **chain reaction** which is accelerating. This is illustrated in Figure 12.18. If this type of reaction continues uncontrolled, a great deal of energy is released in a short time, and a nuclear explosion results.

If the number of neutrons which take part in the chain reaction is controlled so that the number of fissions per unit time is constant, rather than increasing, the rate of release of energy can be controlled. This situation is illustrated in Figure 12.19. These conditions apply in the reactor of a modern nuclear power station, where some of the neutrons released in fission reactions are absorbed by control rods in order to limit the rate of fission reactions.

Figure 12.18 *Acceleration chain reaction.*

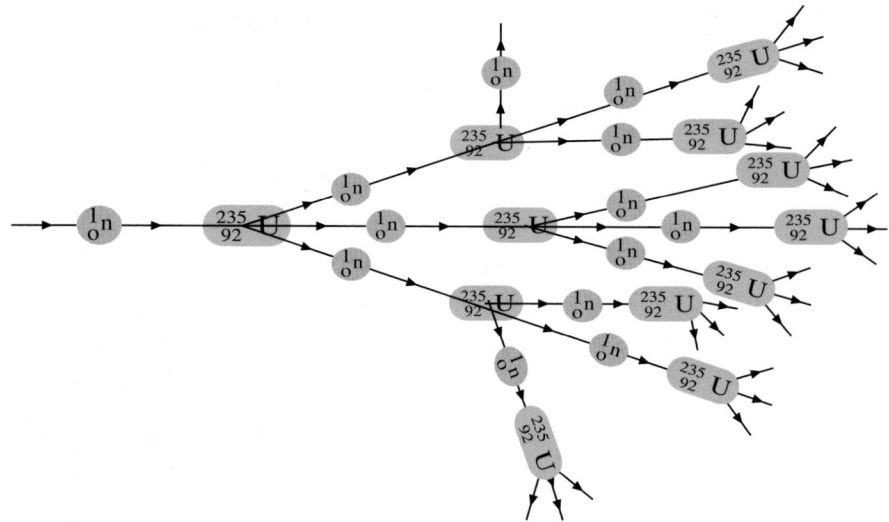

Figure 12.19 *Controlled chain reaction. Two of the neutrons produced by fission are absorbed by control rods. The third neutron induces further nuclear fission.*

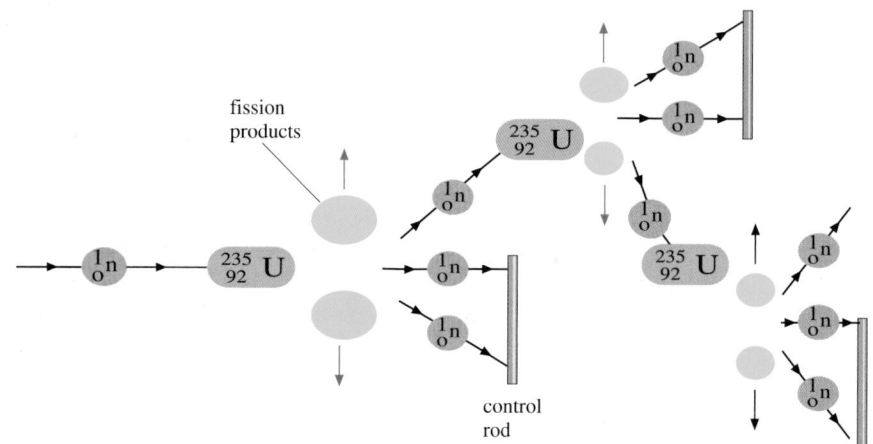

fission products

control rod

Nuclear fission reactor

There are several possible fission reactions which can be made use of in a nuclear reactor. Most are represented by the general equation

$$^{235}_{92}U + {}^{1}_{0}n \rightarrow {}^{236}_{92}U \rightarrow 2 \text{ new nuclides} + 2 \text{ or } 3 \text{ neutrons} + \text{energy}$$

The nuclides formed in this reaction are called **fission fragments**.

Figure 12.20 illustrates a pressurised water-cooled reactor (PWR). Inside the reactor the uranium-235 is placed in long fuel rods which are surrounded by heavy water (water in which deuterium atoms, hydrogen-2, take the place of the normal hydrogen-1 atoms) or graphite. The role of these materials is to slow down the neutrons released during fission. This is important because slow-moving neutrons are more readily absorbed by uranium-235. Materials

12.11 Nuclear fission

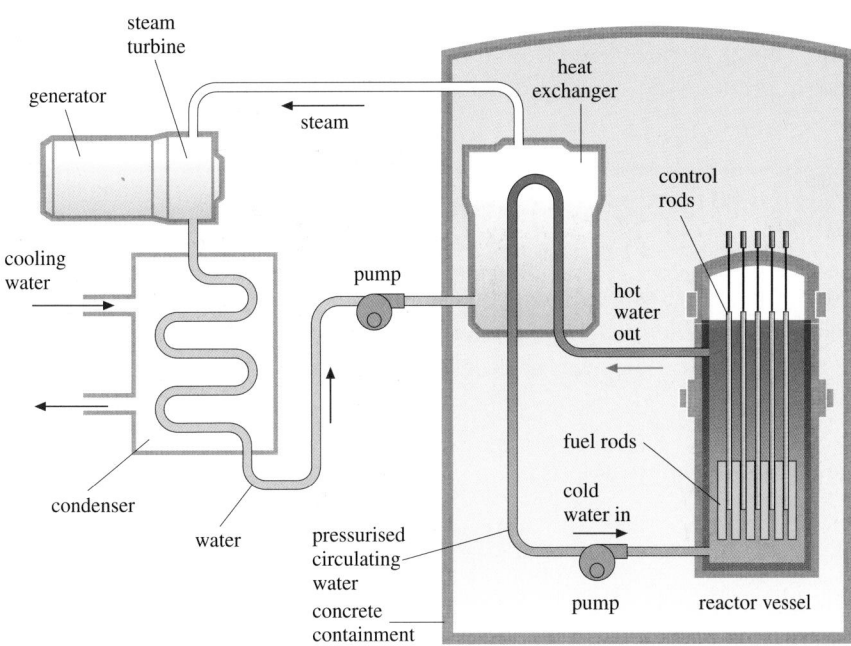

Figure 12.20 *Pressurised water-cooled reactor.*

such as heavy water or graphite which slow down the neutrons are called **moderators**.

The number of neutrons available to trigger further fission reactions is controlled using **control rods** of boron or cadmium. These substances readily absorb neutrons. By pushing the rods further into the reactor core, the number of neutrons absorbed increases and the rate of fission reactions decreases. If the rods are pulled out a little, fewer neutrons are absorbed and the rate of fissioning increases. Ideally, the rods should be adjusted so that the fission reaction proceeds at a constant rate, that is, it neither accelerates nor decelerates. The energy released during fission reactions is seen mainly as the kinetic energy of the resulting lighter nuclei and neutrons (the fission fragments). When the fission fragments are slowed down, the heat energy produced is removed from the reactor by a circulating coolant. This heat energy is then used to produce high temperature steam in a heat exchanger. The steam is used to drive turbine generators.

There are two main types of nuclear reactor used for the commercial generation of electricity. These are the pressurised water-cooled reactor (PWR) and the advanced gas-cooled reactor (AGR).

In the PWR, water circulates through the reactor vessel at a high pressure, typically 150 atmospheres. At such a high pressure the water does not boil, even though its temperature may be as high as 300 °C. This super-heated water passes into a heat exchanger through a series of pipes surrounded by water at a much lower pressure. The low-pressure water absorbs heat energy and changes into steam, which then drives the turbines and electricity generators.

In AGR reactors, helium gas at a temperature of about 600 °C is the coolant. Again, energy is transferred from the core to water to a heat exchanger in order to produce the steam to drive the turbines.

12.12 Nuclear fusion

Most of the energy on Earth comes from the Sun, where it is produced by nuclear fusion reactions. Light nuclei, such as isotopes of hydrogen, join together to produce heavier, more stable nuclei, and in doing so release energy.

One of these fusion reactions is

$$^2_1\text{H} + {}^2_1\text{H} \rightarrow {}^3_2\text{H} + {}^1_0\text{n} + \text{energy}$$

From the binding energy per nucleon curve (Figure 12.17) we see that the binding energy per nucleon for light nuclei, such as hydrogen, is low. But if two light nuclei are made to fuse together, they may form a new heavier nucleus which has a higher binding energy per nucleon. It will be more stable than the two lighter nuclei from which it was formed. Because of this difference in stability, a fusion reaction such as this will release energy.

Although fusion reactions are the source of solar energy, we are at present unable to duplicate this reaction in a controlled manner on Earth. This is because the nuclei involved in fusion have to be brought very close together, and conditions of extremely high temperature and pressure, similar to those found at the centre of the Sun, are required. Reactions requiring these conditions are called **thermonuclear reactions**. Some fusion reactions involving hydrogen isotopes have been made to work in the Joint European Torus (JET), although not in a controlled, sustainable manner.

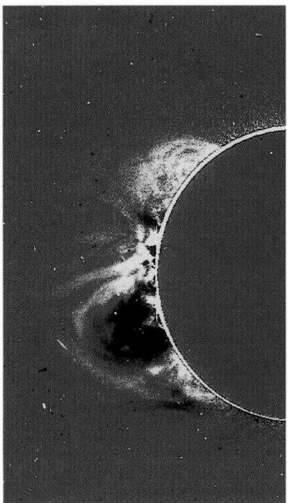

Most of our energy on Earth comes from the Sun. It is produced by nuclear fusion reactions. Light nuclei such as isotopes of hydrogen join together to produce heavier, more stable nuclei.

Sections 12.8–12.12 summary

★ The mass defect of a nucleus is the difference between the total mass of the separate nucleons and the mass of the nucleus.

★ Einstein's mass-energy equivalence relation: $E = \Delta mc^2$.

★ The binding energy of a nucleus is the energy needed to separate completely all its constituent nucleons.

★ The binding energy per nucleon is a measure of the stability of a nucleus: a high binding energy per nucleon means that the nucleus is stable.

★ Nuclear fission is the splitting of a heavy nucleus into two smaller, lighter nuclei of approximately equal mass.

★ Nuclear fusion is the joining together of light nuclei to form a larger, heavier nucleus.

Sections 12.8–12.12 questions

1 Calculate the mass defect, the binding energy of the nucleus, and the binding energy per nucleon for the following nuclei:

   (a) hydrogen-3, $^3_1$H; nuclear mass 3.01605 u

   (b) zirconium-97, $^{97}_{40}$Zr; nuclear mass 97.09801 u

   (c) radon-222, $^{222}_{86}$Rn; nuclear mass 222.01754 u

(proton mass = 1.00728 u; neutron mass = 1.00867 u.)

2 One possible reaction taking place in the core of a reactor is

$$^{235}_{92}U + ^1_0n \rightarrow ^{95}_{42}Mo + ^{139}_{57}La + 2^1_0n + 7^{\ 0}_{-1}e$$

For this reaction, calculate:

   (a) the mass on each side of the equation,

   (b) the change in mass after fission has taken place,

   (c) the energy released per fission of uranium-235,

   (d) the energy available from the complete fission of 1.00 g of uranium-235,

   (e) the mass of uranium-235 used by a 500 MW nuclear power station in one hour, assuming 30% efficiency.

(Masses: $^{235}_{92}U$, 235.123 u; $^{95}_{42}Mo$, 94.945 u; $^{139}_{57}La$, 138.955 u; proton, 1.007 u; neutron, 1.009 u. 0.235 kg of uranium-235 contains 6.0×10^{23} atoms.)

3 Two fusion reactions which take place in the Sun are described below.

   (a) A hydrogen-2 (deuterium) nucleus absorbs a proton to form a helium-3 nucleus.

   (b) Two helium-3 nuclei fuse to form a helium-4 nucleus plus two free protons.

For each reaction, write down the appropriate nuclear equation and calculate the energy released.

(Masses: $^2_1$H, 2.01410 u; $^3_2$He, 3.01605 u; $^4_2$He, 4.00260 u; $^1_1$p, 1.00728 u; $^1_0$n, 1.00867.)

12.13 Probing matter

This photograph was taken with an ion microscope, a device which makes use of the de Broglie wavelength of gas ions (see Chapter 9). It shows a sample of iridium at a magnification of about five million. The positions of individual iridium atoms can be seen.

Photographs like this reinforce the idea that all matter is made of very small particles we call atoms. Advances in science at the end of the nineteenth and the beginning of the twentieth century led most physicists to believe that atoms themselves are made from even smaller particles, some of which have positive or negative charges. Unfortunately even the most powerful microscopes cannot show us the internal structure of the atom. Many

Ion microscope photograph of iridium.

theories were put forward about the structure of the atom, but it was a series of experiments carried out by Ernest Rutherford and his colleagues around 1910 that led to the birth of the model we now know as the **nuclear atom**.

Probing matter using α-particles

In 1911 Rutherford and two of his associates, Geiger and Marsden, fired a beam of α-particles at a very thin piece of gold foil. A zinc sulphide detector was moved around the foil to see the directions in which the α-particles travelled after striking the foil (Figure 12.21).

Figure 12.21 *α-scattering experiment.*

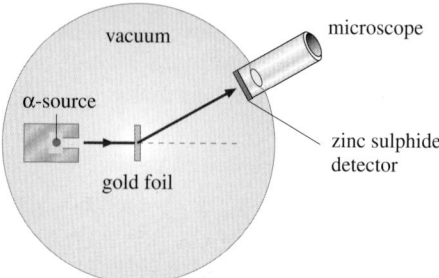

They discovered that

- the vast majority of the α-particles passed straight through the foil with very little or no deviation from their original path

- a small number of particles were deviated through an angle of more than about 10°

- an extremely small number of particles (one in ten thousand) were deflected through an angle greater than 90°

From these observations, the following conclusions could be drawn.

- The majority of the mass of an atom is concentrated in a very small volume at the centre of the atom. Most α-particles would therefore pass through the foil undeviated.

- The centre (or nucleus) of an atom is positively charged. α-particles, which are also positively charged, passing close to the nucleus will experience a repulsive force causing them to deviate.

- Only α-particles that pass very close to the nucleus, almost striking it head-on, will experience large enough repulsive forces to cause them to deviate through angles greater than 90°. The fact that so few particles did so confirms that the nucleus is very small, and that most of the atom is empty space.

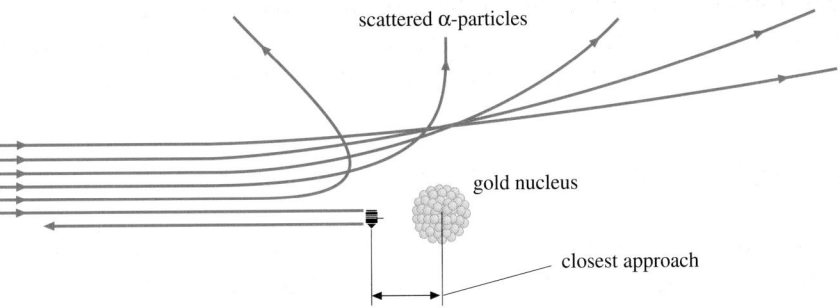

Figure 12.22

> • Because atoms are neutral, the atom must contain negative particles. These travel around, or orbit, the nucleus.

Figure 12.22 shows some of the possible trajectories of the α-particles. Using the nuclear model of the atom and equations to describe the force between charged particles, Rutherford calculated the fraction of α-particles that he would expect to be deviated through various angles. The calculations agreed with the results from the experiment. This confirmed the nuclear model of the atom. Rutherford calculated that the diameter of the nucleus is about 10^{-15} m, and the diameter of the whole atom about 10^{-10} m. Figure 12.23 shows the features of the nuclear model of a nitrogen atom.

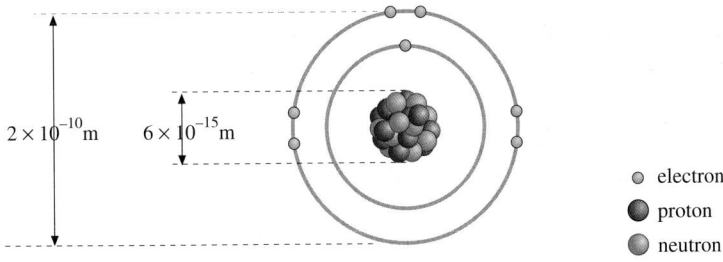

Figure 12.23 *The diameter of a nitrogen atom is more than 30 000 times bigger than the diameter of its nucleus.*

Some years later the α-particle scattering experiment was repeated using α-particles with higher energies. Some discrepancies between the experimental results and Rutherford's scattering formula were observed. These seemed to be occurring because the high-energy α-particles were passing very close to the nucleus, and were experiencing not only the repulsive electrostatic force but also a strong attractive force which appears to act over only a very short range. This became known as the **strong nuclear force**. This is the force that holds the nucleus together.

Probing matter using electrons

Electrons are not affected by the strong nuclear force. It was suggested that they might therefore be a more effective tool with which to investigate the structure of the atom. As we saw in Chapter 9, moving electrons have a wavelike property, and can be diffracted.

If a beam of electrons is directed at a sample of powdered crystal and the electron wavelength is comparable with the interatomic spacing in the crystal, the electron waves are scattered from planes of atoms in the tiny crystals, creating a diffraction pattern. The fact that a diffraction pattern is obtained confirms the regular arrangement of the atoms in a crystalline solid. Measurements of the angles at which strong scattering is obtained can be used to calculate the distances between planes of atoms.

If the energy of the electron beam is increased, the wavelength decreases. Eventually the electron wavelength may be of the same order of magnitude as the diameter of the nucleus. High-energy electron diffraction can be used to determine nuclear diameters. Probing the nucleus with high-energy electrons rather than α-particles gives a further insight into the dimensions of the nucleus, and also gives information about the distribution of charge in the nucleus itself.

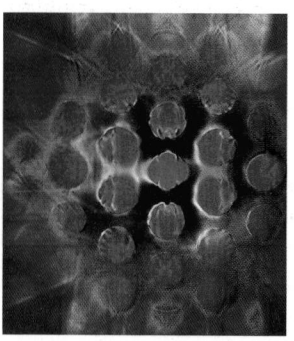

Electron diffraction pattern of a sample of pure titanium.

Section 12.13 summary

★ The Rutherford α-particle experiment confirmed the nuclear model of the atom: the atom consists of a small, heavy, positively-charged nucleus, surrounded by negatively-charged electrons in orbit about the nucleus.

★ The diameter of the nucleus is about 10^{-15} m; the diameter of the atom is about 10^{-11} m.

★ Electron diffraction gives evidence for the regular arrangement of atoms in crystals, and allows the measurement of the distance between planes of atoms in solids.

Exam Questions

1 Cobalt-60 is often used as a radioactive source in medical physics. It has a half-life of 5.25 years. Calculate how long after a new sample is delivered the activity will have decreased to one-eighth of its value when delivered.

2 Iodine-131 is a beta-emitter with half-life of 194 hours. It decays to xenon (Xe). The decay is represented by

$$^{131}_{53}I \rightarrow {}^P_Q Xe + \beta$$

(a) Obtain the values of P and Q in the decay equation.
(b) A sample of iodine-131 of mass 0.20 g is prepared in a reactor.
   i) Given that 0.131 kg of iodine-131 contains 6.02×10^{23} atoms, calculate the number of iodine-131 atoms contained in the sample at the time of preparation.
   ii) Calculate the mass of iodine-131 which will have decayed 48 hours after preparation.

3 The half-life of a radioactive isotope of sodium used in medicine is 15 hours.

(a) Determine the decay constant for this nuclide.
(b) A small volume of a solution containing this nuclide has an activity of 1.2×10^4 disintegrations per minute when it is injected into the bloodstream of a patient. After 30 hours, the activity of 1.0 cm^3 of blood taken from the patient is 0.50 disintegrations per minute. Estimate the volume of blood in the patient. Assume that the solution is uniformly diluted in the blood, that it is not taken up by the body tissues, and that there is no loss by excretion.

4 A fusion reaction that takes place in the Sun is represented by the equation

$$^2_1H + {}^2_1H \rightarrow {}^3_2He + {}^1_0n + \text{energy.}$$

(a) Calculate the energy released in this reaction. (Masses: ^2_1H, 2.0141 u; ^3_2He, 3.0160 u; ^1_0n, 1.0087 u.)
(b) Suggest why it is difficult to achieve controlled reactions of this type.

5 A stationary radium nucleus ($^{224}_{88}Ra$) of mass 224 u spontaneously emits an α-particle. The α-particle is emitted with an energy of 9.2×10^{-13} J, and the reaction gives rise to a nucleus of radon (Rn).

(a) Write down a nuclear equation to represent α-decay of the radium nucleus.
(b) Show that the speed with which the α-particle is ejected from the radium nucleus is 1.7×10^7 m s^{-1}.
(c) Calculate the speed of the radon nucleus on emission of the α-particle. Explain how the principle of conservation of momentum is applied in your calculation.

6 When an α-particle travels through air, it loses energy by ionisation of air molecules. For every air molecule ionised, approximately 5.6×10^{-18} J of energy is lost by the α-particle.

(a) Suggest a typical value for the range of an α-particle in air. Hence estimate the number of air molecules ionised per millimetre of the path of the α-particle, given that the α-particle has initial energy 9.2×10^{-13} J.
(b) It has been discovered that the number of ionisations per unit length of the path of an α-particle suddenly increases just before the α-particle stops. State, with a reason, the effect that this observation will have on your estimate.

CHAPTER THIRTEEN

Synoptic assessment

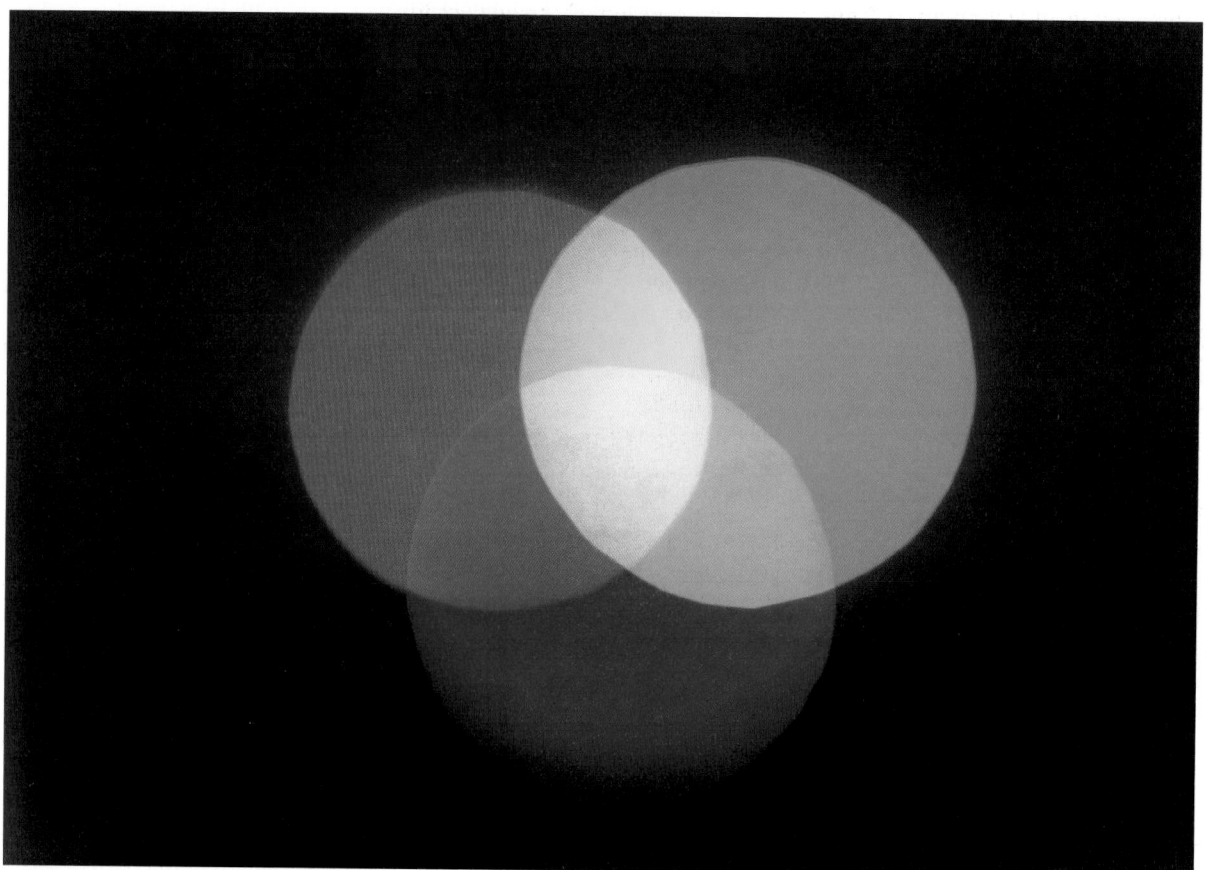

Synoptic assessment requires the student to bring together concepts and ideas from many different areas of Physics.

If you look through the physics specification (syllabus) which you are taking, you will notice that reference is made to **synoptic assessment**. At Advanced level, the examining of all specifications must include synoptic assessment.

Basically, synoptic assessment is included so that students do not study a module, take an examination and then forget what they have learned when they go on to take further modules.

Synoptic assessment is in two parts.

1. Students must show that they have knowledge and understanding of physics across the modules they have studied and that they can draw together ideas from different areas of the subject. Really, this means that it is necessary to know and understand the work in *all* the modules (both AS and A2) when taking the synoptic part of the examination.

2. Students must demonstrate the skills they have acquired whilst studying Advanced level physics. These skills include the presentation, processing, analysis and evaluation of data. The skill assessment may take the form of data analysis and/or comprehension test. The assessment may also be included as part of the assessment of coursework.

Synoptic assessment takes place only at the end of the course. The weighting attached to this assessment is a minimum of 20% of the total marks for the Advanced level examination. Thus, it is very important that you should prepare for the synoptic parts of the examination.

13.1 Knowledge and understanding

For the convenience of being able to examine the contents of the physics specification in a number of separate parts, the specification has been divided into modules. The contents of each module are examined separately. In synoptic assessment, questions are set with the deliberate intention of making students use knowledge from at least two modules. It is important that you practise answering questions of this type, but it is not always necessary to write out your answers. A good means by which many different questions can be covered is by group discussion. Having selected a question, each member of the group should have time – perhaps overnight – to think about what should be included and how it should be presented. At a later time, ideas are pooled.

It is important, however, that you practise on an individual basis. You need to develop the ability to gather together your thoughts and to write them down within the time allowed in the examination. Remember that, in synoptic assessment, you are to bring together ideas from different modules. You should think about what to include within an answer and then plan the presentation before you begin to write.

The planning of experimental work is one element of the assessment of coursework. Planning involves thinking about different ways by which a task may be accomplished and thus involves bringing together ideas from various areas of study. It may well be that synoptic assessment is included in the assessment of planning techniques.

Synoptic questions may look different from those in examination papers testing the content of individual modules. In papers for separate modules, questions are likely to relate directly to subject matter within that module. For synoptic assessment, questions will be much more general and are unlikely to refer to a single specific topic.

Example

1   In some systems, a quantity gradually approaches zero as another
    increases. For example, the pressure of a fixed mass of gas at constant
    temperature becomes smaller as its volume increases, but the pressure
    never becomes zero.
    (a) Give three examples from physics where this type of change
        occurs. For each of the examples, the exact form of the
        relationship between the quantities should be different.
    (b) For each of the changes you have named in (a):
        • describe the change in words and give a mathematical relation
          between the quantities involved,
        • state the conditions necessary for the relation in i) to apply,
        • sketch a graph to show the relation. On your graph, point out
          any special features which distinguish the relation from any
          other.

    (a) Three examples are:
        i)   radioactive decay where the number of nuclei of a
              radioactive isotope decreases as time increases,
        ii)   gravitational field strength decreases as the distance from a
              body increases,
        iii) the resistance of a wire decreases as its area of cross-
              section increases.
    (b) i)   The number N of undecayed nuclei decreases
              exponentially with time t. The change is unaffected by
              environmental conditions.

              $$N = N_o e^{-\lambda t},$$

              where N_o is the number of undecayed nuclei at time $t = 0$
              and λ is the disintegration constant.
              Figure 13.1 shows the variation with time t of the number N
              of undecayed nuclei. It can be seen that the time taken to
              halve the number of undecayed nuclei is constant. This
              time is known as the half-life $T_{\frac{1}{2}}$ of the radioactive isotope.
              The disintegration constant λ and the half-life $T_{\frac{1}{2}}$ are related
              by the expression

              $$T_{\frac{1}{2}} = 0.693/\lambda.$$

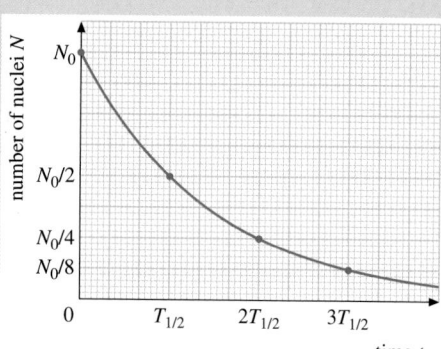

Figure 13.1

13.1 Knowledge and understanding

ii)   The gravitational field strength g due to a point mass decreases inversely with the square of the distance d from the mass. This is also true for a uniform sphere when the distance from the centre of the sphere is greater than its radius. The mass must be isolated. The relation is given by

$$g = GM/d^2,$$

where G is the gravitational constant and M is the mass of the object. Near the Earth's surface, g is known as the acceleration of free fall. Figure 13.2 shows the variation of g with d. Doubling the distance from the mass means that the field strength has been reduced by a factor of four.

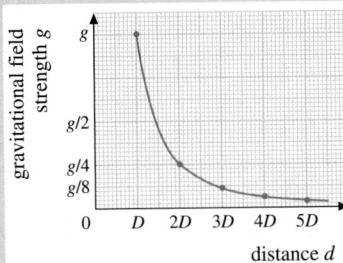

Figure 13.2

iii)   The resistance R of a wire varies inversely with its area of cross-section A. The material of the wire must be homogeneous and the wire must be of constant length and at constant temperature.

$$R = \rho l/A,$$

where ρ is a constant known as the resistivity of the material of the wire and l is the length of the wire. Resistivity is different for different materials. Figure 13.3 shows the variation with A of the resistance R. Doubling the area of cross-section halves the resistance.

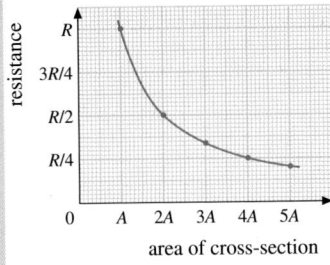

Figure 13.3

CHAPTER 13

13.2 Physics skills

Physics is concerned with the collection of data, its presentation, analysis and the conclusions which can be drawn from it. Synoptic assessment of presentation of data and its subsequent analysis may occur as part of the assessment of coursework but also in the assessment of modules taken at the end of the Advanced level course.

Presentation of data

Whenever possible, data should be presented in the form of a table. There are some basic rules which should be followed when constructing a table for raw and processed data.

1   Each column must have a heading and the appropriate units must be given.

2   There should be a progression from left to right. Data should be given in columns on the left, with processed data on the right.

3   Data must be given to an appropriate number of places of decimals. Processed data must have significant figures consistent with those of the raw data (see Chapter 1).

4   The number of significant figures in any one column must be consistent.

Processing of data

Very often, the purpose of the processing of data is to determine the relationship between two quantities. Since experimental data are not completely reliable, it is usual to present the processed data in the form of a graph. This enables trends to be observed and conclusions to be drawn based on all the data.

One of the easiest means by which average behaviour can be determined is to plot a graph which should result in a straight line. It is relatively easy to draw a straight line with the graph points scattered equally on both sides of the line. The trick is to arrange that the quantities plotted on the two axes will give rise to a straight line!

Figure 13.4 shows a straight line graph where the quantity y is plotted on the y-axis (the vertical axis) and the quantity x is shown on the x-axis (the horizontal axis). The graph has a gradient m and an intercept c. (c is the value of y when $x = 0$.) The equation of the graph line is

$$y = mx + c$$

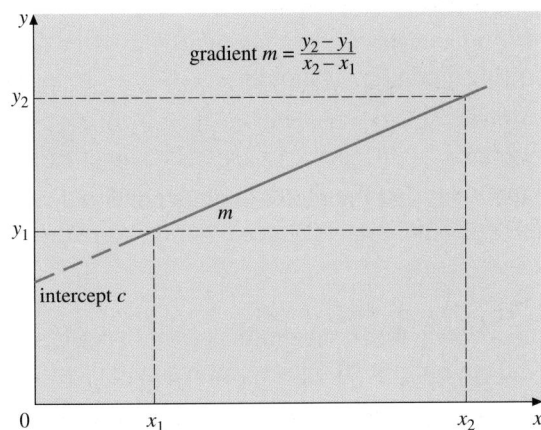

gradient $m = \dfrac{y_2 - y_1}{x_2 - x_1}$

Figure 13.4

We try to arrange the data so that a straight line graph is obtained.

Suppose we wish to find the relation between two quantities p and q. We suspect that the relation is of the form

$$p = aq^n,$$

where a and n are constants. Note that there may be other variables, but in order to find the relation between p and q, they must be kept constant.

Taking logarithms of both sides,

$$\lg p = \lg a + n \lg q.$$

Plotting a graph of $\lg p$ on the vertical axis against $\lg q$ on the horizontal axis will give a straight line graph, since, by comparing

$$\lg p = n \lg q + \lg a$$

and

$$y = mx + c$$

the gradient m of the line is equal to n and the intercept c is $\lg a$.

If the relation between the two quantities is suspected to involve an exponential function, a similar technique may be used. Suppose the relation is thought to be of the form

$$y = ae^{nx}.$$

Taking natural logarithms,

$$\ln y = \ln a + nx$$

In this case, a graph of $\ln y$ should be plotted against x. The gradient of the straight line is n and the intercept is $\ln a$.

Graph plotting

The presentation and analysis of data frequently involves the plotting of a graph. When plotting a graph you should

1   choose scales such that the plotted points cover at least half the area of the graph grid (there may be an exception were the origin is required),

2   choose scales which mean the graph is easy to read (e.g. 10 squares represent 1.0 cm but not 10 squares represent 3.0 cm),

3   label the axes with units,

4   show values on the axes at regular intervals,

5   plot points accurately,

6   draw a best straight line such that the scatter of points on each side of the line is the same.

Example

1   A wooden rod is clamped at one end so that the rod is horizontal. A mass is attached to the other end, a distance l from the clamp. The end of the rule is displaced and then released so that it performs small vertical oscillations of period T.
A student wishes to determine the relation between l and T, and makes the following rough notes of the measurements he takes.

Distance = 0.950 m, 10 oscillations 22.6, 22.4, 22.6 s
Distance = 0.843 m, 10 oscillations 18.8, 18.9, 18.6 s
Distance = 0.73 m, 10 oscillations 15, 15.2, 15.4 s
Distance = 0.652 m, 10 oscillations 12.6, 12.8, 13 s
Distance = 0.54 m, 10 oscillations 9.7, 9.4, 10 s
Distance = 0.462 m, 10 oscillations 8, 8.4, 7.8 s

Tabulate the measurements and, by drawing a suitable graph, determine the relation between T and l.

Assume the relation is of the form

$$T = al^n.$$

Then, taking logarithms to the base 10,

$$\lg T = \lg a + n \lg l.$$

Plotting a graph of lg T against lg l, the gradient is n and the intercept on the lg T axis is equal to lg a.

Tabulating raw and processed data:

l/m	10T/s			T/s	lg (T/s)	lg (l/m)
0.950	22.6	22.4	22.6	2.25	0.353	−0.0222
0.843	18.8	18.9	18.6	1.88	0.274	−0.0742
0.730	15.0	15.2	15.4	1.52	0.182	−0.137
0.652	12.6	12.8	13.0	1.28	0.107	−0.186
0.540	9.7	9.4	10.0	0.97	−0.013	−0.268
0.462	8.0	8.4	7.8	0.81	−0.093	−0.335

Note: The distance l can be measured to the nearest mm, i.e. to the third place of decimals when distance is measured in metres.

The time for ten oscillations can be measured to the nearest 0.1 s which means that the period can be given to the second place of decimals.

The values of the logarithms are given to the same number of significant figures as are found in the data.

The graph is plotted in Figure 13.5.

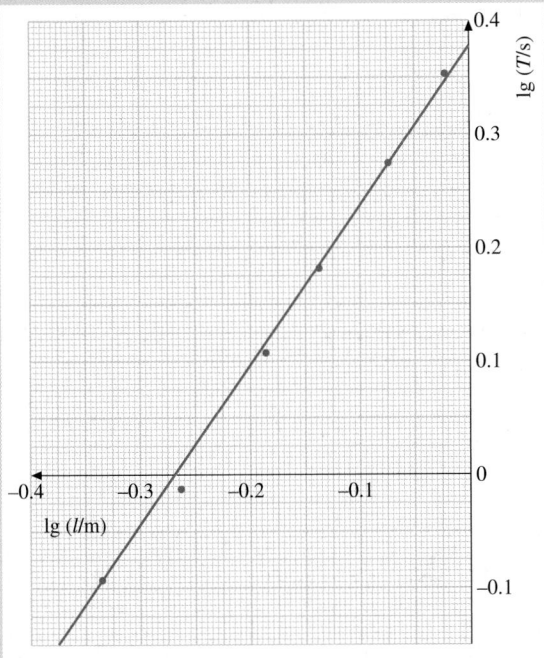

Figure 13.5

From the graph, gradient $= (+0.377 - 0)/(0 - -0.271)$
$= 1.39$
intercept $= +0.377$
$a = 10^{0.377} = 2.38$

The relation is $T = 2.4l^{1.4}$.

Comprehension tests

In general, a comprehension test consists of a passage which is to be read
and then questions are to be answered, based on the passage.

Unless a question asks for direct quotes from the passage, you should not
copy sections from the passage for your answers. Rather, you should use
your own words so that you can show the examiner that you understand
what has been said.

Example

1 Read the passage below and then answer the following questions.

Measuring temperature using a thermistor

*Thermistors are semiconductors whose resistance R varies with
thermodynamic temperature T according to the relation*

$$R = Ae^{b/T},$$

*where A and b are constants. Two main disadvantages of thermistor
thermometers are the non-linear relation between resistance and
temperature and also, the very wide manufacturing tolerance limits
on thermistor characteristics.*

*A direct reading thermistor thermometer was designed by Beakley
(1951). He showed that if a fixed resistor is connected in series with
the thermistor and a source of constant e.m.f., the curve relating
current I in the circuit to temperature T of the thermistor is as
shown on the left.*

*The curve shows a point of inflection at some temperature T_L. Near
this temperature, the graph can be taken as being linear and thus a
linear temperature scale may be established.*

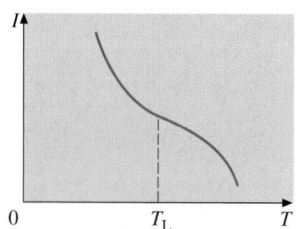

1. Describe the form of the non-linear relation shown by a thermistor.
2. Suggest what is meant in the passage by 'thermistor characteristics'.
3. Draw a diagram of the circuit proposed by Beakley.
4. By reference to the graph, explain what is meant by a point of inflection.
5. Explain why it is desirable for a temperature measuring device to have a linear scale.

1. As temperature increases, the factor b/T decreases. This means that the number $e^{b/T}$ will also decrease. Hence, the resistance of the thermistor will decrease according to the exponential of b/T as thermodynamic temperature increases.
2. The characteristics of an electrical component can be either the form of the variation with voltage of the current in the component, or the variation with temperature of its resistance. The resistance is equal to the ratio of voltage/current at that particular temperature. In the case of a thermistor, it is, in general, the variation of resistance with temperature which is of importance.
3.

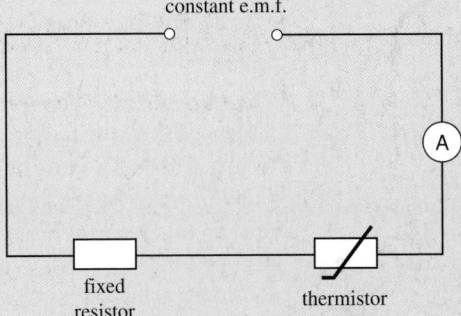

4. At low values of temperature, the graph line is 'concave up'. At higher temperatures, it is 'concave down'. At the point of inflection, the graph changes from one curvature to the other. There is a region on each side of this point where the gradient of the curve is constant.
5. When the scale is linear, equal changes in reading are produced by equal changes in temperature. If the scale is not linear, then it may be that a large change in temperature is required before any appreciable change in reading is detected. The sensitivity of the thermometer will not be constant.

1 Models are frequently used in physics to assist with the understanding of concepts. Choose three models you have studied, each from a different area of physics. For each model,

(**a**) identify and describe the concept which is being modelled,

(**b**) describe the model and how it illustrates the physical concept,

(**c**) discuss the limits to which the model accurately represents the concept.

Hint: models could include
- water circuit board for electric current
- the Bohr model of the atom
- magnetic pucks for α-particle scattering
- lines of force to illustrate force fields
- kinetic theory model of gases
- ripple tank for two-source interference of light.

2 Physics students are frequently discouraged from using vague terms such as 'large' or 'small'. In any topic in physics, what is large to one physicist may be very small to another. Write an account to give reasons to students as to why such vague terms should be avoided. Illustrate your answer by reference to two distinct topics in physics, giving magnitudes of quantities where appropriate.

Hint: Topics could be energy, distances and forces comparing the view of a nuclear physicist with, perhaps, a cosmologist.

3 Outline the features which may be used to describe a wave motion.

By reference to three different types of wave, explain how differences in these features give rise to the different properties of the wave.

Hint: Properties include speed, wavelength/frequency, amplitude/intensity, reflection, refraction, diffraction, interference, polarisation, absorption.
Comparison could be made between, for example, sound, ultrasound, visible light, γ-rays, and radio waves.

4 Discuss the forces, if any, which act on a stationary charge and on a moving charge when each is situated in:

(**a**) a gravitational field,

(**b**) an electric field,

(**c**) a magnetic field.

Hint: For each charge and field, you should consider the magnitude and direction of any force, giving equations and rules where appropriate.

5 Exponential changes are frequently met in the study of physics. By reference to two different areas of study, explain what is meant by an exponential change. You should choose one example of an exponential decrease and one of an exponential increase. Illustrate your answer with sketch graphs, where appropriate.

Hint: Examples may include decay of current in a *CR* circuit, radioactive decay, and a chain fission reaction.

6 Quantitative expressions for energy often have the form $\frac{1}{2}xy^2$, where x is one physical quantity and y is another. Give three examples from different areas of physics where expressions of this form are used. Explain why the expression for energy is this form. For one of the examples you have quoted, derive the expression.

Hint: Examples include kinetic energy, strain energy, and the energy stored in a charged capacitor.

7 The rate at which heat radiation is given off from a furnace is known to depend on the temperature of its surface. The rate of emission of this radiation may be measured with an instrument that gives a reading R which is directly proportional to the rate of emission. Use the following data to determine the dependence of rate of emission on surface temperature T.

T/K	*R* (arbitrary units)
620	8.00
740	17.0
950	50.0
1260	135
1450	275
2090	1070

Hint: Plot a graph of lg *T* against lg *R*.

8 Several planets in the Solar System have moons. Data from some of the moons of Uranus are given below.

moon	mean distance from centre of Uranus/km	period of orbit/days
Cordelia	49500	0.330
Belinda	75300	0.622
Miranda	130000	1.41
Ariel	191000	2.52
Umbriel	266300	4.14
Titania	435000	8.71

(a) Determine the relation between the period T of the orbit and the mean distance d of the moon from the centre of Uranus.

(b) A moon has been discovered which has an orbital period of 0.372 days. Determine its mean distance from the centre of Uranus.

Hint: Plot a graph of lg T against lg d.

9 In a crystal of sodium chloride there are contributions to the electrical conduction process both from the ions and from crystal defects. For temperatures up to about 800 K the defect contribution dominates. In this temperature range, the resistivity ρ varies with temperature T according to the relation $\rho = B\,e^{E/kT}$, where B is a constant, E is the activation energy for defect motion and k is the Boltzmann constant. Experimental data for sodium chloride are listed below.

T/K	ρ/Ω m
570	5.0×10^9
625	1.2×10^9
685	2.9×10^8
715	1.8×10^8
770	6.7×10^7

Find the value of E for sodium chloride. Give your answer in electronvolts (eV).

Hint: Plot a graph of ln ρ against $1/T$.

10 A theory due to Debye suggests that, at very low temperatures, the molar heat capacity C_m of a solid obeys the relation $C_m = AT^3$, where A is a constant and T is the thermodynamic temperature. Data for silver are listed below.

T/K	C_m/J K^{-1} mol^{-1}
4	0.0127
6	0.0437
10	0.195
14	0.525
20	1.58

Draw a suitable linear graph to test whether or not these data conform to the Debye theory. Obtain an estimate of the value of A for silver.

Hint: Plot a graph of lg C_m against lg T.

11 For many types of optical glass, the refractive index n is given by the expression $n = A + B/\lambda^2$, where λ is the wavelength of the light. Data for one type of glass are listed below.

λ/nm	n
365	1.5401
405	1.5335
486	1.5250
707	1.5149
1014	1.5100

Draw a suitable linear graph and use it to find the values of A and B for this glass.

Hint: Plot a graph of n against $1/\lambda^2$.

12 Read the passage carefully and answer the questions which follow.

The helium-neon laser

In a helium-neon laser the lasing material is a mixture of helium and neon gases. An electric discharge is set up in the gas. By this means some of the helium atoms are raised to a metastable state, 20.61 eV above the ground state for helium. Neon atoms have an excited metastable state 20.66 eV above the ground state for neon, i.e. almost exactly the same energy step as for the helium metastable state. The excited helium atoms do not quickly return to their ground state by spontaneous emission, but instead often give their excess energy to a neon atom when they collide. In such a collision, the helium atom returns to its ground state and the neon atom is excited to its metastable state, 20.66 eV above the ground state for neon. This causes population inversion between the neon metastable level and another level, 1.96 eV below the neon metastable level. Stimulated emission may take place to this lower level, causing laser action.

(a) Explain what is meant by the terms *excited state, ground state, spontaneous emission.*

(b) The passage gives details of some of the energy levels in helium and in neon. Sketch two labelled energy level diagrams showing the relevant state of helium and of neon. Mark on the appropriate diagram the transition corresponding to the excitation of the helium atom, and the transition corresponding to laser action in the neon atom.

(c) Calculate the wavelength of the laser light emitted due to the transition in the neon atom.

(d) The passage states that helium atoms in their metastable state may give their energy to neon atoms. However, the excitation energy of the neon excited state is slightly *greater* than the energy available from the helium atoms. Suggest where the required extra energy may come from.

13 Read the passage carefully and answer the questions which follow.

Brownian movement and diffusion

Brownian movement is a term now used to describe many phenomena in which some quantity is constantly undergoing small, unpredictable fluctuations. The effect was discovered in 1827 by Robert Brown, who used a microscope to observe pollen grains suspended in water. He found that the grains moved at random, with frequent changes of speed and direction, further motion being totally uncorrelated with past motion, and that the movement never ceased.

Linked with Brownian movement is the phenomenon of diffusion, in which particles in a medium tend to migrate from regions of higher number density to regions of lower

number density. Consider two adjacent regions Y and Z in a vessel containing water, separated by a shutter. The volumes of the two regions are equal. Both regions contain pollen grains subject to Brownian movement. Initially, suppose that Y contains twice as many grains as Z. At t = 0 the shutter is opened. At that instant, the probability of a grain leaving Y to enter Z is twice as great as the probability that a grain will leave Z to enter Y. Because of this net difference of probabilities, there is a net movement of pollen grains from Y to Z, which continues until the numbers of grains in the two regions are equal. This process of diffusion will continue as long as there is a concentration gradient between the different regions of the medium.

(a) Explain the meaning of the terms *fluctuations, totally uncorrelated with, number density, concentration gradient.*

(b) Before a kinetic theory explanation for Brownian motion was accepted, several other theories were put forward.

  i) Brown at first believed that the motion was due to the fact that the pollen under investigation had originally come from a living plant. From your knowledge of the effect, quote one experimental observation against this theory.

  ii) Another early theory suggested that the movement was due to convection currents in the fluid. Explain how the observed characteristics make this an unlikely idea.

(c) In the second paragraph of the passage there is a statement in words about the probability of particle motion in the diffusion process. In this part of the question you are asked to express in mathematical terms the situation described.

  i) If n_Y and n_Z are the numbers of pollen grains in regions Y and Z respectively at time $t = 0$, write down the relation between n_Y and n_Z.

  ii) If P_{YZ} is the probability of a grain moving from Y to Z, and P_{ZY} is the probability of a grain moving from Z to Y, write down the relation between P_{YZ} and P_{ZY} at $t = 0$.

  iii) Hence deduce the general relation between the probability P of a grain leaving a region and the number n of grains in that region.

  iv) Name an example of another phenomenon in physics which obeys the same general relation between the probability of an event and the number of particles present.

14 Read the passage carefully and answer the questions which follow.

Atmospheric electricity

The Earth's atmosphere is an enormous electrical machine. Thunderstorms act as giant electrostatic generators, delivering electric charge to the Earth and positive charge to the upper level of the atmosphere (the ionosphere). The ionosphere and the surface of the Earth are good conductors, and charge delivered to them spreads over their entire spherical areas. In fair-weather regions, there is a leakage current from the ionosphere to the Earth, completing a huge atmospheric electric circuit. The leakage current is carried by ions which are always present in the atmosphere. The time-averaged leakage current generated as a result of a single thunderstorm is about 1 A. At any time, about 2000 thunderstorms are in action all over the Earth.

The potential difference between the ionosphere and the Earth's surface is, on average, about 3×10^5 V. Most of the electrical resistance in the atmosphere is concentrated in the dense, lower regions of the atmosphere. Thus, most of the 3×10^5 V drop occurs within a few km of the Earth's surface. In fair-weather conditions, this potential difference creates an electric field in the atmosphere which is directed vertically downwards and is of magnitude about 130 V m^{-1} at the Earth's surface. This field maintains the gradual leakage of charge between the ionosphere and the Earth.

(a) Explain what is meant by the terms *a good conductor, ions, time-averaged current, electric field.*

(b) The passage describes how the Earth's atmosphere may be considered as an electrical machine.

   i)   Draw a diagram showing a simple electric circuit equivalent to the atmospheric circuit. Represent the generator (equivalent to all simultaneous thunderstorms) by the symbol for a battery, and the atmosphere throughout all fair-weather regions by the symbol for a resistor. Indicate which pole of the battery corresponds to the ionosphere. Use information from the passage to insert numerical values for the components of your diagram.

   ii) Hence estimate the average total power delivered to the atmospheric circuit by thunderstorm activity.

(c) The passage mentions that ions are always present in the atmosphere. Suggest two processes by which they are produced in the atmosphere.

(d) The fair-weather electric field at the Earth's surface is of magnitude about 130 V m^{-1} and is directed vertically downwards.

   i)   What *point* charge would produce an electric field of this magnitude and direction at a distance of 6.4×10^6 m (the Earth's radius)?

   ii) If a charge of this magnitude were distributed uniformly over the Earth's surface, how much charge would there be on each square metre of surface?

Answers

Section 1.1
1 (a) $6.8 \times 10^{-12}\,\text{F}$ (b) $3.2 \times 10^{-5}\,\text{C}$
(c) $6.0 \times 10^{10}\,\text{W}$
2 800
3 4.6×10^{4}
4 (a) $\text{kg}\,\text{m}^2\,\text{s}^{-2}$ (b) $\text{m}^2\,\text{s}^{-2}\,\text{K}^{-1}$
5 $\text{kg}\,\text{m}^{-1}\,\text{s}^{-2}$
6 $\text{kg}\,\text{m}^2$

Section 1.3
1 systematic: measurement not from centre of lens, poorly calibrated rule
random: poor focusing of image, parallax error
2 57

Section 1.4
1 (a) vector (b) scalar (c) vector
2 Velocity has direction, speed does not. Velocity is speed in a certain direction.
3 Student is correct. Weight is a force which acts vertically downwards.
4 Direction of arrow gives direction of vector. Length of arrow drawn to scale represents magnitude of vector quantity.

Section 1.5
1 (a) 690 N (b) 210 N (c) 510 N at an angle of 28° to the 450 N force
2 Upstream at 78° to the bank.
3 120 N at an angle of 25° to the 50 N force in an anticlockwise direction.

Exam Questions
3 (b) i) 92 N ii) 77 N
(c) i) 59 N ii) 59 N
4 (a) i) $18.3\,\text{m}\,\text{s}^{-1}$ ii) 29 ° above horizontal
(b) i) $10\,\text{m}\,\text{s}^{-1}$ ii) 33 °

Sections 2.1–2.5
1 30 km
2 $180\,\text{m}\,\text{s}^{-1}$
4 3.6 hr; $610\,\text{km}\,\text{hr}^{-1}$
5 $-5.0\,\text{m}\,\text{s}^{-2}$

Sections 2.6–2.8
1 75 m
2 $140\,\text{m}\,\text{s}^{-1}$
3 $47\,\text{m}\,\text{s}^{-1}$
4 A; B
5 3.5°

Exam Questions
2 8
3 61°; 2.8 m; $3.9\,\text{m}\,\text{s}^{-1}$; 1.4 s
4 $9.4\,\text{m}\,\text{s}^{-1}$; $+10\%$

Section 3.1
2 (a) 7.9 N (b) 0.19 J
3 $8.66 \times 10^{4}\,\text{J}$

Section 3.2
1 (a) $500\,\text{N}\,\text{m}^{-1}$ (b) 13.1 cm
2 7.9 J
3 0.36 mm
4 (a) chemical (b) nuclear (c) gravitational potential (d) wind/kinetic (e) sound, light, kinetic energy of gases, potential energy (f) potential
5 (a) 0.51 m (b) $180\,\text{J}\,\text{s}^{-1}$
6 (a) 11 J (b) $14\,\text{m}\,\text{s}^{-1}$
7 (a) 16 MJ (b) $1.1 \times 10^{2}\,\text{MJ}$ (c) $2.1 \times 10^{2}\,\text{MJ}$
8 15%

Section 3.3
1 $-20\,°\text{C}$
2 (a) $40\,\text{J}\,\text{K}^{-1}$ (b) $38\,°\text{C}$
3 22.6 g
4 (a) $2020\,\text{J}\,\text{kg}^{-1}\,\text{K}^{-1}$ (b) 91 s

Section 3.4
1 £21
2 (a) 16.5 kW; 3.2 kN; 96 kW (b) 1.32 MJ; 3.84 MJ

Section 3.5
1 (a) 4.2 Nm (b) 9.1 N
2 (a) 2.9 Nm (b) 8.0 N

3 **(a)** 90 N **(b)** 110 N
4 67 N

Exam Questions
1 **(c)** 1950 J; 780 kW
2 **(c)** 152 °C
3 **(c)** 147 s; 1.02 g s^{-1} **(d)** £2.80
4 **(a)** 60 W **(b)** 900 J kg^{-1} K^{-1}

Sections 4.1–4.4
1 50 kg
2 i) 7.7 m s^{-1} ii) 45.2 m s^{-2} iii) 2040 N
4 $T/2$

Sections 4.5–4.9
1 3.3×10^4 kg m s^{-1}
2 3.6×10^7 N
3 0.27 N
4 1.03×10^5 m s^{-1}
5 Heavy particle's speed is practically unchanged; light particle moves with speed $2u$, in same direction as the incident heavy particle.
6 The heavier body

Exam Questions
1 **(a)** 2.16 kg m s^{-1} **(b)** 0.93 m s^{-1} **(c)** 0.87 J
 (d) 0.044 m
2 3080 m s^{-1}
4 **(a)** $3u$ to left **(b)** i) $3u$ to right ii) $3t_1$
 (c) $3u/2$
5 **(a)** $3m$ **(b)** 0.25

Sections 5.1–5.2
1 −43.9 °C
2 133

Section 5.3
1 212 kPa
2 1.28
3 5.8×10^3 N

Section 5.4
1 6100 m s^{-1}

Section 5.6
1 0; 250 J
2 20 J increase

Exam Questions
1 **(a)** 15.8 Ω **(b)** 149.0 °C
2 **(a)** i) 2.69×10^{25} m^{-3} ii) 3.4×10^{-9} m
 (b) 2.2×10^{-4}
4 **(a)** 939 °C (1212 K) **(b)** 2.39×10^{-4} kg
5 2.4×10^5 Pa
7 **(a)** 16 **(b)** 1 **(c)** 16 (in each part, the ratio is hydrogen:oxygen)
8 **(a)** 0 **(b)** 250 J **(c)** temperature rises

Section 6.1
1 **(a)** 760 C **(b)** 1000 W **(c)** 57 Ω
3 **(a)** 0.20 A **(b)** 0.60 W **(c)** 5400 J
3 3.5×10^6 J
4 2.0 kW

Section 6.2
1 6.7 m
2 **(b)** I/A   0.20 0.40 0.60 0.80 1.00 1.20 1.40
    R/Ω   0.85 1.20 2.45 3.65 4.56 5.47 6.21
3 **(a)** 0.62 Ω **(b)** 4.3×10^{-7} Ω m
 (c) 9.2×10^{28} m^{-3}

Section 6.3
1 0.05 Ω; 0.3 Ω
2 **(a)** 0.25 A **(b)** 1.6 Ω **(c)** 12 J

Section 6.4
1 **(a)** 169 Ω **(b)** 13 Ω
2 5 Ω; 3.0 A
3 **(a)** 25 Ω

Exam Questions
1 **(a)** 4.5 V **(b)** i) 50 Ω ii) 0.090 A
 iii) 0.90 V
3 Ten resistors, each of resistance 12 kΩ and power rating 0.5 W, connected in parallel.
4 **(a)** 4 Ω **(b)** 8 Ω **(c)** 3 Ω **(d)** 1.0 A
5 **(a)** i) 1.6×10^{-2} Ω ii) 1.1×10^{-3} Ω
 iii) 28 W **(b)** i) 4.4 s ii) 4.4×10^{21}
 (c) i) 11.7 V ii) 307 W
6 **(a)** 1.02 V, 1.22 W **(b)** ii) 7.53 m
 iii) 1.41 W

Section 7.1
1 **(a)** 200 Hz **(b)** 5.0×10^{-3} s
2 **(a)** 7.5×10^{14} Hz to 4.3×10^{14} Hz **(b)** 1.2 m
3 1/4
4 **(a)** 5.7×10^{-8} J **(b)** 4.9×10^{-5} W

ANSWERS

Section 7.2–7.3
3 (c) 3.5 mm
5 (a) 5.0 m

Section 7.4
2 (b) 8.5 cm
3 (a) $45 \, \text{m s}^{-1}$ **(b)** 18.8 Hz

Section 7.5
1 (a) $2.5 \times 10^{-6} \, \text{m}$ **(b)** $10.2°$ **(c)** 5, 3
3 $0.72°$ $(1.26 \times 10^{-2} \, \text{rad})$

Exam Questions
1 $2.0 \times 10^{9} \, \text{W m}^{-2}$
2 (b) $\pi \, \text{rad}$ $(180°)$ **(c)** 5:1 **(d)** $224 \, \text{m s}^{-1}$
4 26.6 cm
6 (b) 212 mm **(b)** 449 mm

Section 8.1
1 3.1 eV to 1.8 eV
2 (a) 340 nm **(b)** $2.5 \times 10^{-19} \, \text{J}$
3 $5.6 \times 10^{-19} \, \text{J}$ (3.5 eV)

Section 8.2
1 $3.9 \times 10^{-11} \, \text{m}$
2 $2.6 \times 10^{3} \, \text{m s}^{-1}$

Section 8.3
1 430 mm
2 91 nm

Exam Questions
2 8.0×10^{14}
3 (a) 500 nm **(b) i)** $3.9 \times 10^{-19} \, \text{J}$
ii) $9.3 \times 10^{5} \, \text{m s}^{-1}$
4 (a) $4.4 \times 10^{-35} \, \text{m}$ **(b)** $3.8 \times 10^{-34} \, \text{m}$
(c) $2.4 \times 10^{-11} \, \text{m}$ **(d)** $1.8 \times 10^{-13} \, \text{m}$
6 A: 480 nm; B: 1.8 μm; C: 670 nm

Section 9.1
2 (a) $3.3 \, \text{rad s}^{-1}$ **(b)** 0.53 revolutions per second

Section 9.3
1 (a) 220 Hz **(b)** $1.4 \times 10^{3} \, \text{rad s}^{-1}$
(c) $42 \, \text{m s}^{-1}$ **(d)** $5.9 \times 10^{4} \, \text{m s}^{-2}$
(e) $7.1 \, \text{m s}^{-1}$

2 $2.8 \, \text{m s}^{-1}$

Section 9.4
1 $9.826 \, \text{m s}^{-2}$
2 (a) $5.8 \, \text{N m}^{-1}$ **(b)** 15 mm **(c)** 0.59 s
(d) -8.2 mm

Section 9.5
1 3.2:1
2 (a) $0.71A$ **(b)** $0.71A$

Exam Questions
1 (b) $2.72 \times 10^{-3} \, \text{m s}^{-2}$
2 (a) i) $25 \, \text{rad s}^{-1}$ **ii)** 4.0 revolutions per second **(b) i)** 0.84 N
3 (b) i) $2A\rho g\Delta x$ **ii)** $2g\Delta x/L$
iv) $(1/2\pi)\sqrt{(2g/L)}$
4 (a) 1.3 s **(b)** 15 N
5 (a) $1.28 \times 10^{14} \, \text{Hz}$

Section 10.1
3 (a) $2.1 \times 10^{21} \, \text{N C}^{-1}$ **(b)** 660 N

Section 10.2
1 (a) $1.1 \times 10^{-10} \, \text{F}$ **(b) i)** $1.3 \times 10^{-9} \, \text{C}$
ii) $2.4 \times 10^{4} \, \text{V m}^{-1}$
2 (a) 2.0 μF **(b)** 1.2 μF
3 (a) i) $4.1 \times 10^{-3} \, \text{J}$ **ii)** $9.0 \times 10^{-5} \, \text{C}$
(b) i) $6.0 \times 10^{-5} \, \text{A}$ **ii)** 1.5 s **iii)** 0.75 s

Section 10.3
1 $5.5 \times 10^{3} \, \text{kg m}^{-3}$
2 -0.50%
3 $7.78 \times 10^{8} \, \text{km}$

Exam Questions
1 (a) $1.0 \times 10^{5} \, \text{V m}^{-1}$ **(c)** $5.0 \times 10^{-4} \, \text{N}$
2 $2.2 \times 10^{39}; 2.2 \times 10^{39}$
3 (a) $2.8 \times 10^{-2} \, \text{N}$ **(c)** $1.9 \times 10^{-7} \, \text{C}$
4 (a) $4.4 \times 10^{-7} \, \text{F}$
5 (a) $5.7 \times 10^{-11} \, \text{N}$
6 (c) 1.5 m to left of A
7 (b) $8.86 \times 10^{4} \, \text{km}$

Section 11.2
1 (a) East–West; $2.5 \times 10^{3} \, \text{A}$ **(b)** No, large current would melt the wire.
2 (a) $1.5 \times 10^{-4} \, \text{T}; 2.3 \times 10^{-3} \, \text{N m}^{-1}$ **(b)** Forces not large enough to move copper wire

3 **(a)** 86 mT **(b)** uniform field, no field outside poles, wire normal to the field

4 **(a)** left to right using $r = mv/BQ$ **(b)** positive with argument based on Fleming's left-hand rule.

Section 11.3

1 **(a)** 1.4 T; 4.0 mWb; 1.6 Wb **(b)** 150 V across switch **(c)** Large voltage induced which could be dangerous.

2 **(a)** 0 **(b)** 50 mV **(c)** 29 mV

3 **(a)** 0.034 **(b)** 4100 **(c)** Thicker wire reduces heating effect caused by the larger current.

Exam Questions

1 **(a)** North–South **(b)** East–West; 26 μT; 55°
(c) 2.0×10^{-4} N m^{-1}

2 **(a)** 1.6×10^{-5} N m **(b)** 32°; 18 mA

3 **(a)** 4.5 mT; 1.1×10^{-4} Wb **(b)** 7.1 mV

4 $E = 0$ when current constant. Spikes in opposite directions at times t_{on} and t_{off}. E at $t_{off} > E$ at t_{on}.

Section 12.1–12.7

3 2.5×10^9 Bq

4 **(a)** X: 3.5×10^5 Bq; Y: 2.0×10^5 Bq

5 1.2×10^{-10} kg

6 **(a)** 2.2 hr **(b)** 1.3×10^{21}

Section 12.8–12.12

1 **(a)** 0.00857 u, 7.98 MeV, 2.66 MeV/nucleon
(b) 0.687 u, 640 MeV, 6.60 MeV/nucleon
(c) 1.79 u, 1660 MeV, 7.50 MeV/nucleon

2 **(a)** 1.808 u, 2.008 u **(b)** 0.200 u **(c)** 186 MeV
(d) 7.62×10^{10} J **(e)** 0.11 g

3 3.7 MeV; 14 MeV

Exam Questions

1 15.8 years

2 **(a)** 131; 54 **(b)** i) 9.2×10^{20} ii) 0.032 g

3 **(a)** 0.046 hr^{-1} **(b)** 6000 cm^3

4 **(a)** 3.27 MeV

5 **(c)** 3.0×10^5 m s^{-1}

6 **(a)** About 50 mm; 3300 mm^{-1}

Section 13 Exam Questions

7 $R \propto T^4$

8 **(a)** $T^2 \propto d^3$ **(b)** 53 800 km

9 0.85 eV

10 2.0×10^{-4} J K^{-4} mol^{-1}

11 $A = 1.5065$; $B = 4.58 \times 10^{-15}$ m^2

12 **(c)** 633 mm

13 **(c)** i) $n_Y = 2n_Z$ ii) $P_{YZ} = 2P_{ZY}$ iii) $P \propto n$

14 **(b)** ii) 6×10^8 W **(d)** i) 5.9×10^5 C
ii) 1.1×10^{-9} C

Index

The publishers would like to thank the following individuals, institutions and companies for permission to reproduce photographs in this book. Every effort has been made to trace ownership of copyright. The publishers would be happy to make arrangements with any copyright holder whom it has not been possible to contact.

ActionPlus 15, 21, 24, 46, 51; Alex Bartel/Science Photo Library 48 (all three); Alfred Pasieka/Science Photo Library 3; Andrew Lambert 2,(bottom left), 10, 52, 75, 112, 155 (both) 179, 180, 183 (all four), 190, 237, 238, 287 (both), 295; © Archivo Iconografico, S.A./CORBIS 34 (top), 105; Barry Mayes/Life File 39 (bottom); Bill Sanderson/Science Photo Library 86; Bob Winsett/CORBIS 182; CERN/P Loiez/Science Photo Library 309; © CORBIS 87, 243 (inset), 277; DaimlerChrysler Rail Systems Ltd 274 (inset); Dan Schechter/Science Photo Library 130; Dr David Wexler/Science Photo Library 340; European Space Agency/Science Photo Library 336; Eye of Science/Photo Library 274 (main); EMPICS 39 (top); Professor Erwin Mueller/Science Photo Library 337; Foto-Unip/Still Pictures 160; Geoff Tompkinson/Science Photo Library 280; Geoscience Features Picture Library 305; Professor Harold Edgerton/Science Photo Library 98; Imperial College/Science Photo Library 207, 209, 210 (top); Dr Jeremy Burgess/Science Photo Library 54; J-L Charmet/Science Photo Library 62; © Jim Sugar Photography/CORBIS 214; © John McAnulty/CORBIS 34 (bottom left); Kit Kittle/CORBIS 239; Lucien Aigner/CORBIS 201; M & C Denis-Hoot/Still Pictures 2 (bottom middle); Martin F Chillmaid/Science Photo Library 319; Mehau Kulyk/Science Photo Library 290; Mike Guidry 210 (bottom); Mike Maidment/Life File 79; © Museum of Flight/CORBIS 2(bottom right); National Grid 130 (graph); N Feather/Science Photo Library 313; Nigel Dickinson/Still Pictures 45; Norman 'Nobby' Clarke 205; Novosti/Science Photo Library 221 (right); © Patrick Ward/CORBIS 47; © Paul A Sauders/CORBIS 243 (main); Peter Aprahamian/Sharples Stress Engineers/Science Photo Library 173; Peter Johnson/CORBIS 197; Peter Menzel/Science Photo Library 84; PING Europe Ltd 99; Polaroid 172 (both); R Maisonneuve/Publiphoto/Science Photo Library 291 (bottom); Rosenfeld Images Ltd/Science Photo Library 2 (middle left); Royal Observatory, Edinburgh/Science Photo Library 2 (top left); Science Museum 318; Science Photo Library 113, 138; Tek Image/Science Photo Library 221 (left); Teltron Limited 291 (top); Tony Craddock/Science Photo Library 34 (top left); Tony Hallas/Science Photo Library 1; W A Steer/Daresbury Laboratory/University College London 204; Wendy Brown 135; Yves Lefevre/Still Pictures 8 (top)

The illustrations were drawn by: Jeff Edwards; Hardlines Illustration and Design; Richard Duszczak.

With thanks to Hertfordshire Science Teaching Scholarship.

Orders: please contact Bookpoint Ltd, 130 Milton Park, Abingdon, Oxon OX14 4SB. Telephone: (44) 01235 827720, Fax: (44) 01235 400454. Lines are open from 9.00–6.00, Monday to Saturday, with a 24 hour message answering service. Visit our website www.hoddereducation.co.uk

British Library Cataloguing in Publication Data
A catalogue record for this title is available from The British Library

ISBN (10): 0 340 75779 5
ISBN (13): 978 0340 757796

First published 2000
Impression number   10 9 8 7 6
Year   2006

Cover photo: Digital Stock (top right); Photodisc (top left, bottom left, bottom right)
Typeset by Wearset, Boldon, Tyne and Wear
Designed by Lynda King
Printed in Dubai for Hodder Educational, a division of Hodder Headline group, 338 Euston Road, London NW1 3BH